MACHINE TOOLS

Dedication

Dedicated to my Parents

MACHINE TOOLS

[WORKSHOP TECHNOLOGY]

[For Diploma in Mechanical Engineering and also for Basic Knowledge of Degree Students]

R N Datta MIE (I), C Engg (I), TT (Dip)

Formerly Lecturer in Engineering [Mechanical]

Jagadish Chandra Polytechnic and The Calcutta Technical School, Kolkata

New Central Book Agency (P) Ltd

Co-Published by

Bijay Publications

E-mail : bijaypublications@gmail.com

Exclusively Distributed by

Bijay Pustak Bhandar (P) Ltd

90/6A M. G. Road, Bhawani Dutta Lane, Kolkata-700007

✆ : (033) 4605-7856, 8777898576, 8274916104

E-mail : bijaypustakbhandar@gmail.com

NCBA

REGD OFFICE
8/1 Chintamoni Das Lane, Kolkata 700 009, India
email: ncbakolkata@gmail.com

CORPORATE OFFICE
501A/1 Right Side [Upper Ground Floor]
Govind Puri, Kalkaji, New Delhi 110 019
email: delhincba@gmail.com

BIJAY PUBLICATIONS
90/6A M. G. Road, Bhawani Dutta Lane, Kol-7
E-mail : bijaypublications@gmail.com

BIJAY PUSTAK BHANDAR (P) LTD
90/6A M. G. Road, Bhawani Dutta Lane, Kol-7
E-mail : bijaypustakbhandar@gmail.com

MACHINE TOOLS • R N Datta

First Published: 1990
Second Edition: 2001
Third Central Edition: 2004
Reprinted: 2005, 2010, 2011, 2012, 2013, 2015, 2025

PUBLISHER • PRINTER
New Central Book Agency (P) Ltd
8/1 Chintamoni Das Lane, Kolkata 700 009

CO-PUBLISHED BY
Bijay Publications
90/6A M G Road, Bhawani Dutta Lane, Kolkata 700007

TYPESETTER
Anin, Kolkata

TECHNICAL EDITORS
Pradip Kumar Chowdhury, Prabhat Jas

ISBN: 978 81 7381 733 5

Price ₹ 435.00

Preface to the Third CENTRAL Edition

The reception accorded to the First and Second Editions of this book encouraged the Author to believe that it has been found highly useful for the students of Diploma as well as for preliminary knowledge of Degree Students.

This edition differs much from previous editions and the opportunity has been taken for correcting few slips and incorporation of addition and modification and also addition of more miscellaneous objective type questions.

In the second edition, two main chapters, i.e., **N.C. and C.N.C. Machine Tools** and **Installation of Machine Tools**, are incorporated. More than **500 objective type questions with answers** are also incorporated covering all chapters.

It is hoped that the book will prove useful as a textbook for students of Diploma, Degree and also for Machinist, Turner Trade Courses.

Author is thankful to his beloved students for their help in the preparation of this book.

Lastly, the author shall remain ever grateful to the readers for constructive suggestions for further improvement in the subsequent editions.

Kolkata
16th February, 2004

R. N. Datta

Preface to the First Edition

I am attached with the subject for over twenty-two years at Bengal Engineering College, Shibpur, Howrah, and at Jagadish Chandra Polytechnic, Berachampa, 24 Parganas (N). I find there is a dearth of suitable books on the subject mainly at the Polytechnic level and similar courses in Mechanical Engineering.

The Machine Tools is one of the principal subjects in Mechanical Engineering, whose study prepares the student for subsequent work in the shop-floor. This book has been prepared in accordance with the Diploma Syllabus in different states in India and comprises four chapters in Volume I and ten chapters in Volume II.

Volume I primarily deals with the chapters on "The Fundamentals of Machine Tools", "The Engine Lathe", "The Capstan and Turret Lathes" and "The Drilling Machines". The subject matter are arranged in a very simple, systematic manner on comparative basis with simple diagrams which are easily understandable and reproducible in the examinations. Topics are question-oriented with varieties of examples and problems.

This book is intended for the students of Diploma or Licentiate in Mechanical Engineering, Certificate or trade courses or similar courses in Mechanical Engineering, for the technicians or apprentices engaged in workshops and industries.

Volume II deals with Milling Machine, Shaping-Planning-Slotting Machines, Grinding Machines, Jigs and Fixture, Press-work, Horizontal and Vertical Boring Machines, Broaching Machines, Principles of non-traditional Machines, Safety Measures — Type of Guard — Safety Devices, and Different Finishing Methods and Processes.

I am thankful to Sri Amit Nag and Sri Pradip Kumar Bhattacharya of Barasat, 24 Paraganas (N), for their help in the preparation of this book.

I welcome the comments from the readers regarding presentation of subject matter, diagrams, etc.

I should also be pleased to receive suggestions from the readers for the improvement in the subsequent publications.

R. N. Datta

CONTENTS

VOL. I

VOL. II

Volume I

1

FUNDAMENTALS

A1.11 : Introduction

Intension of modern industry is to create appropriate number of useful products at reasonable price. To produce the products require a large number of metal-cutting and metal-forming machine tools, differing in purpose, processing, capacities and sizes. Metal-cutting and metal-forming processes are the basis of engineering industry and are related either directly or indirectly in the production of nearly every product of our modem civilization. The study of metal-cutting and metal-forming machine tools is a basic knowledge of the fundamentals of machine parts, mechanism, method of machining of materials, operations, etc. To operate or run any machine tools require electric motor in which electrical energy will be converted to rotational energy and this mechanical output, i.e., rotational energy will again transform to different mechanical motion, like translatory, intermittent translatory, etc., needed for various machines of doing the job.

A1.12 : Machine

Machine is a device for transforming or transferring energy. Example : Internal Combustion Engines, where heat energy in diesel or petrol oil supplied to this engine is converted to mechanical energy at the crank-shaft end.

A1.13 : Tools

The parts in contact with the job for doing different operations or generation of surfaces of products are known as tools. Tools may be classified according to the operation performed by the source of power.

A1.14 : Hand Tools

The tools which are operated or controlled by hand power in performing different operations in fitting, assembly, lay-out, etc., known as hand tools. These tools are cutting tools, measuring tools, hammers, files, wrenches, etc.

A1.15 : Machine Tools

In performing metal cutting we need the help of tools, driven by power (in which hand power is insufficient to operate), in the form of rotational energy either from a motor or any prime mover through transmission system are known as machine tools. Machine tools are those machines that make machines and are the products of our industrial era. The main components of machine tools are a motor, a transmission system and a job and cutting tool(s). Examples of such tools are : Engine Lathe, Drilling Machine, Shaping Machine, Milling Machine, etc.

A1.16 : Comparison Between Hand Tools And Machine Tools

Points	Hand Tools	Machine Tools
1. Definition	Tools used for removing metal to attend desired shape and size and operated by hand power	Tools used for removing metal to attend desired shape and size and operated by machine power
2. Example	Hammers, Files, Wrenches, etc.	Engine Lathe, Capstan Lathe, Drilling machine, etc.
3. Amount of Force/ Torque/Power required	Less	More
4. Suitable for Jobs	Small and light jobs	Small, medium and heavy jobs
5. Accuracy and Efficiency	Less	More
6. Suitable for productive /non-productive jobs	Mainly non-productive jobs	Mainly productive jobs
7. In case of power failure	Does not effect	Will not work
8. Fatigue to worker	More	Less
9. Time required to machine the surface	More	Less

A1.17 : Principles Of Metal Cutting

Most important principles of metal cutting are :

1. Job material must be softer than the tool material.
2. Tool shape and angles should be so designed that the same will produce maximum metal removal.
3. There must be two relative motions in any metal cutting, two relative motions are either in the tool or in the job or one in job and other in the tool.
4. Proper selection of speed, feed, depth of cut, cutting fluid used and also machine's rigidity, operator's alertness, etc.

A1.18 : Purposes Of Machine Tools

Machine tools remove metal from a job to produce a desired size and shape. All machine tools should have the following purposes :

1. To have the machine tool of a high degree of perfection in construction, ability of high precision work and tremendous advancement in the field of machine-tool building.

2. To hold and support the job in proper position on a table or on any accessories.

3. To hold the required tool on a tool post or on a spindle.

4. To arrange the required speed, feed and depth of cut with respect to the job-tool materials and required operation.

5. To give the desired two relative movements on the tool or on the job or one in job and another in tool depending on the machine tool and operation to be performed.

A1.21 : Types Of Machine Tools

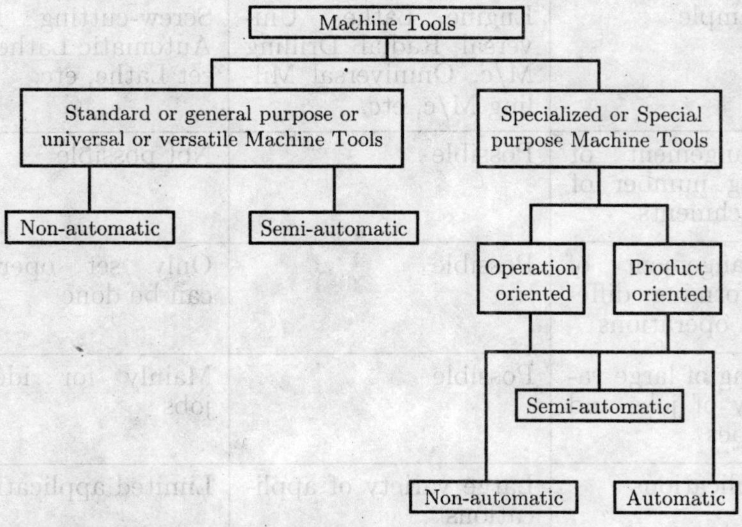

A1.22 : Classification Of Machine Tools

Machine tools can be classified to one of the following groups :

(a) *General Purpose or Universal or Versatile or Standard Machine Tools*—These types of machine tools are designed to perform a great variety of machining operations with a variety of work-pieces by using number of attachments suitable for tool room and maintenance works. Examples of this type machine tools are Engine Lathe, Universal Radial Drilling Machine, Omniversal Milling Machine, etc.

(b) *Single Purpose Machine Tools*—These types of machine tools are designed to perform a single definite machining operation in machining number of identical workpieces, suitable for productive work. Example : for turning the cam contours on cam-shafts, finishing operation of a riffle barrel, machining of crankpins of crank-shafts, etc.

(c) *Special Purpose Machine Tools*—These type of machine tools are manufactured individually and intended for performing a certain specified operations in machining a large number of identical workpieces. Applications of these machines are in large lots and mass production.

A1.23 : Comparison Between Standard M/C Tools And Special Purpose M/C Tools

Points 1	Standard Machine Tools 2	Special Purpose Machine Tools 3
1. Definition	To perform a great variety of machining operations with variety jobs by using number of attachments	To perform certain definite operations in machining large number of identical jobs
2. Example	Engine Lathe, Universal Radial Drilling M/c, Omniversal Milling M/c, etc.	Screw-cutting Lathe, Automatic Lathe, Turret Lathe, etc.
3. Arrangement of using number of attachments	Possible	Not possible
4. Arrangement of performing different operations	Possible	Only set operations can be done
5. Using of large variety of jobs and shapes	Possible	Mainly for identical jobs
6. Applications	Large variety of applications	Limited applications
7. Location	Maintenance and tool room shops	Production shop
8. Operation and Setting time	Slower method	Faster method

A1.31 : Types Of Drive In Machine Tools

Power from the source (i.e., motor mostly used as a power source) is transmitted to the machine spindle by the following methods: (a) Group drive or Line shaft drive, (b) Individual drive or Self drive.

A1.32 : Group Drive (Line Shaft Drive)

In this drive, a powerful motor drives an overhead shaft known as line shaft or main shaft and the line shaft drives many counter-shafts according to the number of machines in the shop floor which in turn, drives the different machine spindles. Line shaft gets power from motor and is always rotating as the motor rotates. The desired counter-shaft may/may not be driven from the line shaft through fast and loose pulley arrangements, there having a cone pulley and by belting, power from countershaft transferred to machine spindle through another cone pulley on machine shaft. Group drive possesses low initial cost, economical in fixed charges on power consumption and maintenance, and having overloading capacities and thus one motor can run different machine tools at a time.

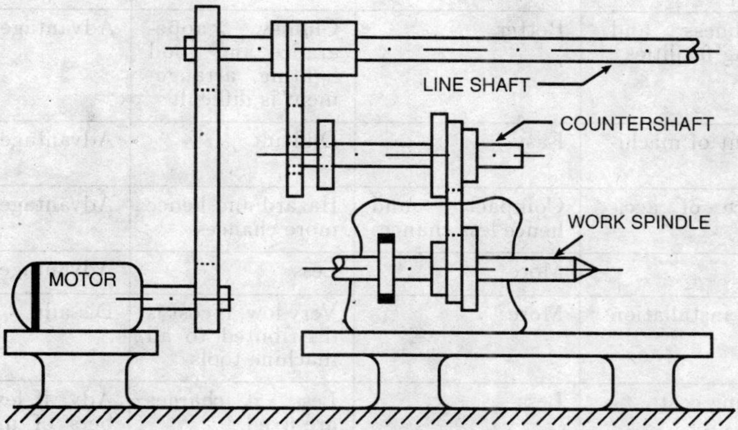

S-1.32 GROUP DRIVE

A1.33 : Individual Drive (Self Drive)

Each machine required its own electric motor and starter, generally mounted on single base or encased. Power from motor is transferred to the machine spindle by either of belting, coupling, chaining, gearing, etc.

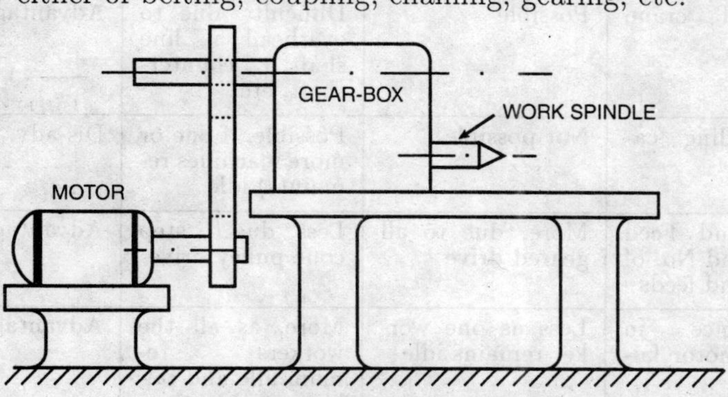

S-1.33 INDIVIDUAL DRIVE

Modern industries are using individual drive due to its easy lay-out, cleanliness, less chance of accident, adaptation of overhead crane, less annoyance during motor failure, better lighting, wide speed variations, etc.

A1.34 : Comparison Between Individual Drive And Group Drive

Points 1	Individual Drive (I.D.) 2	Group Drive (G.D.) 3	Advantages of I.D. over G.D. 4
1. Definition	In an individual drive system, each machine tool is driven by its own motor	In a group drive system, several machine tools are driven by a single motor of having high power	—
2. One motor is responsible for	One machine tool	A group of machine tools	Dis-adv.
3. Cleanliness and lighting facilities	Better	Clumsy appearance and good lighting arrangement is difficult	Advantage
4. Lay-out of machines	Easy	Difficult	Advantage
5. Chance of accident	Compact and hence less chances	Hazard and hence more chances	Advantage
6. Safety	More	Less	Advantage
7. Initial installation cost	More	Very low, if cost is distributed to all machine tools	Dis-adv.
8. Running cost	Less	Less, if charges are fixed	Adv. if few numbers of machines are used at a time
9. Power consumption	Less, if few machines are used	Less, if all machines are worked at a time	Same as column 8
10. Replacement of belt and maintenance cost	Takes little time and less costly	Takes more time and more costly	Advantage
11. Overhead crane facilities	Possible	Difficult due to overhead line shaft, counter-shaft, etc.	Advantage
12. Over-loading capacity	Not possible	Possible, if one or more machines remain(s) idle	Dis-adv.
13. Speed and Feed Range and No. of speeds and feeds	More, due to all geared drive	Less, due to step cone pulley drive	Advantage
14. Disturbance in case of motor failure	Less as one worker remains idle	More, as all the workers remain idle and talking each other, makes noise	Advantage
15. Frictional power losses	Less	More (shafts, belts, pulleys absorbs greater power)	Advantage

Points	Individual Drive	Group Drive	Advantages of I.D. over G.D.
1	2	3	4
16. Suitability for variable power consumption	Not suitable	Most suitable	Dis-adv.
17. Economical	When running only a particular machine	When charges are fixed on power consumption and maintenance	Same as column 8
18. Type of spindle drive	Mostly all geared head drive	Mostly cone-pulley drive	Advantage

A1.35 : Compare Flat Belt & V-Belt

Sl. No.	Points	Flat Belt	V-Belt
1.	Cross-section	Rectangular	Trapezoidal
2.	Manufactured from material	Leather, Rubber, Canvas, Steel, etc.	Endless belt of strong canvas or fabric chords, vulcanized with rubber, synthetic rubber, etc.
3.	Transmission of power	Up to a distance of 10 m with a maximum surface speed of 1400 m/min.	Short centre distance but not less than larger pulley diameter and not more than the sum of the two pulley diameters. Industrial medium and heavy power demands, while a maximum surface speed of 1500 m/min. Belt runs in 40 vee grooves
4.	Velocity ratio	For open belt 5, for flat belt with an idler pulley 10	For V-belt of 8 to 15
5.	Number of belts used at a time	Only one belt	If any one of the belts fails, the other belts will transmit power. But 2 or more belts are used at a time
6.	In case of belt failure	No transmission of power	If any one of the belts fails the other belts will transmit power
7.	Chances of slip	Avoids oil contact, more slip but can be decreased by using resin, etc.	Avoid any oily substance contact. More surface contact and less slip
8.	Maintenance in case of failure	Easy to repair or connect the two ends	Repairing of belt is not possible but will be replaced by a new one

Sl. No.	Points	Flat Belt	V-Belt
8.	Maintenance in case of failure	Easy to repair or connect the two ends	Repairing of belt is not possible but will be replaced by a new one
9.	Application	In mills, factories, etc.	Cooling fan and the engine, in automobile like trucks, buses, cars, etc., modern self drive machine tools, etc.

A1.41 : Cutting Tool

Cutting tool is required in machine tools for machining processes in metal-cutting or metal-forming to attain the desired shape and size of a product.

A1.42 : Classification Of Cutting Tools

Cutting tools used in different machine tools may be classified as follows :

(a) With respect to the shape of the tool—

- (i) Solid (Lathe turning tool, H.C.S.)
- (ii) Tool bit (Lathe turning tool fixed on holder, H.S.S., cast non-ferrous satellite cobalt)
- (iii) Tipped tool (Different machining processes, brazed on holder, cemented carbide, ceramic, etc.)
- (iv) Pointed tool (Diamond pointed on holder end and costly)
- (v) Grain size (Abrasives type tool, mostly used in grinding wheel(s), etc.)

(b) With respect to the type of cutting—

- (i) Orthogonal cutting (Two-dimensional cutting, for square shoulder operation, etc.)
- (ii) Oblique cutting (Three-dimensional cutting, mostly used in turning operation)

(c) With respect to materials used in tool—

- (i) High Carbon Steel (H.C.S.)
- (ii) High Speed Steel (H.S.S.)
- (iii) Cast non-ferrous steel cobalt
- (iv) Cemented Carbide
- (v) Ceramic
- (vi) Diamond
- (vii) Abrasives

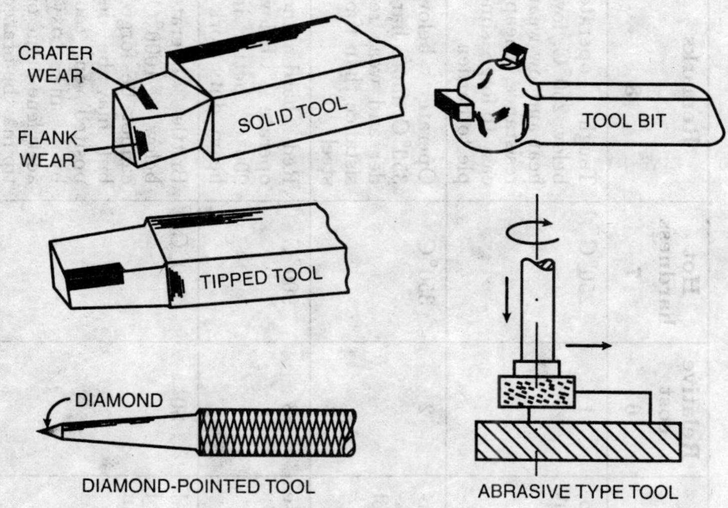

S-1.42 CUTTING TOOLS

(d) With respect to the type of operations—

(i) Turning tool
(ii) Knurling tool
(iii) Parting-off tool
(iv) Thread-cutting tool, etc.

(e) With respect to number of cutting point used in one tool—

(i) Single-point cutting tool (turning, shaping and planing tools)
(ii) Double-point cutting tool (twist drill)
(iii) Multi-point cutting tool (Milling cutter, Reamer, Broaching tool and Power saw, etc.)

(f) With respect to the direction of tool movement during cutting—

(i) Right-hand tool (cutting from right to left)
(ii) Left-hand tool (cutting from left to right)

(g) With respect to the motion of the tool—

(i) Translatory motion—Lathe tools
(ii) Intermittent translatory motion—Planning tools
(iii) Reciprocating motion—Shaping tool
(iv) Rotary motion—Milling tool
(v) Rotary and vertical translatory—Drilling tool, etc.

Single-point cutting tool is used on Lathe, Shaper and Planer Machine Tools. Double-point cutting tool is used in Drilling Machines. Multi-point cutting tool is used in Milling, Broaching, Power saw, etc., Machine Tools.

A1.43 : Comparative Study Of Different Cutting-Tool Materials Used In Different Machine Tools

Tool materials 1	Compositions 2	Heat treatment 3	Cutting speed 4	Normal use 5	Relative cost 6	Hot hardness 7	Remarks 8
1. Carbon Tool Steel or Tool Steel	0.8 to 1.5% Carbon and rest Iron	Heat 900°C, water cooled and drawn at 250°C	12 m/min	Reamer, Tap, Drill, Single point tool	1	250°C	Tough, operate below 250°C, low heat and low wear resistance, cheap, easy to forge, simple to harden
2. Medium Alloy Steel	Carbon, 1 to 1.5% Tungsten, Chromium, Vanadium and Molybdenum 4 to 6% rest Iron	Heat 1000°C to 1100°C, oil cooled and drawn at 350°C	18 m/min	Band saw, circular saw, hacksaw, cutting tool at slower speed	2	350°C	Operate below 350°C, more harder and wear resistance than tool steel
3. High Speed Steel	Tungsten, 18% Chromium, 4% Vanadium, 1% Rest Iron	Heat 1300°C, oil cooled and drawn at 900°C	30 m/min	General purpose tool	8	900°C	Red hard and operate below 900°C, better in heavy duty work
4. Cast Non-ferrous Stellite Cobalt	Cobalt, 40% Chromium, 30% Tungsten, 20% Carbon, 4% Iron, 6%	Cast at 1000°C	60 m/min	Tool bits, inserted tools, blades of cutter, suitable in Rubber, Plastics	30	1000°C	Brittle operate below 1000°C, cannot be forged but may be deposited on tool by an oxy-acetylene flame or tip may be brazed

Tool materials 1	Compositions 2	Heat treatment 3	Cutting speed 4	Normal use 5	Relative cost 6	Hot hardness 7	Remarks 8
5. Cemented carbide or sintered carbide	5 to 15% Cobalt as binder, 95 to 85% Tungsten carbide (94: 6)	Pre-sintered at 900°C and Sintered at 1500°C	150 m/min	Tipped tool, Die and punch, , Drop forging tool	Rs 500 per normal tip (approx.)	1500°C	Very brittle, operate below 1500°C. Tips are brazed on tool holder
6. Ceramic	Aluminium oxide	Sintered at 1200°C	600 m/min	Tipped tool for cutting C.I., plastics	Rs. 700 per normal tip (approx.)	1200°C	Operate below 1200°C, used generally without coolant, low heat conductivity. Excellent compressive strength, brittle. Ceramics are also brazed on tool holder
7. Diamond	Ore	Hardest	1000 m/mim	For cutting glass, plastics, ceramics, etc., copying Milling and lathe tools		1650°C	Operate below 1650°C, hardest, incompressible, quick heat transfer, low coefficient of friction
8. Abrasives mainly of Aluminium oxide and Silicon carbide	Abrasives grains in natural and artificial form			Aluminium oxide for cutting high tensile material, Silicon carbide for low tensile material and non-ferrous materials			Used for harder material and for super finish, loose, bonded into wheels and stone, embedded on paper and cloths

A1.44 Comparative Study Of Right Hand Tool & Left Hand Tool

<table>
<tr><th>Left-hand tool</th><th>Right-hand tool</th></tr>
<tr><td>1. Left-hand tool moves from the left-hand to the right-hand end, i.e., to move the tool from the headstock end to the tail-stock end.</td><td>1. Right-hand tool moves from right to the left-hand end of the lathe bed, i.e., to move the tool from the tailstock end to the headstock end.</td></tr>
<tr><td colspan="2">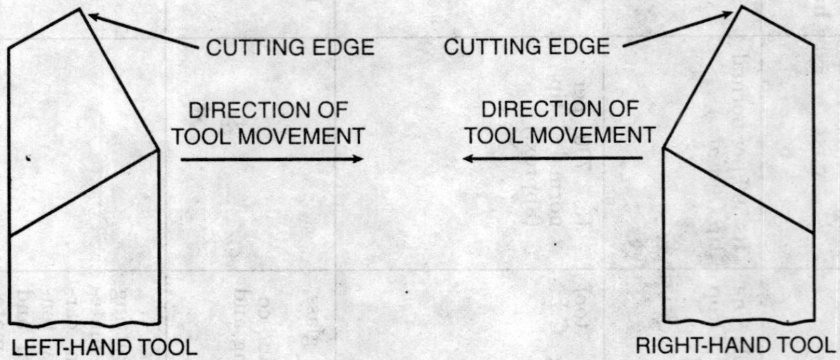

S-1.44</td></tr>
<tr><td>2. Left-hand tools generally used for turning operations which leave a shoulder on the right-hand end of the job. Left-hand thread cutting, facing operations, etc.</td><td>2. Right-hand tool generally used in most of the operations in lathe machine, e.g., turning, right-hand thread cutting, etc.</td></tr>
<tr><td>3. Cutting edge of left-hand tool is on its right-hand end.</td><td>3. Cutting edge of Right-hand tool is on its left-hand end.</td></tr>
</table>

A1.45 : Properties Of Different Materials Used In Cutting Tool

Nickel	Increases the strength and toughness.
Chromium	Increases the hardness with high strength and high elastic limit and also corrosion resistance.
Tungsten	Excellent hot hardness and forms stability.
Vanadium	Fine grain structure, and increases strength with addition of Nickel and Chromium. Also increases keenness of cutting edges.
Manganese	Improves strength of steel in hot rolled and heat treated condition and wear resistance.
Cobalt	Increases hardness and strength also hot hardness and wear resistance.
Silicon	High elastic limit and corrosion resistance.
Molybdenum	Increases keenness of cutting edges.

A1.46 : Characteristics Of Cutting-Tool Materials

Ideal cutting-tool material should possess the following characteristics:

1. It should retain its strength and hardness at elevated temperature.
2. It should have high toughness to withstand shock and vibration and to prevent breakage.
3. It should be low co-efficient of friction for less wear and better surface finish.
4. Tool material should be cheap, easily manufactured at reasonable price and easily re-grind.
5. Tool material and shank material should be as close as possible to avoid cracking.

A1.47 : Method Of Metal Cutting

Metal cutting may be divided into two main classes :

1. Orthogonal or Two-dimensional cutting.
2. Oblique or Three-dimensional cutting.

A comparative study between the two methods of cutting :

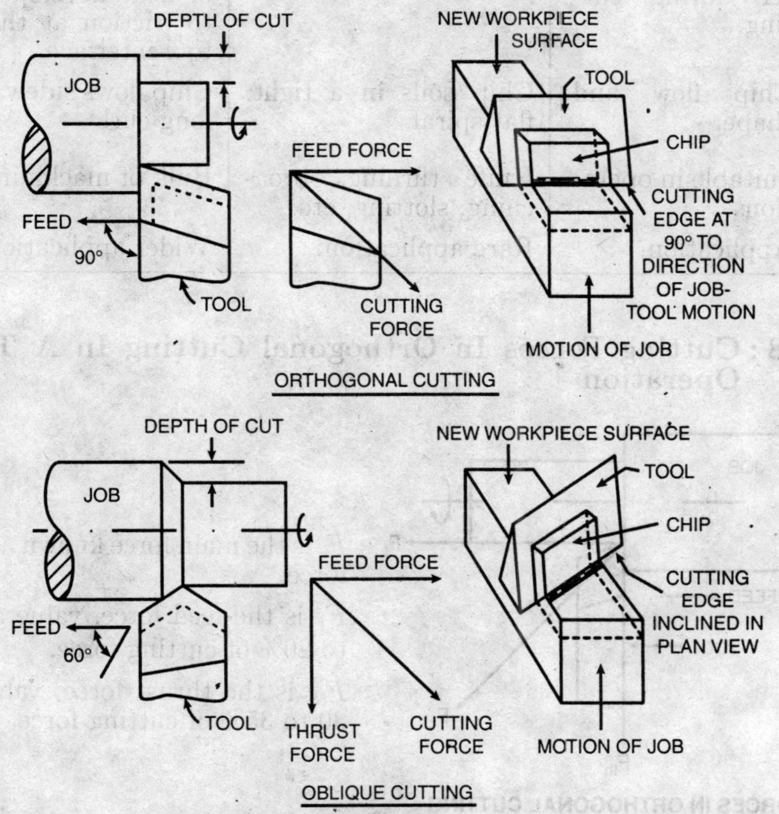

S-1.47 ORTHOGONAL & OBLIQUE CUTTING

A1.47A : Comparison Between Orthogonal And Oblique Cutting

Points	Orthogonal cutting	Oblique cutting
1. Definition.	In orthogonal cutting, the cutting edge of the tool is 90° (degree) to the direction of relative motion between the work and tool.	In oblique cutting, the cutting edge of the tool is inclined at an angle less than 90° (degree) to the direction of relative motion between the work and tool.
2. Alternative name.	It is also named as two-dimensional cutting.	It is also named as three-dimensional cutting.
3. Volume of metal removal for a set of conditions.	Less metal removal due to square cutting.	More metal removal due to oblique cutting, removes greater area of chip for same depth of cut and other conditions.
4. Tool life.	Shorter life.	Longer life.
5. Friction and chatter during cutting.	More.	Less, as small amount of heat developed due to friction at the job-tool interface.
6. Chip flow and shape.	Chip coils in a tight, flat spiral.	Chip flows sideway in a long curl.
7. Suitable in operations.	Knife turning, broaching, slotting, etc.	Bulk of machining.
8. Application.	Rare application.	Wider application.

A1.48 : Cutting Forces In Orthogonal Cutting In A Turning Operation

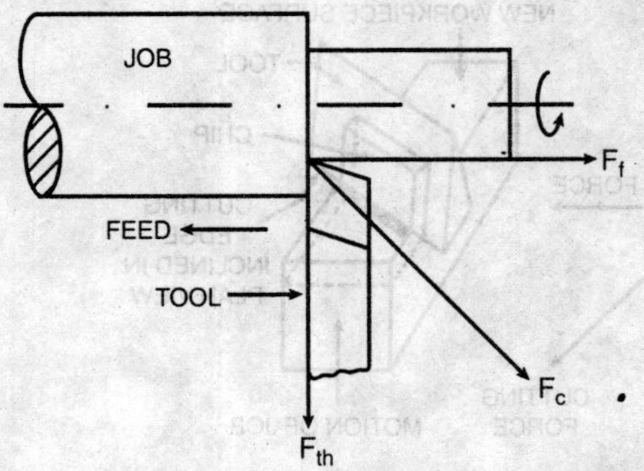

FORCES IN ORTHOGONAL CUTTING
S-1.48

F_c is the main force known as cutting force.

F_f is the feed force, value about 15 to 20% of cutting force.

F_{th} is the thrust force, value about 30 to 35% of cutting force.

A1.49 : Application Of Oblique Cutting In Machining Operation

Oblique cutting have wider application due to the following advantages :

1. More metal removal due to more contact area of job-tool for the same depth of cut.
2. Longer tool life.
3. Less heat generated per unit area.
4. Less friction between job-tool interface.
5. Less chatter.
6. Easy removal of chip through sideways in the form of curl.

A1.410 : Tool Life

A particular tool cannot be used in operation for an unlimited period of time. Every tool possesses a definite life. So tool life is the time duration of a tool, will operate satisfactorily until it will blunt; that means it is the continuous service after which the tool becomes dull. It can also be expressed as the maximum number of work-pieces can be machined by a tool.

Tool life depends on the following factors:

(i) Hardness of the material to be cut.

(ii) Quality of tool material, i.e., from which the tool is to be prepared.

(iii) Proper selection of the tool angles with respect to job-tool materials, types of operations, etc.

(iv) Size of the tool.

(v) Cutting speed at which cut is being made by the tool.

(vi) Feed and depth of cut at which cut is being made by the tool.

(vii) Cross-sectional area of the chip.

(viii) Cutting fluid characteristics and its type.

(ix) Rigidity of the machines.

(x) Operator's skill and alertness, etc.

A1.411 : Measurement Of Tool Life

Effective time elapsed between two successive grinding of a cutting tool is called tool life. In several ways the tool life of a cutting tool can measure as:

(i) Number of workpieces can be machined by the tool between re-sharpenings.

(ii) Effective operational time, i.e., the actual time the tool is in contact with the workpiece.

(iii) Equivalent cutting speed.

(iv) Volume of metal removal, or weight of the chip removed by the tool between re-sharpenings.

A1.412 : Determination Of Tool Life In Terms Of Volume Of Metal Removal

In practice, tool life, in terms of volume of metal removal can be measured easily.

Let us assume that the volume of metal removal from a workpiece between re-sharpenings for a definite depth of cut, feed and cutting speed can be calculated as follows:

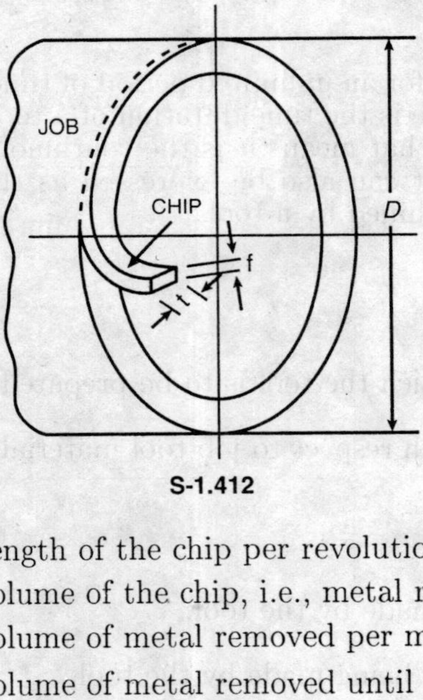

S-1.412

(i) Tool life of the tool in terms of metal removal until tool failure, T_L mm^3.

(ii) Feed, f mm/revolution.

(iii) Depth of cut, t mm.

(iv) Cutting speed, V metre/minute.

(v) R.P.M. of the work, N r.p.m.

(vi) Time of tool failure, T minutes.

Therefore, area of cross-section of chip $= f \times t$ mm$^2 = ft$ mm^2.

Length of the chip per revolution $= \pi D$ mm.

Volume of the chip, i.e., metal removed per revolution $= ft \times \pi\ D$ mm^3.

Volume of metal removed per minute $= ft\ \pi\ D \times N$ mm^3/ min .

Volume of metal removed until tool failure $ft\pi DN \times T$ mm^3.

$$T_L = ft\pi DNT \text{ mm}^3 = \pi DNftT \text{ mm}^3 \qquad (1.1)$$

Again, we know cutting speed, $V = \pi DN/1000$ metre/minute or $1000\ V = \pi DN$. By substituting the equation of cutting speed, 'V' in equation (1.1) we get, Tool Life, $T_L = 1000\ VftT$ mm^3/min.

A1.413 : Tailor's Tool Life Equation

Tailor has shown a relationship between the cutting speed (V) and tool life (T_L), can be expressed in the form : $VT_L^n = k$, where, $V =$ Cutting speed in m/min, $T_L =$ Tool life in minute, $k =$ Constant, depending on the tool and work materials, tool shape, cutting conditions and environment, etc.

It is clear from the above equation, tool life is inversely proportional with the cutting speed.

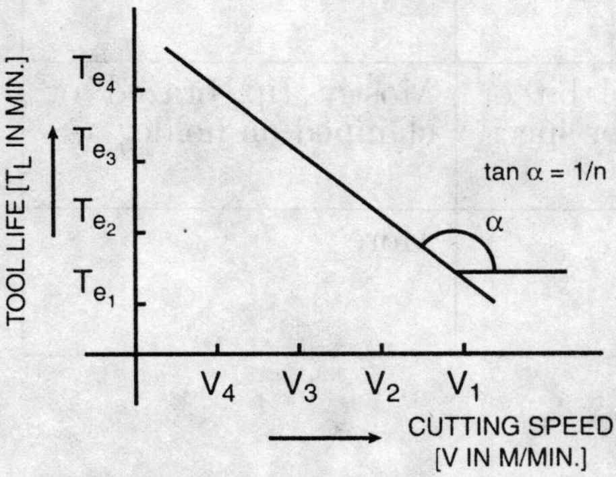

(PLOTTED IN LOG-LOG PAPER)

S-1.413

Value of exponent (n) may vary with the tool materials

$n = 0.08$ to 0.18 for H.S.S. tools.

$= 0.18$ to 0.4 for carbide tools.

$= 0.4$ to 0.8 for ceramic tools.

From the above equation, it is clear that, "tool life is the function of cutting speed." If we plot, different values of tool life and cutting speed on a log-log graph paper, shows a straight line relationship.

A1.414 : Compare Tool Material High Speed Steel With Cemented Carbide

Sl. No.	Points	H.S.S.	Cemented Carbide
1.	Composition	Generally, Tungsten 18% Chromium 4% Vanadium 1% Rest Iron	5 to 15% Cobalt as binder, 95 to 85% Tungsten Carbide (94:6)
2.	Cutting speed	About 30 m/min.	About 150 m/min.
3.	Hot hardness	900°C	1500°C
4.	Properties	Red hard and better in heavy duty work	Very brittle and better in production work
5.	Heat treatment	Heat to 1300°C, oil cooled and drawn at 900°C	Presintered at 900°C and sintered at 1500°C
6.	Suitable in	Maintenance, Tool room and Jobbing work	Production work even can be used in Tool room
7.	Grinding facility	Easily machined and can be given any desired shape with respect to operation	Tool shape cannot be machined but it will take the shape of the holder. Different tools for different operations
8.	Cost	Less	More

Sl. No.	Points	H.S.S.	Cemented Carbide
9.	Tool shape	Mostly tool bit or solid tool for specific purpose	Mostly tip brazed or clamped on holder
10.	Tool life for same amount of metal removal	Less	More

A1.51 : Machinability

The term machinability is defined as the property of the material by virtue of which it can be machined easily with a cutting tool under a given set of conditions.

Machinability is influenced by the following variables (or, the machinability depends on the following factors) :

1. Work Material

(a) Chemical composition.

(b) Microstructure.

(c) Rigidity and dimensions of work.

(d) Properties of the material.

2. Tool Material

(a) Tool signature (or tool geometry).

(b) Tool material.

(c) Nature of movement of the tool with respect to work.

3. Cutting Conditions

(a) Cutting speed during machining.

(b) Cross-sectional area of chip (i.e., feed and depth of cut).

(c) Cutting fluid used.

4. Machines Parameters

(a) Rigidity of the machine.

(b) Work holding arrangements.

(c) Machine's capacity (power, cutting forces, accuracy, etc.)

5. Operator's Skill and Alertness.

A1.52 : Criteria Of Machinability

The following are the criteria to determine the machinability:

1. *Tool life :* Higher the value of tool life, better is the machinability and again, if the cutting speed is increased, tool life will decrease and vice-versa. Hence machinability based on tool life or cutting speed.
2. *Surface finish :* Higher the surface finish on a material under a given set of conditions, the better is its machinability.
3. *Cutting forces :* Higher the cutting forces during machining of a material, the lower is its machinability.
4. *Other specific criteria :* Like, thrust, drilling torque, rate of metal removal, shear, angles, etc. can also be employed to judge the easy of machining.

A1.53 : Machinability Rating (Or Index)

$$\text{Machinability rating of the material to be tested} = \frac{\text{Cutting speed of test material for } T_L \text{ min . tool life}}{\text{Cutting speed of reference material for } T_L \text{ min . tool life}} \times 100$$

A1.61 : Cutting Fluid

Cutting fluid (either in the form of liquid or gaseous state) is supplied at the job-tool interface to improve the cutting condition and also tool life. It is also named as coolant or lubricant.

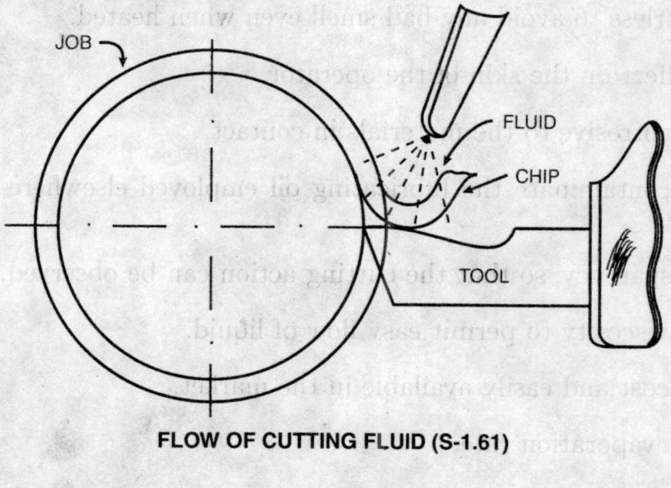

FLOW OF CUTTING FLUID (S-1.61)

A1.62 : Purposes Of Cutting Fluids

1. *To cool the tool*—It decreases friction, increases tool life, good surface finish, increases hot hardness and also preventing metallurgical damage.

2. *To lubricate and reduce friction*—To lubricate or easy flow of chips and movement of tool between job-tool interface, and tool-chip interface respectively. Reduction of friction, means less heat generation and hence less power consumptions and also increases tool life.

3. *To break continuous chips into small parts*—It minimizes working hazard.

4. *To wash the chip away from the tool*—It keeps the cutting zone clear and decreases the working hazard.

5. *To cool the workpiece*—To prevent excessive thermal distortion.

6. To improve surface finish and decrease tool wear.

7. To protect the newly machined surface from corrosion and rusting.

8. To reduce cutting forces and low built-up edge.

A1.63 : Properties Of Cutting Fluid

An ideal cutting fluid should have the following properties:

1. High heat absorption during machining operation.

2. Good lubricating effect to reduce co-efficient of friction.

3. High flash point to eliminate the fire hazard.

4. Stability and non-foaming tendency.

5. Chemically inactive so as not to react.

6. Odourless to avoid any bad smell even when heated.

7. No effect on the skin of the operator.

8. Non-corrosive to the materials in contact.

9. Non-contaminate the lubricating oil employed elsewhere on the machine.

10. Transparency, so that the cutting action can be observed.

11. Low viscosity to permit easy flow of liquid.

12. Low cost and easily available in the market.

13. Low evaporation rate.

A1.64 : Cutting Fluid With Main Functions

The fluid which can serve either only as coolant or as lubricant or both, when a stream of that fluid is directed normally on the job-tool interface, is known as cutting fluid.

Cutting Fluid

As coolant	As lubricant	As both coolant and lubricant	As flashing action
1. To cool the cutting tool, chip, job. To reduce distortion, etc., in cutting.	To easy movement of tool on job, reduce cutting forces.	–	To wash-off chips.
2. Less tool wear, longer tool life.	Less friction, less heat generation, etc.	–	Better surface finish, high feed rate.
3. Air, water, etc.	Mineral oil, vegetable oil, animal oil, etc.	Emulsion of water and oil, mixed oil, compounded oil, etc. At various proportion depending on job materials, operational type, cutting conditions, etc.	

A1.65 : Choice Of Cutting Fluid

Cutting fluid mainly depends on the friction at the job-tool interface. Also depends on job materials, type of operations, cutting conditions, machine conditions, etc.

A1.66 : Prevention Of Rust On Job & Machine

Water alone can form rust and to avoid rust, cutting fluid is mixed thoroughly with water and oil. Sometimes, sodium nitrate or similar chemical compound can be used with the fluid as rust inhibitor (or preventor).

A1.67 : Types Of Cutting Fluid

Soluble oil or water emulsion—Water emulsion or soluble oil is a blend of water, mineral oil and emulsifying agents or coupling agents. The soap acts as an emulsifying agent or coupling agent. Sometimes, additives, like antifoam agents, water softening agents, rust inhibitors are added to soluble oil. Emulsion of mineral oil in water in the ratio of 1 : 10 to 1 : 100 according to the needs for the operations and job materials.

Mostly used in the operations of lathe work, milling, grinding, shaping, planning, power sawing, drilling, etc.

Cutting Oils : A mixture of mineral oil and fatty oil/Lard oil with chemical additive with addition of Sulphur or Chloride, known as chemical additive oil. Mostly suitable for automatic machines, automatic screw-cutting machines, turret lathes, etc., for production work.

Synthetic Coolants :	Like soluble oil, it is also blending of many chemical agents with water in the ratio 1 : 50 to 1 : 250 part of water but without/with petroleum products and mostly used in grinding operation.
Solid Lubricants :	Soap bar and wax stick are used sometimes as lubricants to the cutting tool.

A1.68 : Comparative Study Of Different Cutting Fluids With Their Main Functions

Cutting Fluid 1	As Coolant 2	As Lubricant 3	Rust Inhibitor 4	Use 5	Remark 6
Water	Better	very poor	No, but can be improved by using rust inhibitors.	Where coolant only necessary.	Very low cost.
Soluble oils (water + oil + soap)	Good	Good	Yes, if its emulsion is perfect.	In most of the machining operations in Lathe, Drilling, Milling, etc.	Clean atmosphere, less costly percentage of oil varies with operations.
Mineral oil	Good	Good	Yes	Where more contact between job-tool like screw-cutting, automatic machines, etc.	Risk of skin cancer, cost moderate.
Fatty oil or vegetable oil	Better	Better	Yes	Do	Risk of skin cancer, cost more.
Animal oil	Best	Best	Yes	Do	Risk of skin cancer, cost most.
Mixed or compounded oil (Mineral oil + Fatty oil or, Mineral oil + Animal oil or, Fatty oil + Animal oil)	Excellent both for coolant and lubricant at reasonable price. Percentage of different oil in the mixture varies with the type of operations.		Yes	In most of the automatic machine.	

Cutting Fluid 1		As Coolant 2	As Lubricant 3	Rust Inhibitor 4	Use 5	Remark 6
Chemical additive oil with mixed oil/straight oil.	Chloride base	Better	Better	Yes		Anti-weld characteristic, cost reasonable.
	Sulphur base	Better	Better	Yes	Machining Alloy Steel	Tough and stringy material, machining cost reasonable.
Solid lubricant (wax, soap, etc.)		Less	Better	Yes	Generally used at cutting point for easy movement of the cutting tool.	Cost cheap.
Synthetic coolant		Better	Poor	Possible	Grinding operations	Life better than soluble oil, clean atmosphere, may be toxic, moderate cost.

A.1.69 : Selection Of Cutting Fluid

Selection of cutting fluid mainly depending on job materials and type of operation performed. However, the following cutting fluid may be used for the respective job materials for most of the operations.

Cast Iron, Brass	: Dry (means air blow)
Hard steel, Bronze	: Mineral lard oil
Copper	: Turpentine
Aluminium	: Kerosene
Soft steel	: Cutting oil or cutting compound

A1.71 : Single Point Cutting Tool With Nomenclature & Tool Angles

Cutting tool nomenclature means systematic naming of the various parts and angles of a single point cutting tool.

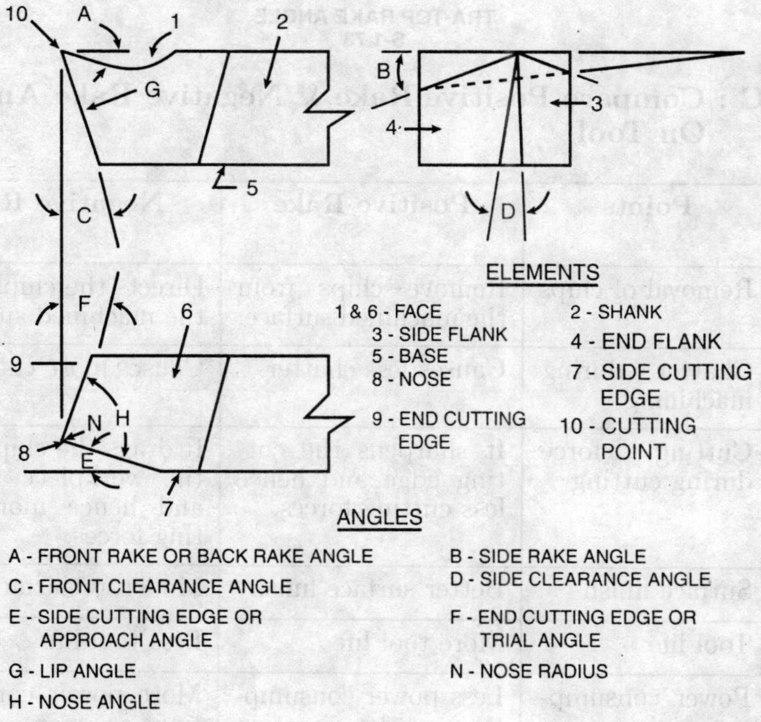

S-1.71(2)

A1.72 : Tool Geometry Or Tool Signature

Systematic namings of different single point cutting tool angles are known as Tool geometry or Tool signature for machining M.S., M.C.S. and alloy steel by H.S.S. cutting tool.

10	-	12	-	5	-	6	-	16	-	15	-	1.2
A	-	B	-	C	-	D	-	E	-	F	-	N R

i.e.,	Front-Rake or Back-Rake (A)	Side-Rake (B)	Front-clearance or end clearance (C)	Side-clearance (D)	Side-cutting edge angle or (90-approach angle) (E)	End-cutting edge (F) or trial angle (F)	Nose radius 1.2 mm (NR)

A1.73 : Top Rake Or Back Rake Or Front Rake Angle

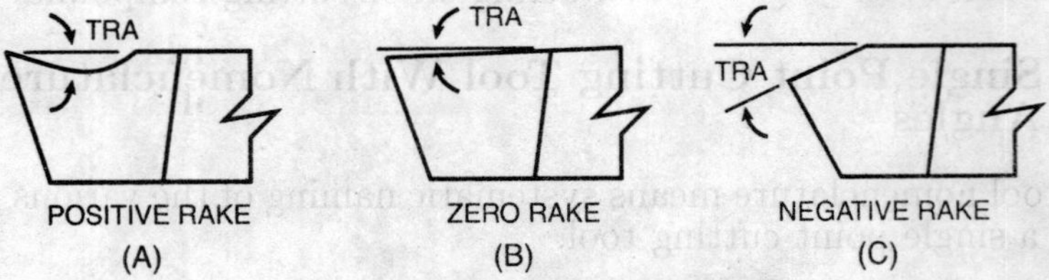

TRA-TOP RAKE ANGLE
S-1.73

A1.73C : Compare Positive Rake & Negative Rake Angles On Tool

Sl. No.	Points	Positive Rake	Negative Rake
1.	Removal of chips	Remove chips from the machined surface.	Directs the chips on to the machined surface.
2.	Chatter during machining	Causes less chatter.	Causes more chatter.
3.	Cutting force during cutting	It sharpens the cutting edge and hence less cutting forces.	It drags the chips from the workpiece surface and hence more cutting forces.
4.	Surface finish	Better surface finish	Moderate surface finish
5.	Tool life	More tool life	Less tool life
6.	Power consumption	Less power consumption as less cutting forces during machining.	More power consumption as more cutting forces during machining.
7.	Suitability in Machine Tools	Specially in Lathe tools also in milling cutter	Specially in reciprocating type of machines like shaping, planning, slotting tools also in milling cutters.
8.	Pressure between Job-Tool	Less	More

Sl. No.	Points	Positive Rake	Negative Rake
9.	Strength of the tool	Cross-section of the tools is not so robust	Robust or strengthen tool is required.
10.	Rigidity of the set-up	Rigid set-up is essential	More rigid set-up is essential.
11.	Rake Angle on Tools	Sketch Sl.73(A)	Sketch S1.73(C)

A1.73Q : Reasons For Using Cemented Carbide Tool In Production And H.S.S. Tool In Tool Room & Maintenance Work

Cemented carbide tool having high cutting speed, high hot hardness, minimum tool wear and can be used as tip which will incur less cost of the tool and also having long tool life, high depth-of-cut can be provided for quick metal removal which are essentials for high production and for all these above reasons it is mostly used in production work.

H.S.S. tool can be used as a bit which can easily be ground for different operations, having better cutting speed, less tool wear, comparatively high hot hardness w.r.t. alloy steel and can easily be machined. So, H.S.S. tool most be suitable in jobbing work or training purpose or tool room works or maintenance work.

Positive rake angle

1. Remove chips from the machined surface.
2. Sharpens the cutting edge and hence less cutting forces.
3. Produces better surface finish.
4. Increases tool life and decreases power consumptions.
5. May be used in Lathe and Milling operations.

Negative rake angle

1. Directs the chips on to the machined surface.
2. Increases pressure between Job-Tool.
3. Causes more chatter.
4. Decreases the rigidity of Machines.
5. Strengthens the tool.
6. Preferred on rigid set-up.
7. Used in Shaping, Planning, Milling, etc.

Zero rake angle

1. Difficult to penetrate into the job.
2. More cutting forces and chatter.
3. Poor surface finish.
4. Rare use, may be used in Thread Cutting and Milling operations,

A1.74 : Side Rake Angle (P.S.R.)

1. Negative side rake makes the tool stronger.
2. Positive side rake gives better cutting action.
3. P.S.R. lowers cutting forces.
4. P.S.R. less power consumption.

A1.75 : Front Clearance Angle & Side Clearance Angle

1. Higher value reduces wear.
2. Higher value reduces strength and rigidity of the tool.
3. Lower value gives better support and suitable for machining tougher metals.

A1.76(A) : End Cutting Edge Angle

1. Reduces the dragging action which will cause chatter.
2. Improves surface finish.
3. Angle may vary 0 to 30°.

A1.76(B) : Approach Angle Or Side Cutting Angle

1. Chip thickness depends on approach angle.
2. Increases tool life by decreasing approach angle.
3. Small approach angle will promote chatter.
4. Approach angle 90° for slender shaft, 70° for general turning, can vary 0 to 90°.

A1.77 : Nose Radius

1. Higher tool life.
2. Better surface finish.
3. Too large value will lead to chatter.
4. Its value varies from 0.4 to 1.5 mm.

A1.78 : Setting Of Cutting Tool (Turning Tool)

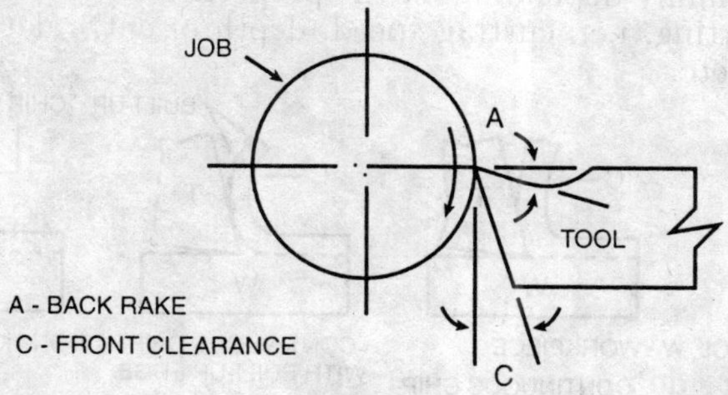

S-1.78

Tool is set, just at right angle to the axis of revolution or above or below the centre-line of the job.

Effects of Tool setting above the centre-line

1. Front clearance angle becomes smaller and will create more friction.
2. Back rake angle becomes larger and chip will separate easily.

Effects of Tool setting below the centre-line

1. Front clearance angle becomes larger and will create less friction.
2. Back rake angle becomes smaller and chips are difficult to separate.

On conclusion

Ideal condition of setting of tool should be on centre-line during operation, which will maintain both front clearance and back rake angle. This setting will also make easy flow of chips and will reduce friction.

A1.81 : Chips

In metal cutting, the cutting tool removes a layer from the workpiece to obtain a desired shape and dimensions of the workpiece. The layer separated (deformed layer) near the cutting edge of the tool is known as chip.

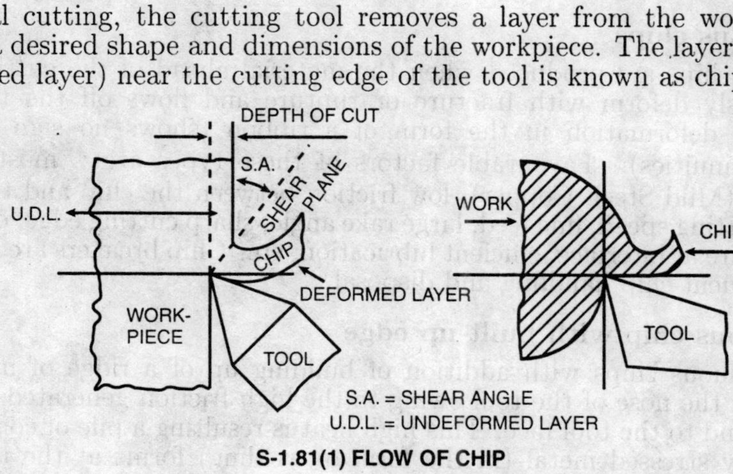

S-1.81(1) FLOW OF CHIP

Types of Chips

Chip types mainly depending on the properties of work material and the conditions of cutting, i.e., cutting speed, depth of cut, rake angle, feed, use of cutting fluid, etc.

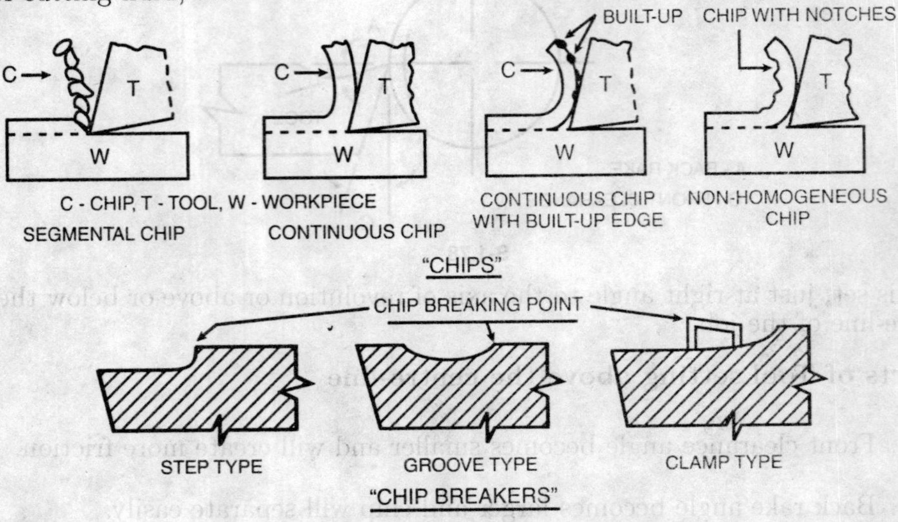

S-1.81(1) DIFFERENT TYPES OF CHIPS & CHIP BREAKERS

(i) Discontinuous or segmental chip

(ii) Continuous or ribbon type chip

(iii) Continuous chip with built-up edge

(iv) Non-homogeneous chip

Discontinuous or Segmental chip

Segmental chips are produced by the actual fracture into small pieces ahead of the cutting edge. The fracture or rupture of the workpiece material takes place during plastic deformation. These chips are produced while machining brittle material (Cast Iron) or while cutting ductile material (Mild Steel) at very low speed, greater depth of cut and small rake angle.

Discontinuous chips are not desirable due to the poor surface finish and excessive tool wear.

Continuous chips

These chips are produced when the material ahead of the cutting edge continuously deform with fracture or rupture and flows off the tool face by plastic deformation in the form of a ribbon (shows no sign of tears or discontinuities). Favourable factors of these types are : most ductile materials (Mild Steel, Copper), low friction between the chip and tool face, normal cutting speed, fine feed, large rake angle, sharp cutting edge, optimum temperature at interface, efficient lubrication, etc. Chip breakers are essential for convenient chip handling and disposal.

Continuous chip with built-up edge

Continuous chips with addition of building up of a ridge of metal are formed on the nose of the tool owing to the high friction generated between the chip and to the tool face. This high heat is resulting a pile of compressed and highly stressed metal (in the form of welding) forms at the tool nose (of cutting tip) and this creates a false cutting edge which breaks a portion of chip into fragments or deposited on the workpiece material and rest are carried away by the chips. Continuous chip with built-up edge will occur while machining ductile materials at low cutting speed, high feed, low rake angle, without the use of cutting fluid, etc. This type is not desirable owing to high power consumptions, poor surface finish and higher tool wear.

Non-homogeneous chip

These chips are produced due to non-uniform strain in workpiece material during the chip formation. Temperature rise on the shear plane during shear deformation causes notches on the free side of the chip, while the side adjoining the tool face is smooth. Non-homogeneous chips are produced while machining typical material in which the yield strength decreases with rise in temperature like steel, titanium alloys, etc., at medium cutting speeds.

A1.82 : Chip Breakers

While turning ductile material producing long unbroken continuous chip will be extremely difficult to handle or to remove from the swarf tray requires breaking into comparatively small pieces for easy handling or to minimize the working hazard. The device used to break the long ribbon is known as chip breaker. It is either a groove ground along the cutting edge or a small step on the top face of the tool. During turning the hardened chip flows on the tool face, strikes the leading edge of the tool causes the chip to break into small pieces. Chip breakers are of groove, step and clamp types.

A1.83 : Tool Wear

Wear means worn on the tool flanks (side and top) due to high temperature and pressure developed at the rubbing surfaces in metal cutting, i.e., tool-workpiece interface at the side flank of the tool and chip-tool interface at the top flank of the tool. Wear on the side flank of the tool is known as flank wear and wear on the rake face (or top flank) is known as crater (or a depression) wear.

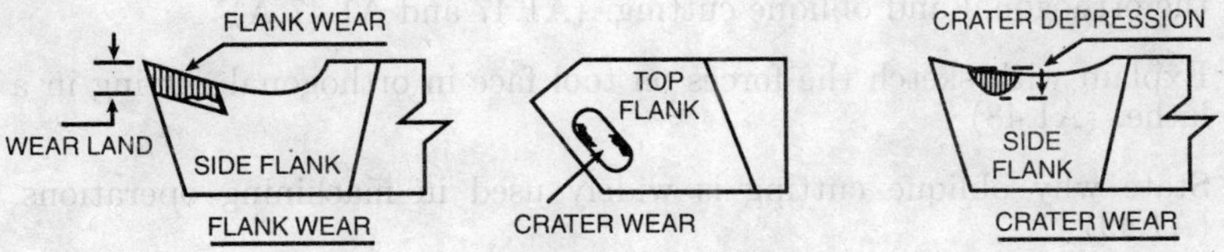

S-1.83 TOOL WEAR

Questions

1. State the requirements of Machine Tool. (A1.11)
2. Define the terms (with example): Machine, Tools, Hand Tools, Machine Tools. (A1.12 to A1.15)
3. Distinguish between Hand Tools and Machine Tools. (A1.16)
4. Write the principles of Metal Cutting. (A1.17)

5. State the purpose of Machine Tools. (A1.18)

6. Write the types of Machine Tools in tabulated form. (A1.21)

7. Explain with example the different types of Machine tools. (A1.22)

8. Compare between the Standard Machine Tools and Special Purpose Machine Tools. (A1.23)

9. Explain in brief the types of drives commonly used in running the Machine Tools. (A1.31 to A1.33)

10. Compare between the Individual drive and Group drive. (A1.34)

11. Define the term Cutting Tool. (A1.41)

12. Classify different types of cutting tools with example and sketch wherever possible. (A1.42)

13. Make a comparative study of different cutting-tool materials such as Carbon tool steel, Medium alloy steel, High speed steel, Cast non-ferrous stellite cobalt, Cemented carbide, Ceramic, Diamond and Abrasives with respect to their composition, heat treatment, cutting speed, normal use, relative cost, hot hardness and other properties. (A1.43)

14. Define and compare with sketch of Right-hand tool and Left-hand tool. (A1.44)

15. Write the properties of different materials mostly used in cutting tool. (A1.45)

16. State the characteristics of cutting-tool materials. (A1.46)

17. Write the methods of metal cutting. Define and compare with sketch the orthogonal and oblique cutting. (A1.47 and A1.47 A)

18. Explain with sketch the forces on tool face in orthogonal cutting in a lathe. (A1.48)

19. State why oblique cutting is widely used in machining operations. (A1.49)

20. State the term tool life. Write the factors on which it depends. (A1.410)

21. How many ways the tool life can be measured? (A1.411)

22. Determine the tool life in terms of volume of metal removal. (A1.412)

23. Write the expression used by the scientist 'Tailor', the relation between cutting speed and tool life. (A1.413)

24. State the term machinability. Write the points on which the machinability influenced or depends. (A1.51)

25. Explain the criteria required to determine the term machinability. (A1.52)

26. How you will determine the machinability rating (or Index)? (A1.53)

27. State the term cutting fluid with sketch. (A1.61)

28. Express the purposes of cutting fluids. (A1.62)

29. Write the ideal properties required for a good cutting fluid. (A1.63)

30. Compare the cutting fluids as coolant, as lubricant, as both coolant and lubricant, as flashing action. (A1.64)

31. How you will choose a suitable cutting fluid in a machining operation? (A1.65)

32. How is the job and the machine to be protected against rust in case of using water as cutting fluid? (A1.66)

33. State the different types of cutting fluids commonly used. (A1.67)

34. Make a comparative chart of different cutting fluids with their main functions. (A1.68)

35. Select a suitable cutting fluid for using different job materials to satisfy most of the operations. (A1.69)

36. Sketch three views of a single point cutting tool and show its nomenclature and angles. (A1.71)

37. What do you mean by tool signature (or geometry) and how it is expressed? (A1.72)

38. Sketch and explain the purposes of positive rake, zero rake and negative rake in case of using as top rake or back rake. (A1.73)

39. Write purposes of different single point tool angles including nose radius. (A1.73 to 1.77)

40. Conclude the effects of the setting of cutting tool point in relation to job axis. (A1.78)

41. Compare V-Belt and Flat Belt. (A1.35)

42. Compare H.S.S. and Cemented Carbide tool materials. (A1.414)

43. Compare Positive Rake and Negative Rake angles. (A1.73C)

44. State different types of Chips, Chip breakers and Tool wear. (A1.81, A1.82, A1.83)

2

LATHE MACHINE

A2.11 : Machine Shop

The machine shop is a shop where materials are cut to the required shape and size according to the necessity for the use of civilization or ready to sell in the market. It is a floor of assembly of several machine tools like Lathe, Planning, Shaping, Milling, Drilling, Power Saw, etc.

A2.12 : Relative Motion Of Job & Tool

In any metal cutting, two relative motions are needed between job and tool. These relative motions are produced by combination of translatory, reciprocating, and rotary movement either in the tool or in the job or in both. This relative motions vary from machine to machine.

Relative Motion for Various Standard Machine Tools

Motion of the Job	Motion of Tool	Name of the Standard Machine Tools
Rotary	Forward translation	Turning in Lathe
Fixed	Rotary as well as translation for feeding	Drilling
Intermittent translation for feeding	Reciprocating	Shaping
Reciprocating	Intermittent translation for feeding	Planing
Translatory	Rotary	Milling
Forward translation	Rotary	Boring

A2.21 : Lathe

The lathe is a most versatile machine tool among all standard machine tools. It is also known as father of all standard machine tools. Functions of Lathe is to remove metal in the form of chip(s) from a piece of work by mounting the same rigidly on a machine spindle and revolving at required speed and the cutting tool is fed against the work either longitudinally or cross-wise to make the work to the required shape and size.

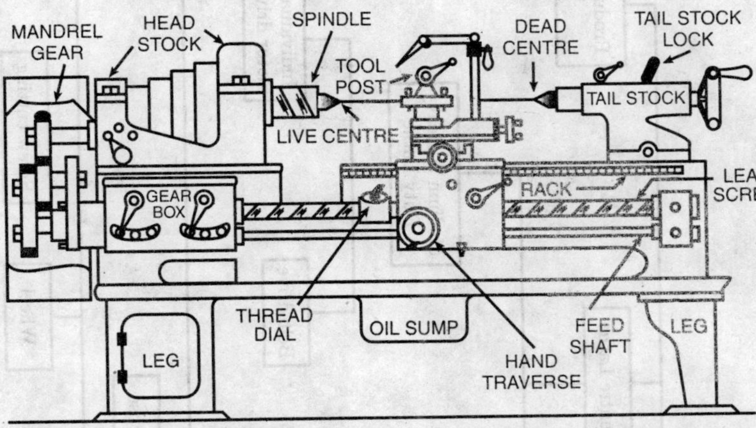

ENGINE LATHE OR CENTRE LATHE (CONE PULLEY DRIVE)
S-2.21(A)

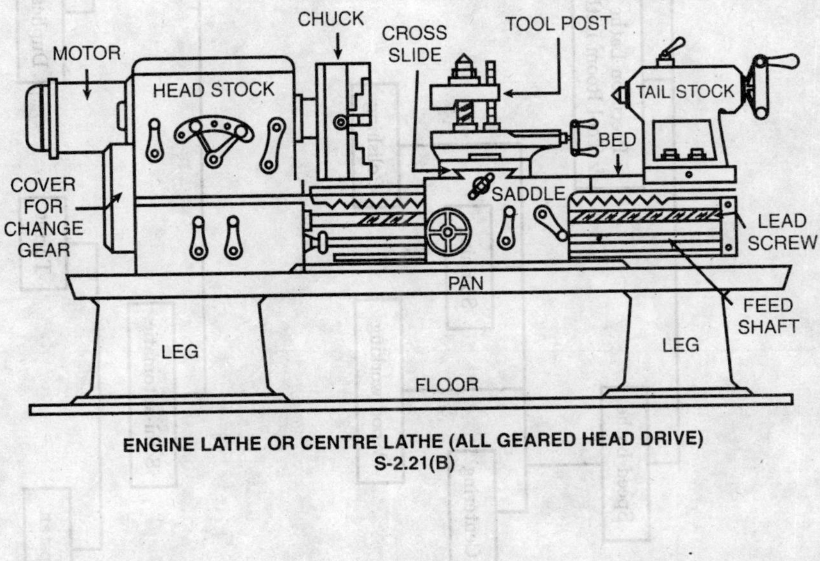

ENGINE LATHE OR CENTRE LATHE (ALL GEARED HEAD DRIVE)
S-2.21(B)

A2.22 : Types Of Lathe Machine

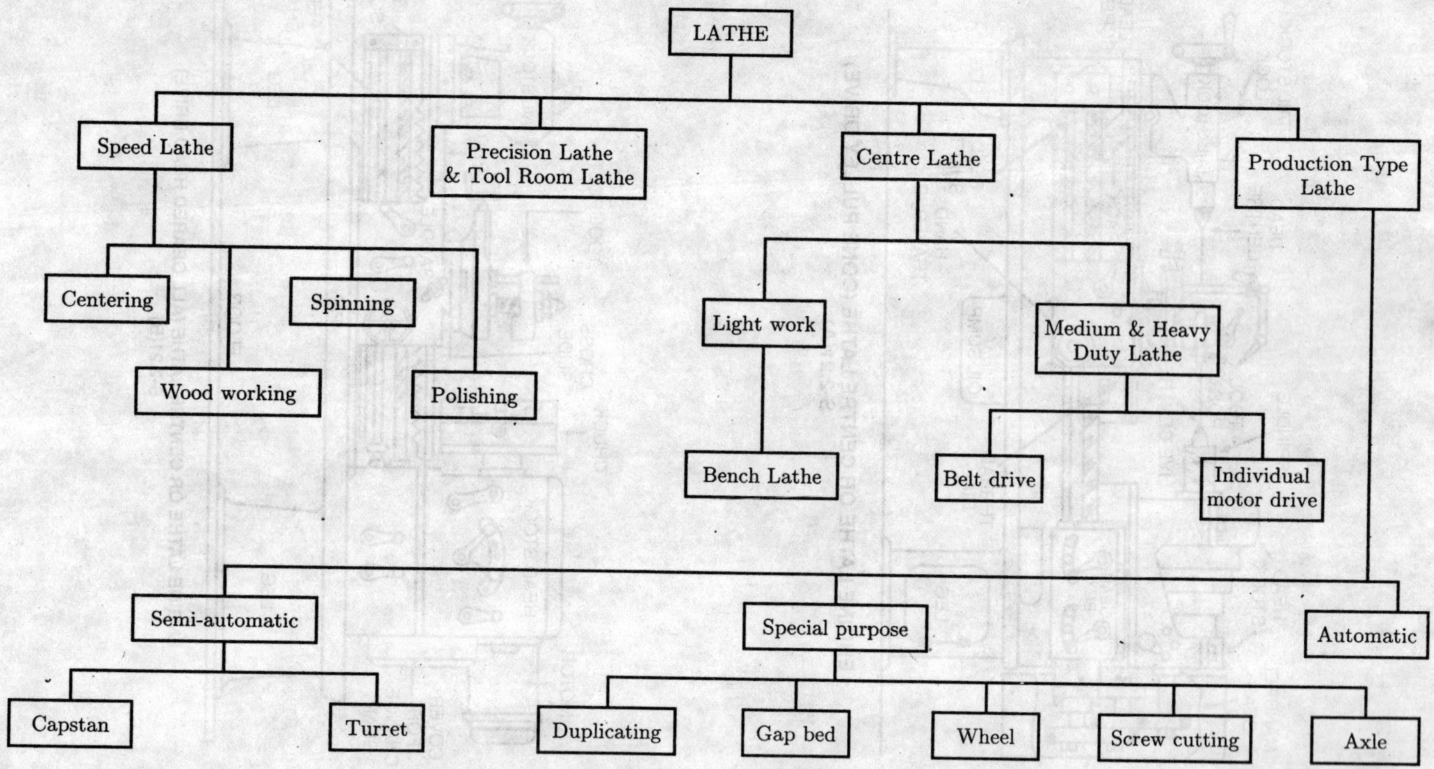

A2.23 : Classification Of Lathe Machines

Lathe machine is classified with respect to the following:

(i) Speed of the machine—

(a) *High speed :* These types are lighter in construction, minimum cutting force, hand feeding and requires high speed in operations, like wood working, centering, polishing, spinning.

(b) *Moderate speed :* These types are heavier in constructions, both power and hand feeding, more cutting forces and hence moderate speed in operations like turning, facing, grooving, drilling, etc.

(c) *Low speed :* These types are heavier in constructions both power and hand feeding, more cutting forces but at low speed in operations like thread cutting, knurling, etc.

(ii) Versatile of operations—

(a) *Light work:* Bench Lathe, mounted on Bench, hand and power feeding, suitable for small precision jobs, Swing diameter up to 300 mm.

(b) *Moderate work :*

(i) With the arrangement of lead Screw in Engine Lathe or Centre Lathe—This type of lathe is the most versatile, can perform most of the operations, both hand and power feeding is available, wide range of speed and feed variables to suit various operations like turning, spinning, knurling, thread cutting, etc.

(ii) Without the arrangement of Lead Screw in Plain Turning Lathe—This type similar to engine lathe but having no arrangement of lead screw and hence screw cutting is not possible. For medium sized lathe swing diameter from 300 to 600 mm and for heavy duty lathe swing diameter over 600 mm.

(c) *Precision work :* Tool room lathe similar to engine lathe but with higher accuracy, more rigidity of the machine having high range of speed, feed and different system of screw-cutting arrangement, with arrangements of fixing most of the attachments, suitable for the production of small tools, dies, gauges, fixtures, similar precision products.

(iii) Axis of rotation of the workpiece—

(a) *Horizontal :* Most of the lathes are horizontal like Centre lathe, Bench lathe, Tool room lathe, etc.

(b) *Vertical :* Vertical turning and boring mills, some jobs are difficult to mount on face plate or chuck and to rotate about horizontal axis, can be fixed/mounted on table/face plate in horizontal plane and rotate about vertical axis for turning, boring, etc., operations.

(iv) Number of spindles in a machine—

(a) *Single-spindle lathe :* Most of the lathes we know having one spindle per machine.

(b) *Multi-spindle lathe :* Generally, suitable for production work having two or more spindles per machine.

(v) Product work—

(a) *Specific product oriented :* Automatic lathe, all the operations of a particular product can be performed by the machine itself without the participations of the operator, i.e., raw material to be fed in one side and finished product will come out at the other end of the automatic lathe machine without the assistance of the operator.

(b) *Specific operation oriented/special purpose lathes:*

(i) Wheel/facing lathe.

(ii) Duplicating/copying lathe.

(iii) T-lathe.

These types of lathes are suitable for performing definite types of machining work. Such as machining or facing of railway wheels and axles, duplicating operations, turning crankshafts, cams, machining of rotor for jet engines, etc.

(c) *Semi-automatic :* Capstan and turret lathes—

Operations in sequence of a product can be performed in production work. These types are the development of engine lathes instead of tailstock, there is a hexagonal turret to hold six or more tools in a operational sequence, there is another square tool post on carriage to hold four tools and a rear tool post for parting off tool in bar type works. Setting of eleven or more tools at a time according to the operational sequence of a particular product helps to minimize the machining time.

Capstan lathes are suitable for bar work. And turret lathes are mostly used for chucking work, heavier products fast production jobs. This machine needs the help of operator partly, not fully like engine lathe or nor absence of operator as in the case of an automatic lathe and in between engine lathe and automatic lathe and so it is named as semi-automatic lathe.

(d) *Automatic lathe :* High speed heavy duty machine suitable for mass production work. Once tools are set according to the job, all the operations relating tool changing, job changing, feed, speed, etc., will be done by the machine itself. Operator's attention is only required for bar feeding, disposal of finished job and frequent checking of job, etc.

A2.31 : Specification Of Centre Lathe (Size of Lathe)

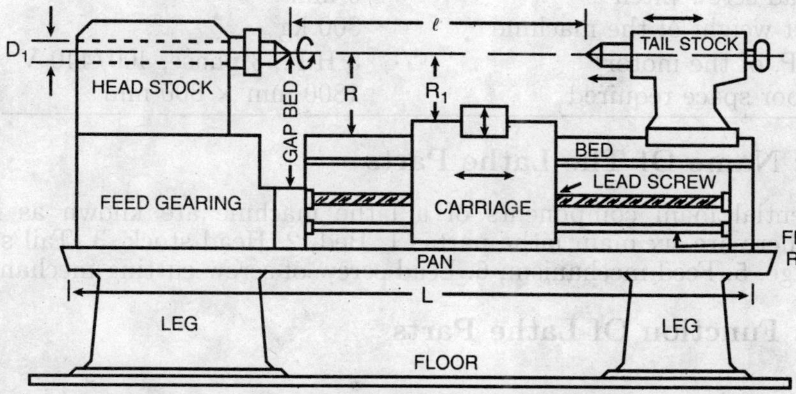

S-2.31 ENGINE OR CENTRE LATHE

1. Name of the machine M/c—Engine Lathe or Centre Lathe.
2. Swing diameter—over bed in mm (2R).
3. Swing diameter over carriage in mm ($2R_1$).
4. Length of the bed in mtr. (L).
5. Diameter of the hole in head stock spindle for bar work in mm (D_1).
6. Minimum length of the job that can be accommodated in between centres in mtr. (l).
7. Pitch of the lead screw.
8. Horse power of the machine.
9. Speed range and no. of speeds of the H.S. spindle.
10. Feed range and no. of feeds of the tool.
11. Floor space required in sq. mtr.
12. Weight of the machine in tonne.

A2.32 : Specifications Of A Particular Lathe Machine

Specifications of an Indian particular Self-centering Screw Cutting (SSSC) Lathe are given below :

Model	···	SSSC Lathe no. 3
Swing diameter	···	300 mm
Bed length	···	1830 mm
Centre to centre distance	···	1070 mm
Swing diameter over carriage	···	180 mm
Spindle hole for bar work	···	35 mm
Spindle speed range	···	30 to 450 r.p.m.
Longitudinal feed	···	0.048 to 1.35 mm
Cross-feed	···	0.0078 to 0.223 mm
Lead screw pitch	···	6 mm
Net weight of the machine		800 kg
H.P. of the motor	···	5 H.P., 3 phase, 400/440 V
Floor space required		1800 mm × 800 mm

A2.41 : Name Of The Lathe Parts

The essential main components of a lathe machine are known as lathe parts. There are six main lathe parts: 1. Bed, 2. Head stock, 3. Tail stock, 4. Carriage, 5. Feed mechanism, 6. Lead screw of screw-cutting mechanism.

A2.42 : Function Of Lathe Parts

Bed

(a) It is sufficiently rigid, good damping qualities to absorb vibration and prevent deflection produced by cutting forces.

(b) It supports head stock and receives both statical, impact and dynamical loading.

(c) It supports both tail stock and carriage and facilitates its movement when necessary.

(d) Bed is made of chilled cast iron.

Head stock

(a) It holds the job on its spindle nose having external screw threads and internally Morse Taper for holding Lathe centre, rotated at different speed either by cone pulley drive or by all geared drive.

(b) It transmits power from the head stock spindle to the feed rod and lead screw for automatic feeding and thread-cutting mechanism.

(c) There is a hole extends through the spindle for long bar work.

Tail stock

(a) It supports the other end of the long job when machined in between centres and minimizes its sagging.

(b) It holds a tool for performing different operations such as drilling, boring, reaming, tapping, etc.

(c) Small amount of taper for a long job may be done by offsetting the tail stock.

Carriage

(a) It holds, moves and controls the cutting tool.

(b) It gives rigid supports to the tool during operations.

(c) It transfers power from feed rod to cutting tool through apron mechanism for longitudinal cross-feeding.

(d) It facilitates the thread cutting with the help of lead screw, half-nut, etc.

Feed mechanism

1. It transmits power from feed rod to cutting tool through the apron gears.

2. It moves the carriage or cross-slide for different operations such as turning, boring, facing, grooves, parting-off, etc., except thread cutting.

3. It gives the tool to move longitudinal, cross and angular by hand feeding.

4. It also facilitates the automatic longitudinal and cross-feeding.

Thread-cutting Mechanism

1. There is a long threaded screwed shaft named as lead screw used for thread-cutting operation.

2. This lead screw is also,used to provide motion to the cutting tool for turning, boring, reaming, etc., when the machine is not equipped with a feed rod.

A2.50 : Transmission Systems In Lathe Machines

A2.51 : Head Stock Spindle Drive

The Head stock spindle is to run at different revolution according to the requirements of job w.r.t. operations. Different revolutions of the spindle may be achieved either by stepped pulley drive or by gear drive.

A2.52 : Stepped Pulley Drive

Belt and pulley are used to transmit power and motion by friction from one shaft to other.

Flat belt and V-belt are generally used as driving belts. V-belts are having better arc of contact and generally used for small centre distances. 1% speed may reduce during driving by flat belt.

In 4-stepped pulleys drive, 4 different speeds of the Head stock spindle can be attained. Spindle speeds are varied in Arithmetic Progression.

Let Driver shaft rotates at N r.p.m. Pulley diameters are D_1, D_2, D_3 and D_4 respectively. Driven shaft having 4 pulleys of same diameter as having in driving shaft but the diameters are in increasing order. We know, the speed is inversely proportional with diameters.

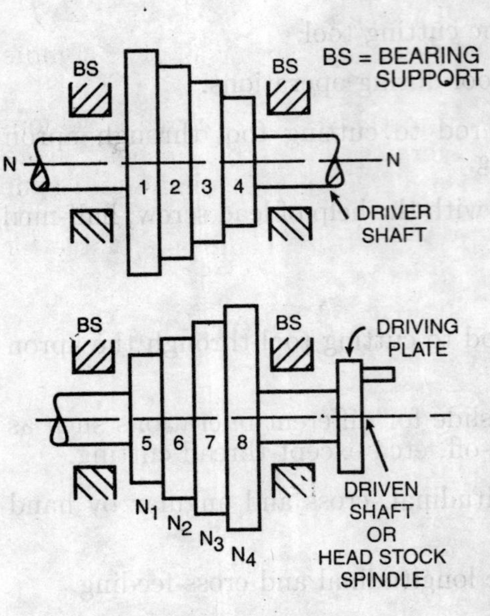

4-STEP PULLEY DRIVE
S-2.52

Therefore, $\dfrac{N}{N_1} = \dfrac{D_5}{D_1}$ $\cdots(1)$

when belt on pulleys 1 & 5.

$\dfrac{N}{N_2} = \dfrac{D_6}{D_2}$ $\cdots(2)$

when belt on pulleys 2 & 6.

$\dfrac{N}{N_3} = \dfrac{D_7}{D_3}$ $\cdots(3)$

when belt on pulleys 3 & 7.

$\dfrac{N}{N_4} = \dfrac{D_8}{D_4}$ $\cdots(4)$

when belt on pulleys 4 & 8.

where, N_1, N_2, N_3 and N_4 are the speeds of driven shaft. We know also, $D_1 = D_8$, $D_2 = D_7$, $D_3 = D_6$ and $D_4 = D_5$. N is the speed of driving shaft.

A2.53 : Stepped Pulley Drives With Back Gearing

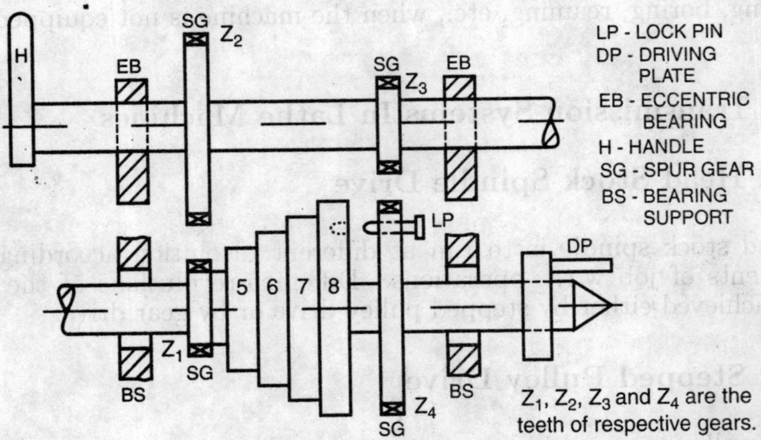

BACK GEAR (S-2.53)

The numbers of speeds will be doubled by using a back gearing.

$$\text{Overall transmission ratios } (TR) = \frac{Z_2}{Z_1} \times \frac{Z_4}{Z_3}$$

The revolutions N_1, N_2, N_3 and N_4, can have on the head stock spindle by disengaging the back gearing and engaging lock-pin with pulley 8.

The following additional numbers of revolutions can be obtained by engaging the back gearing with the help of a handle and eccentric gearing and disengaging the lock-pin from pulley 8. Power transmits from the stepped pulley to spur gears 1-2-3-4 to the H.S. spindle.

$$N_5 = \frac{N_1}{TR},\ N_6 = \frac{N_2}{TR},\ N_7 = \frac{N_3}{TR} \text{ and } N_8 = \frac{N_4}{TR} \quad \because \left(TR = \frac{Z_2}{Z_1} \times \frac{Z_4}{Z_3}\right).$$

The revolutions N_5, N_6, N_7 and N_8 are the slow speeds of H.S. spindle when using back gearing.

This system of drive is simple and less costly. But this drive is seldom used due to the time consuming and dangerous shifting of the belt. Note that, the stepped pulley and spur gear I are free on the spindle and spur gear, 4 is keyed to the head stock spindle. The spur gear 2 and 3 are keyed on the back gear shaft. In back gearing, reduces speed and is suitable for thread cutting, knurling, tapping operations, etc.

A2.54 : All Geared Head Drive

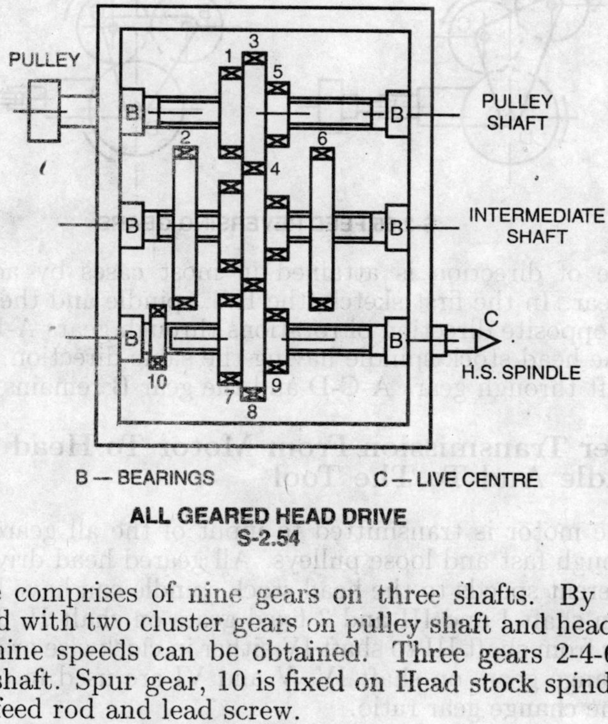

ALL GEARED HEAD DRIVE
S-2.54

This drive comprises of nine gears on three shafts. By operating two levers attached with two cluster gears on pulley shaft and head stock spindle respectively, nine speeds can be obtained. Three gears 2-4-6 are fixed on intermediate shaft. Spur gear, 10 is fixed on Head stock spindle to transmit power to the feed rod and lead screw.

The gear combination for nine different speeds are

$$\frac{T_1}{T_2} \times \frac{T_2}{T_7};\quad \frac{T_1}{T_2} \times \frac{T_4}{T_8};\quad \frac{T_1}{T_2} \times \frac{T_6}{T_9};$$

$$\frac{T_3}{T_4} \times \frac{T_2}{T_7};\quad \frac{T_3}{T_4} \times \frac{T_4}{T_8};\quad \frac{T_3}{T_4} \times \frac{T_6}{T_9};$$

$$\frac{T_5}{T_6} \times \frac{T_2}{T_7};\quad \frac{T_5}{T_6} \times \frac{T_4}{T_8};\quad \frac{T_5}{T_6} \times \frac{T_6}{T_9};$$

This type of drive is mostly suitable in heavy duty machines and where smooth running is required. Spindle speeds are varied in geometric progression.

A2.55 : Feed Reversing Gear (Tumbler Gear)

The tool movement from right to left or from left to right depends on the direction of rotation of feed rod and lead screw. Tumbler gear or feed reversing gear can performed of this task.

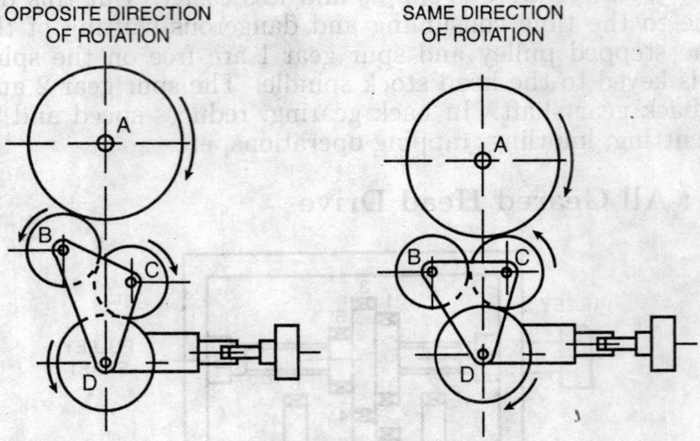

S-2.55 FEED REVERSING GEARS

The change of direction is attained in most cases by addition of an intermediate gear. In the first sketch, the H.S. spindle and the output stud-shaft have the opposite direction of rotations through gears A-B-C-D. In the second case, the head stock spindle having the same direction of rotation to the output shaft through gears A-C-D and the gear B remains idle.

A2.56 : Power Transmission From Motor To Head Stock Spindle And To The Tool

Power from the motor is transmitted to input of the all geared head drive by belting through fast and loose pulleys. All geared head drive consist of 9 gears for 9 different speeds to the head stock spindle can have by sliding two cluster gears on shaft I and III and 3 fixed gears on shaft II. Tumbler gears transmit power from shaft III to shaft IV either in clockwise or anti-clockwise directions. Change gears on shaft IV, V and VI are used in thread cutting according to the change gear ratio.

Gear box-1 having cluster gears on shaft VI and three fixed gears on shaft VII, can transmit. Three different speeds to the output shaft VII. There are ten cone gears fixed on shaft VIII receive power from shaft VII through two gears in a bracket moves on it. The shaft VIII, thus have 30 different speeds to transmit to feed rod by gearing and to lead screw by clutching. The front portion of the carriage is known as apron contains gearing arrangement by which rotary motion of feed rod and lead screw may be converted to translatory motion of tool in longitudinal and transverse direction during turning, facing and thread-cutting operations, etc.

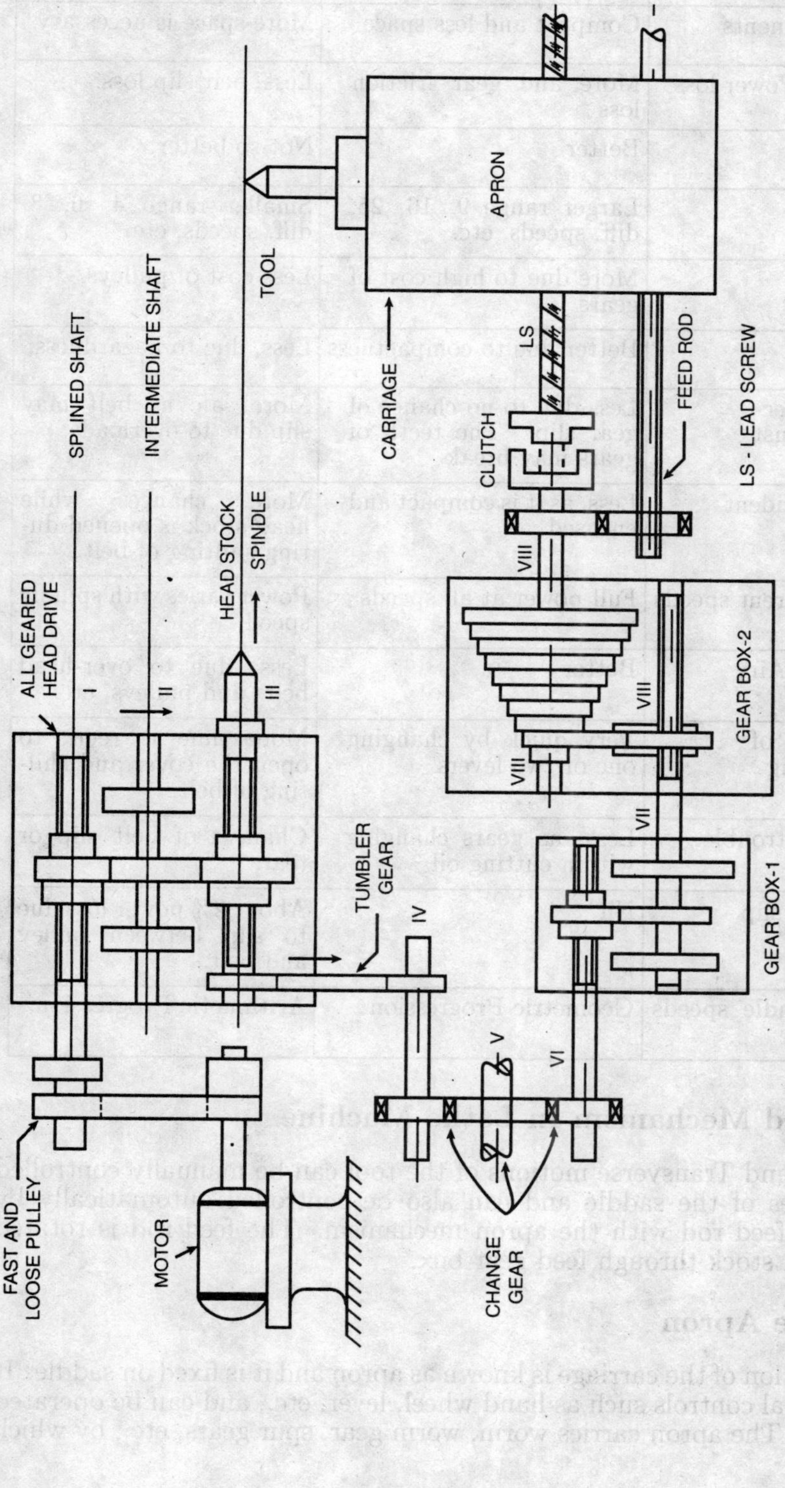

S-2.56 POWER TRANSMISSION FROM MOTOR TO HEAD STOCK SPINDLE

A2.57 : Comparison Between All Geared Head Drive & Cone Pulley Drive

Points	All geared head drive	Cone pulley drive
Space requirements	Compact and less space	More space is necessary.
Friction and Power loss	More and gear friction loss	Less, belt slip loss.
Appearance	Better	Not so better.
Speed range	Larger range 9, 16, 25, diff. speeds, etc.	Smaller range 4, 6, 8, diff. speeds, etc.
Initial cost	More due to high cost of gears	Less cost of pulleys.
Cleanliness	Better due to compactness	Less, due to hazardness.
Machine protection against overloads	Less due to no chance of gear slip. The teeth of gears may break	More safe as belt may slip due to overloads.
Chance of accident	Less, as it is compact and encased	More, chances while head stock is opened during shifting of belt.
Power of different speeds	Full power at all speeds	Power varies with spindle speed.
Lighting and Airy	Better	Less, due to over-head belts and pulleys, etc.
Attainment of speed changing	Very quick by changing one or two levers	More time is reqd. to open the cover and shifting of belt.
Maintenance trouble.	Less, as gears changing within cutting oil	Chances of belt slip or tear.
Power lost in slip	Nil	About 2% power loss due to slip between pulley and belt.
Different spindle speeds on	Geometric Progression	Arithmetic Progression.

A2.58 : Feed Mechanism In Lathe Machine

Longitudinal and Transverse motions of the tool can be manually controlled by the handles of the saddle and can also be controlled automatically by engaging the feed rod with the apron mechanism. The feed rod is rotated from the head stock through feed gear box.

A2.59 : The Apron

The front portion of the carriage is known as apron and it is fixed on saddle. It contains several controls such as hand wheel, lever, etc., and can be operated from outside. The apron carries worm, worm gear, spur gears, etc., by which rotary movement of feed rod can be converted to longitudinal and transverse motion of the tool.

A2.510 : The Lead Screw

It is necessary for thread-cutting operation. Lead screw is driven by head stock-spindle through feed gear box and moves carriage in a longitudinal direction by the help of two half nuts which close around the lead screw when engaged by a hand lever from outside.

A2.511 : Full-Proof Arrangement (Or Self-Locking Mechanism)

It is a safety arrangement against simultaneous engagement (due to carelessness) of half nuts and the longitudinal or cross feed. The arrangement also ensures that when the half nuts are engaged with the lead screw the worm moves side ways disconnecting the feed motion.

A2.512 : Apron Mechanism In Lathe Machine

For longitudinal feed

Automatic	:	Handle Position at L
		Worm-Worm Wheel—SG-1-2-3-4
Hand	:	Hand Wheel—(LFHW)-SG-5-3-4
		When Handle Position at 'N'

For transverse feed

Automatic	:	Handle Position at T
		Worm-Worm Wheel—SG 1-2-6-Cross feed screw
Hand	:	Handle position at 'N'
		Hand wheel (CFHW)—SG-7-6-Cross feed screw

LFHW	=	Longitudinal Feed Hand Wheel, W_1
CFHW	=	Cross Feed Hand Wheel, W_2
SG	=	Spur Gear

Apron Mechanism

Working

The sketch in next page shows the front elevation of apron mechanism. Worm on feed rod is keyed, rotated with the feed rod and slides on it. Worm gear (WH) and gear-1, gears 3 & 4, hand wheel W_1 & gear 5, hand wheel W_2 & gear 7, and gear 6 & cross feed screw all are mounted on the respective shafts.

Longitudinal automatic feed of the tool

The worm 'W' drives worm gear 'WH', connected with gear 1. The Handle 'H' rests on 'L'. The movable gear 2 engages gear 3 and gear 4, is mounted on the same shaft, engages with the rack teeth, which is fixed under the bed and hence having the automatic longitudinal feed of the tool.

Transversal automatic feed of the tool

The handle 'H' is to be set at 'T'. The movable gear 2, then meshes with the gear 6 of cross feed screw and hence having automatic transversal feed of the tool.

Longitudinal hand feed

Handle 'H' is to be set at neutral position 'N'. By rotating the hand wheel W_1, motion of the hand wheel is transferred to gear 4 via gears 5 & 3 and hence the longitudinal feed by hand power.

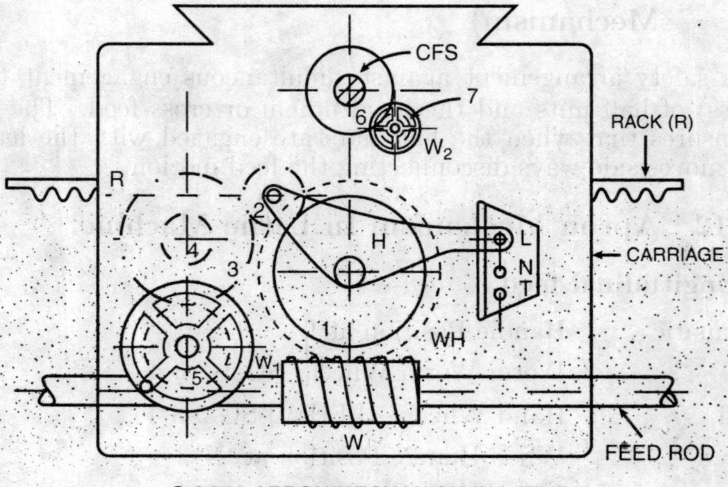

S-2.512 APRON MECHANISM IN LATHE

L - POSITION OF HANDLE FOR LONGITUDINAL FEED
N - NEUTRAL POSITION OF HANDLE, CFS - CROSS FEED SCREW
T - POSITION OF HANDLE FOR TRANSVERSE FEED
WH - WORM WHEEL, W- WORM, H - HANDLE (LEVER), R - RACK

Transversal hand feed

Keeping the handle 'H' at the neutral position 'N', the cross feed screw may be rotated by the hand wheel W_2 via gears 7 & 6 and hence having the cross feed by hand power.

A2.61 : Parts, Accessories & Attachment

Parts of a machine tool are the essential components, without which that particular machine cannot be identified. Example, without tail stock, engine lathe cannot be detected and only it is similar to wheel lathe.

Accessories are required for holding and supporting the working job or the cutting tool conveniently. Example: Accessories are centres, driving plate (catch plate) lathe dog or carriers, chucks, steady rest, follower rest, mandrel, face plate, etc.

Attachments are intended to increase the accuracy and efficiency of the machine for specific purpose and also named as additional equipments. These are taper turning attachment, dividing head attachment, feed stop, grinding attachment, milling attachment, etc.

A2.62 : Lathe Centres

Elements of a centre

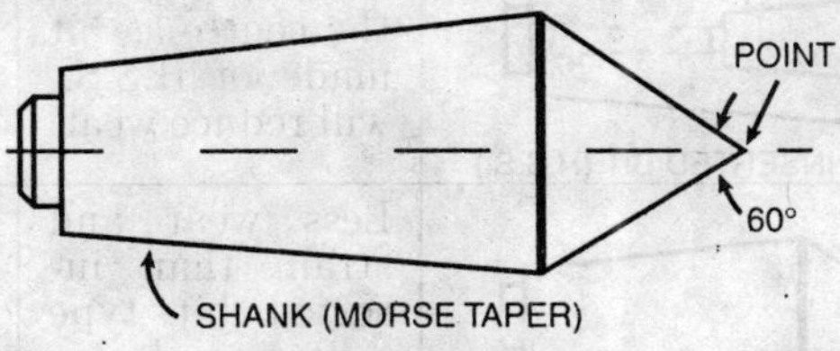

S-2.62 LATHE CENTRE

Included angle for

General work = 60°, Hard work = 75°, Morse taper no. 4 mostly used.

Lathe centres are also named as Live Centre and Dead Centre. If the centre is mounted on head stock and rotated with the head stock is known as Live Centre. The centre which is mounted on Tailstock spindle end and remains stationary during the work rotation is known as Dead Centre.

These are used as supporting the ends of a long job. It takes the thrust during metal cutting and entire load of the job.

It should be made of hard material (generally alloy steel) to resist wear due to friction and deflection due to the weight of the job. Oil, talc, graphite is used at the point to reduce friction in between job and centre.

Dead centre is hardened to reduce friction at job-centre interface.

A2.63 : Types Of Centre With Their Applications

Types	Sketch	Application	Remark
Ordinary	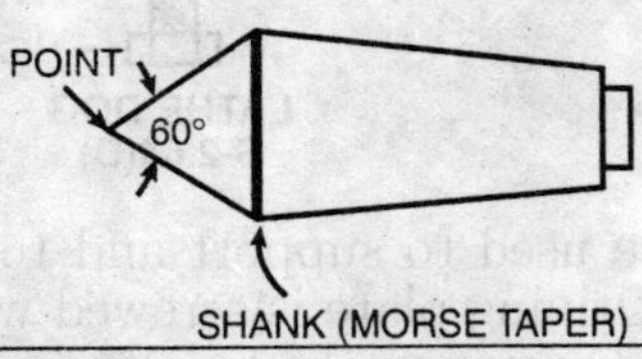	For general purpose	Included angle for general work = 60°. Hard work and heavy loads = 75°
Half Centre	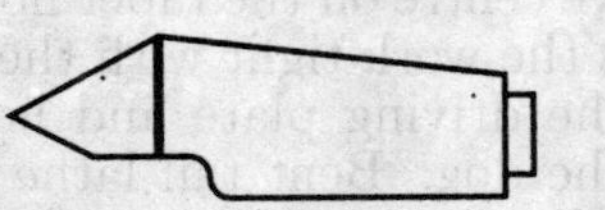	It is used at the tail stock end to allow more space for the tool when turning down the end.	See page–59 for application Fig. S-2.75
Ball Centre	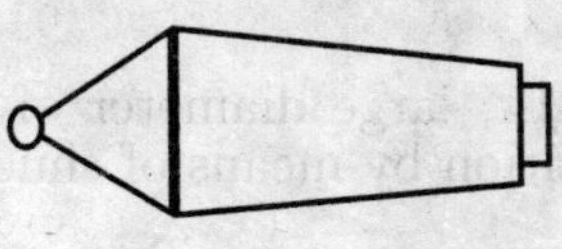	It minimises wear and strain during taper turning by tail stock set-over method.	For application see page–67 Fig. S-2.722
Pipe Centre	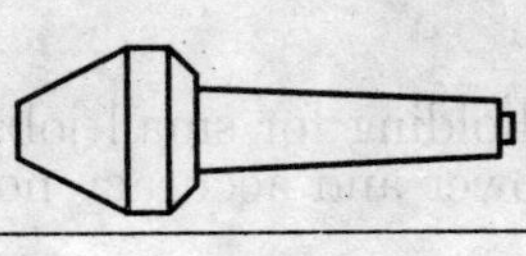	It is used for holding the hollow jobs such as shell, pipes, etc.	

Types	Sketch	Application	Remark
Inserted Bit Type Centre	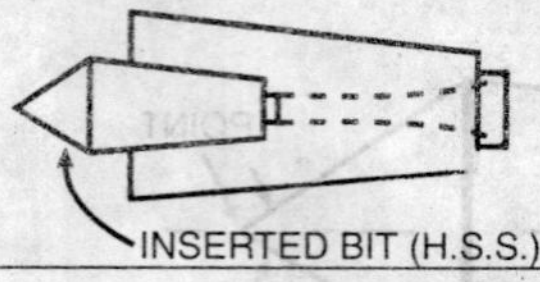	It minimises the running cost of the centre as bit made of H.S.S., will reduce wear.	Material—High speed steel.
Tipped Centre	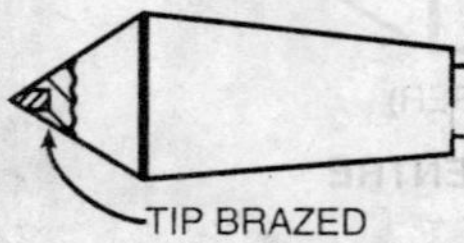	Less wear and strain than inserted bit type centre and expensive may be used in production work.	Tip may be of cemented carbide
Frictionless Centre	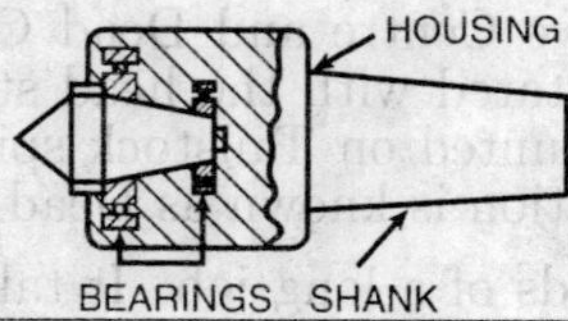	It is used in tail stock end for supporting heavy job rotating at high speed.	Job rotating with H.S.S. centre and shank, H.C.S. Very much expensive

A2.64 : Driving Plate (Catch Plate) & Lathe Dog (Carrier)

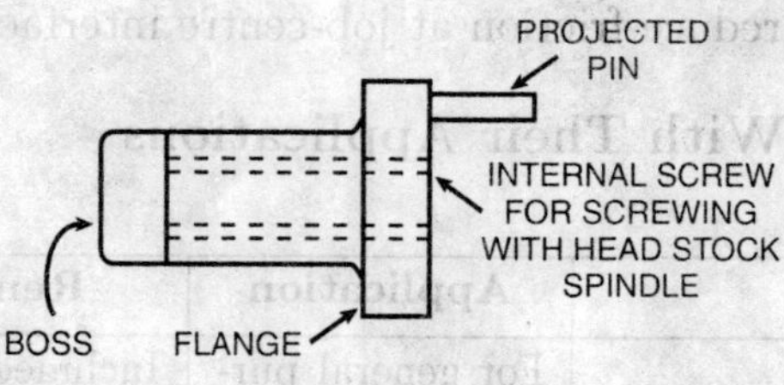

DRIVING PLATE
S-2.64(A)

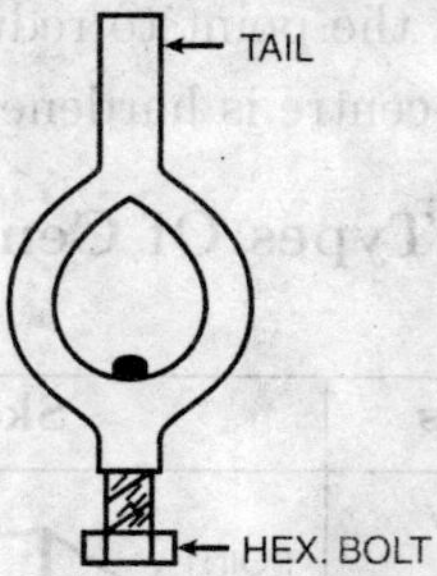

LATHE DOG
S-2.64(B)

Driving plate and Lathe dog are used to support and to drive the job, when it is held in between centres. Driving plate is screwed with the nose of the head stock spindle. There is a live centre on the taper nose of H.S. spindle to support the job. Lathe dog holds the work tight with the help of a bolt(s) is connected with straight tail of,the driving plate and hence job is being driven by the driving plate and lathe dog. Bent tail lathe dog is more safe than the straight tail type. [For application, see page-67 and Fig. S-2.722].

A2.65 : Chuck

Large number of jobs of shorter length, large diameter of irregular shapes are held rigidly and in convenient position by means of chucks. It is screwed with the nose of the machine spindle.

Types of chuck :

Two-jaw

It is a very simple chuck used for holding for small jobs and small stock work, polishing, etc., where holding power and accuracy not so important.

Three-jaw concentric or Universal or self-centering chuck

This type of chuck is suitable for regular shaped job (such as round hexagonal), etc., and particularly in capstan lathe work.

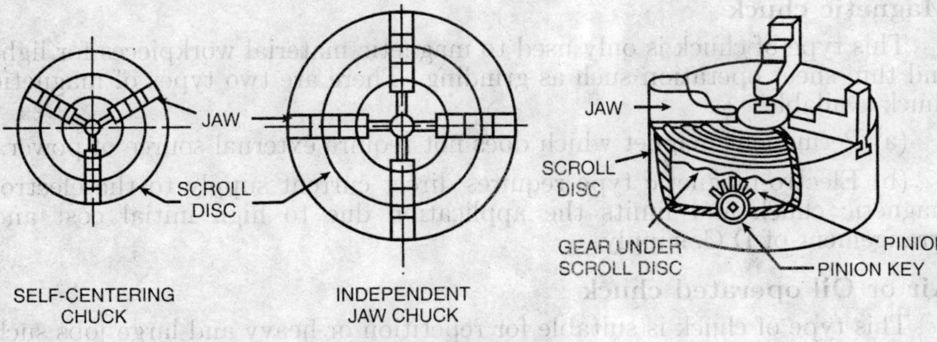

S-2.65(A) UNIVERSAL CHUCK

Three jaws engage with a scroll gear operated by a hand key inserted into the square recess of the pinion heads, which moves the three jaws either forward or backward concentrically when any one of the jaws is moved.

Four-jaw independent

This type of chuck is most adaptable for heavy and irregular shaped work such as casting, etc. According to the name, this chuck having four jaws may be operated independently. It takes more time to hold the job but rigidity and gripping is more than the universal type.

Combination chuck

This type of chuck having both the arrangements of self centering and independent jaw individually is used for both the types, i.e., regular or irregular jobs. The jaws may be operated independently by separate screws and all the jaws can also move at a time by the scroll disc.

Collet chuck

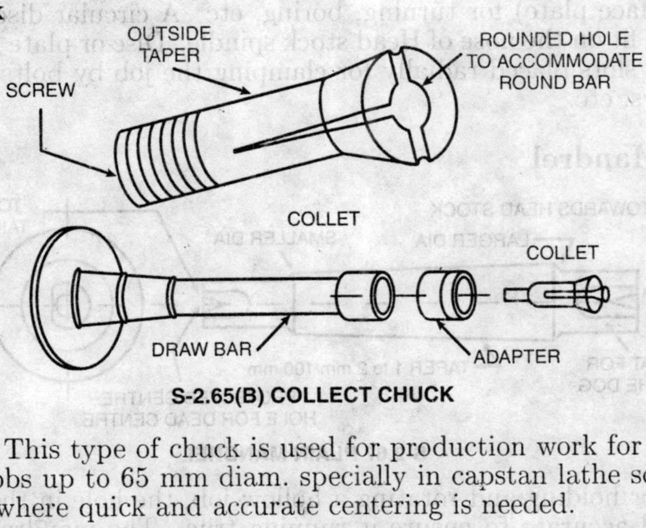

S-2.65(B) COLLECT CHUCK

Functions : This type of chuck is used for production work for holding the bar type of jobs up to 65 mm diam. specially in capstan lathe sometimes in turret lathe, where quick and accurate centering is needed.

The inside bore section of the collet is depending on the cross-section of job to be held and different collets are necessary for different cross-sections of the bar.

The outside surface of collet is tapered to fit in a taper hole of chuck. The tail end of collet is threaded and it meshes with a key. When the key is turned from outside, the collet is drawn in and the split tapered end is to be pushed inward and the job is accurately and securely held in the chuck.

Magnetic chuck

This type of chuck is only used to magnetic material workpieces for light and thin sheet operation such as grinding. There are two types of magnetic chuck available.

(a) Permanent magnet which does not require external source of power.

(b) Electro-magnetic type requires direct current supply to the electro-magnetic chuck and limits the application due to high initial cost and arrangement of D.C. supply.

Air or Oil operated chuck

This type of chuck is suitable for repetition or heavy and large jobs such as roll drum for paper mills, etc., for its fast and effective gripping capacity. For operating this type of chuck requires additional plant for the supply of either compressed air or oil. The jaws are opened or closed by the supply of air or oil pressure and can be operated either by hand or automatically.

A2.66 : Face Plate

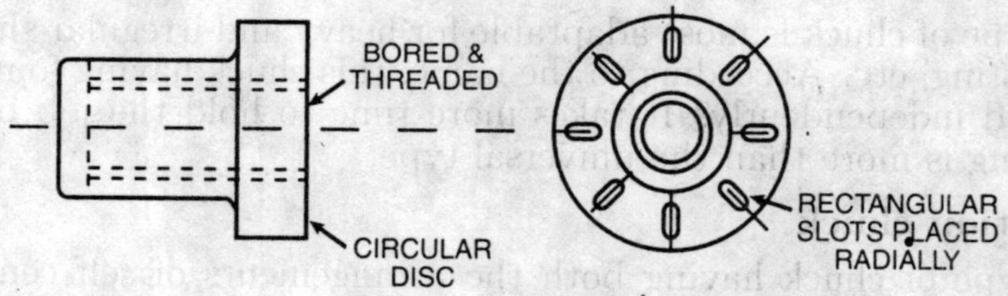

S-2.66 FACE PLATE

A large variety of jobs either from casting or forging, is often difficult to machine either on centres or on chuck, may be fastened on circular disc (known as face plate) for turning, boring, etc. A circular disc is bored and threaded to fit on the nose of Head stock spindle. Disc or plate having several rectangular slots placed radially for clamping the job by bolts, angle plates, clamps, jaws, etc.

A2.67 : Mandrel

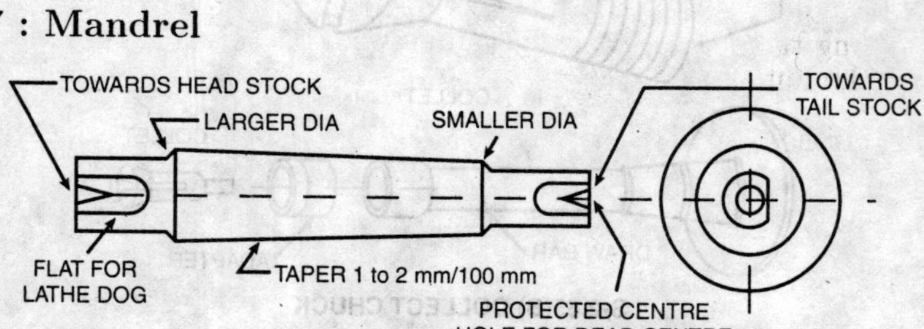

S-2.67 PLAIN MANDREL

It is used for holding and rotating a hollow job, the hole in the job must be cleaned and accurate to ensure it running true. The mandrel is placed on centres and being rotated by driving plate and lathe dog, and it drives the job by friction.

Material : Generally high carbon steel, may also be mild steel case hardened.

A2.68 : Types Of Mandrel

1. Plain Mandrel

This type is most common in the shop floor and wide application in production jobs.

2. Step Mandrel

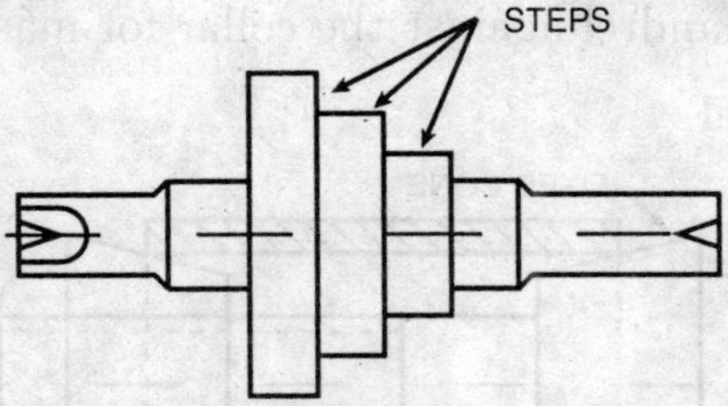

S-2.68(2) STEP MANDREL

In Step mandrel, one mandrel may be used to drive different jobs of different sizes of holes. Suitable for thinner and odd size job such as turning collars, washer, etc.

3. Gang Mandrel

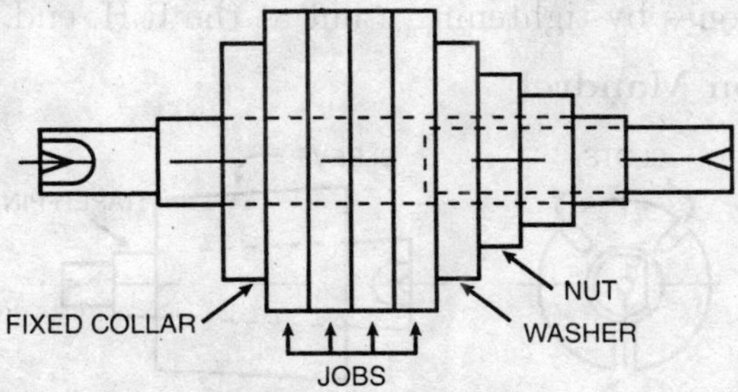

S-2.68(3) GANG MANDREL

This type of mandrel is mostly suitable for production work. A number of identical hollow jobs can be machined at a time which will reduce setting and machining time. Jobs are fixed in between fixed collar and a washer, by tightening the nut.

4. Collar Mandrel

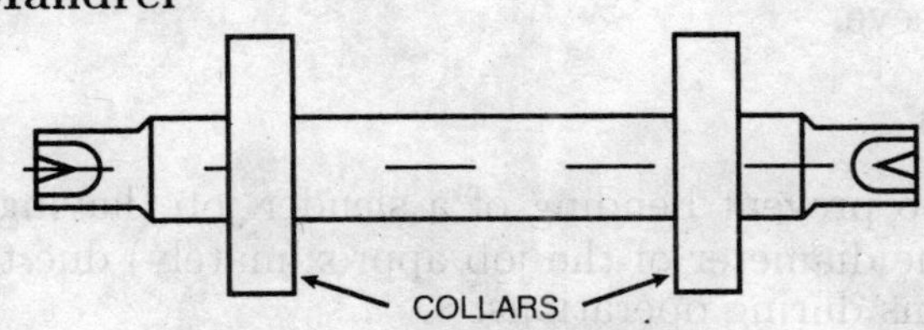

S-2.68(4) COLLAR MANDREL

This type of mandrel is suitable to drive the jobs having larger diameter holes more than 100 mm. Collars are used to hold larger dia. hole jobs instead of using a solid one, which will reduce the weight and cost of the mandrel.

5. Screwed Mandrel

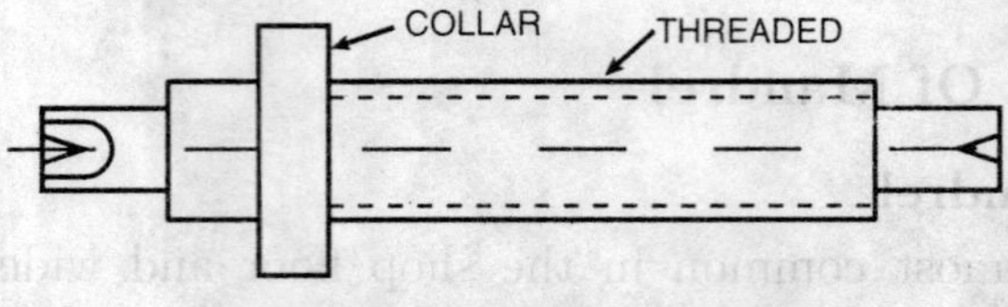

S-2.68(5) SCREWED MANDREL

This mandrel is used for driving the jobs having internal threads. The jobs are screwed on mandrel against the collar for machining operations.

6. Cone Mandrel

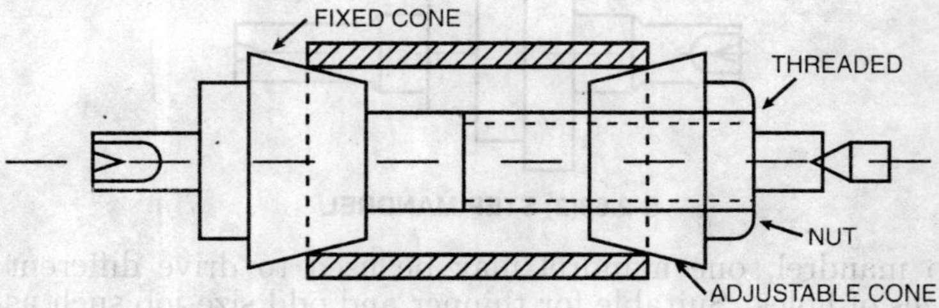

S-2.68(6) CONE MANDREL

It is suitable for the jobs having different hole diameters is placed tightly in between two cones by tightening a nut at the R.H. end.

7. Expansion Mandrel

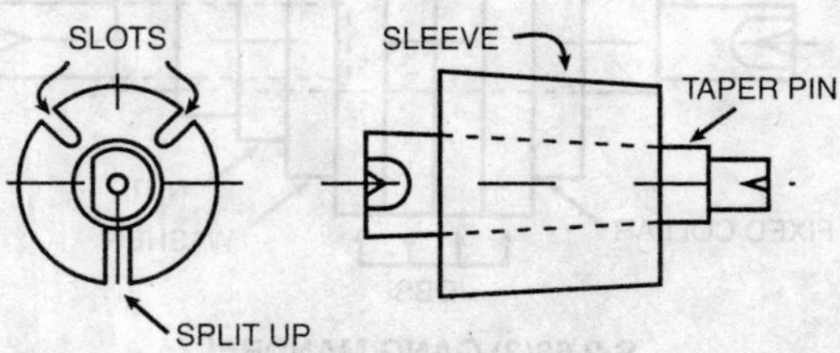

S-2.68(7) EXPANSION MANDREL

Jobs of different diameter hole variation up to 2 mm in dimension can be placed on sleeve and the job can make tight by expanding the sleeve through its two slots and one split-up portion after pushing the pin into the sleeve. Original pin can also be used as a mandrel. There is no taper on the outer surface of the sleeve.

A2.69 : Rests

Rests are used to prevent bending of a slender job (having its length more than 12 times the diameter of the job approximately) due to its own weight and the vibrations during operations.

There are two types of rests generally used in Lathe machine:

1. Steady rest or Fixed steady.

2. Follower rest or Travelling steady.

Steady rest

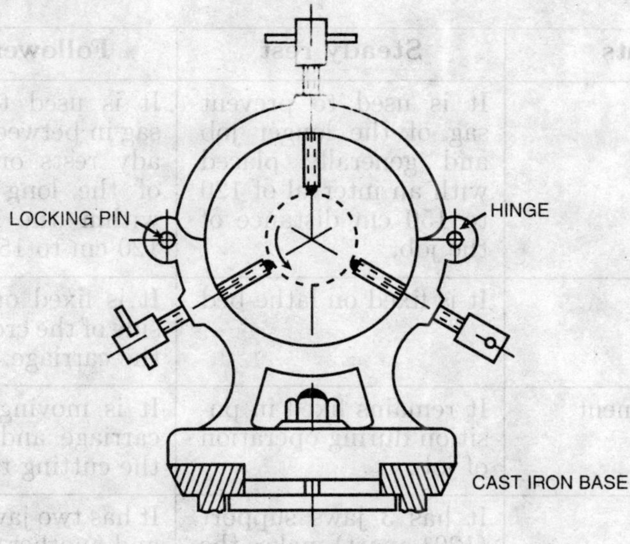

S-2.69(A) STEADY REST

It is fixed on bed during operation. It prevents sag or deflection with an interval of 120 to 150 cm of the job length. There are three jaws 120° apart, two at the lower half and other one at the top of the upper half, can move radially and to run the slender job within three jaws concentrically. One or more rest may be used at a time depending on the job length. It can also be used as tail stock support for drilling, boring, etc., operations.

Follower rest

It is bolted on the rear side of the carriage cross slide. It prevents bending at the job-tool contact point, i.e., within the length of two steady rests or supports. It will follow with the tool and giving the continuous support to the job. There are two jaws and other one is the tool, the three help to rotate the job in a concentric path. Only one follower rest is necessary in particular machine during operations.

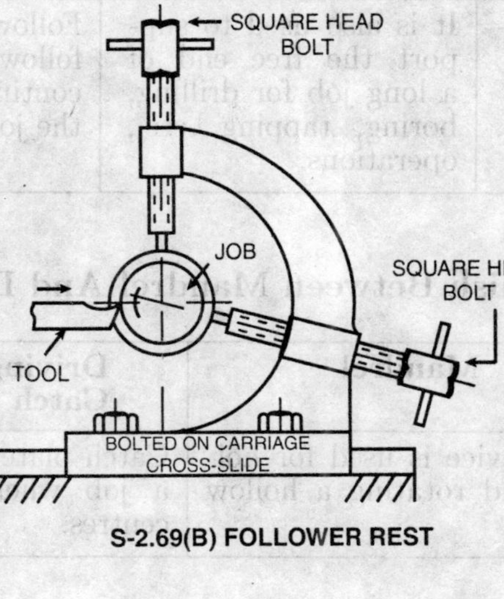

S-2.69(B) FOLLOWER REST

A2.610 : Comparative Study Of Steady Rest & Follower Rest

Points	Steady rest	Follower rest
Function	It is used to prevent sag of the longer job and generally placed with an interval of 120 to 150 cm distance of the job.	It is used to prevent sag in between two steady rests or supports of the long job, i.e., within the length of 120 cm to 150 cm.
Location	It is fixed on lathe-bed.	It is fixed on the rear side of the cross-slide of the carriage.
Rest movement	It remains fixed in position during operation of job.	It is moving with the carriage and following the cutting tool.
No. of jaws	It has 3 jaws support (120° apart) makes the job to move concentric.	It has two jaws support and another is the tool itself makes the job to move concentric.
No. of rests needed at a time	One or more rest(s) is used at a time depending on the length of the job.	Only one rest is used at a time irrespective of the job length.
Rest can be used as tail stock support.	Yes	Not possible.
Machining of the job at the position of the rest.	Not possible, until it changes its position.	It will not disturb the supporting portion as it follows with the tool.
Machining of full length of the job.	Not possible, when it is used.	Possible, without disturbing the setting.
Suitable for operations.	It is also used to support the free end of a long job for drilling, boring, tapping, etc., operations.	Follower rest always follows the tool giving continuous support to the job.

A2.611 : Distinguish Between Mandrel And Driving Plate

Points	Mandrel	Driving plate or Catch plate
Purpose	This device is used for holding and rotating a hollow job.	Catch plate is used to drive a job when held between centres.

Points	Mandrel	Driving plate or Catch plate
Mounting	The job revolves with the mandrel which is mounted between two centres by catch plate and carriers.	Catch plate is either bolted or screwed to the nose of the head stock spindle.
Type of drive	Mandrel drives the job by friction.	Catch plate imparts a positive drive between lathe spindle and job.
Independent rotary movement	Mandrel itself cannot revolve without the help of catch plate and carriers.	Catch plate can revolve the job without taking the help of mandrel.

A2.612 : Compare Between Face Plate & Driving Plate (Or Catch Plate)

Points	Face plate	Driving plate
Location	It consists of a circular disc bored out and threaded to fit the nose on the head stock spindle.	It is either bolted or screwed to the nose of the head stock spindle and having a projected pin which drives the carrier in contact with the pin.
Job holding time	More	Less
Weight	More weight	Less weight.
Purpose	It is used for holding the job which is difficult to mount in between centres or by chuck.	It is used to drive a job when it is held in between centres.
Type of jobs suitabie for holding	It is generally used for the jobs having irregular shapes, bigger diameter, less thickness, etc.	It is mostly used for slender jobs having comparatively smaller diameter.
Parts need to hold the job	Face plate along with angle plates, nuts and bolts, counter weights, etc., are necessary for holding the job.	Catch plate along with carrier or lathe dog is necessary for holding the job.
Job rests on	Job placed on the face of a face plate and bolted with angle plates, bolts and nuts, etc.	Job, supported on pin attached to catch plate and driven by lathe dog or carrier.

A2.613 : Compare Universal & Independent Jaw Chuck

Points	Universal chuck	Independent jaw chuck
Application	It is generally used for concentric or regular jobs.	It is mainly used for irregular jobs. Regular jobs can also be fixed.
Job fixing time	Job fixing is quick and automatically being done.	Job fixing takes more time for correct centering.
Jaw movement	If one key is operated, all the jaws will automatically rotate.	Jaws are independently moved according to the centering desired.
Gripping ability	Gripping capacity is not so good in comparing with independent jaw chuck.	Better gripping capacity as jaws can be tightened independently.

A2.71 : Job Holding Principles

The jobs can be held on head stock spindle by the following methods :

1. Holding and driving the small jobs by chuck.
2. Holding the longer job in between centres and driving by catch plate and carrier.
3. Holding the hollow job on mandrel and the mandrel is driven by the catch plate and carrier.
4. Holding the small thickness and odd job on face plate, clamping by angle plates, nuts and bolts, etc.
5. Holding the production job on fixture.

A2.72 : Principles Of Operation In Lathe Machine

In lathe operation, the head stock spindle holding the job and rotates with the same speed of the spindle, the carriage holding the tool on tool post moves translatory path either longitudinally or cross-wise to give the desired feed, the two relative motions of job-tool make certain operation and removes metal in the form of chips, giving the desired shape of the job.

A 2.73 : Lathe Operations

Lathe is such a versatile machine tool which can produce another lathe, involve several operations. It is useless to mention all the operations, though it will be mentioned some of the important operations : 1. Centering, 2. Facing, 3. Straight turning, 4. Shoulder turning, 5. Taper turning, 6. Eccentric turning, 7. Chamfering, 8. Thread cutting, 9. Knurling, 10. Spinning, 11. Drilling, 12. Boring, 13. Reaming, 14. Tapping, 15. Parting-off.

A2.74 : Centering Operation In Lathe

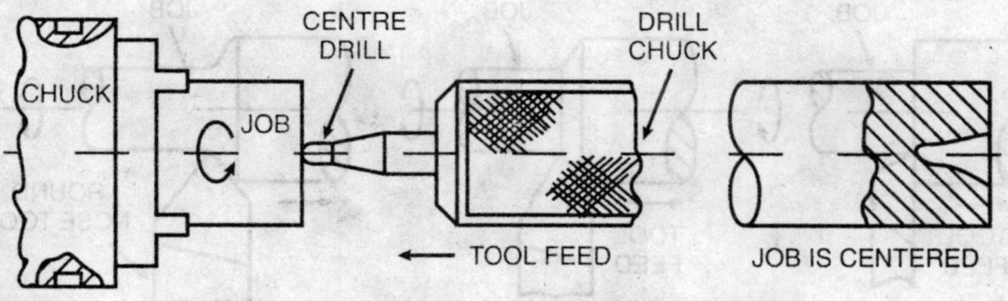

S-2.74 CENTERING OPERATION IN LATHE

It is the operation of producing conical holes on job face(s) to make the bearing support of the lathe centre(s) when the job is held in between centres.

A2.75 : Facing Operation In Lathe

It is the operation of making the ends of a job to produce a flat surface square with axis of operation or to make the desired length of the job. In this operation job rotates and the tool advances in perpendicular direction to the axis of the job rotation.

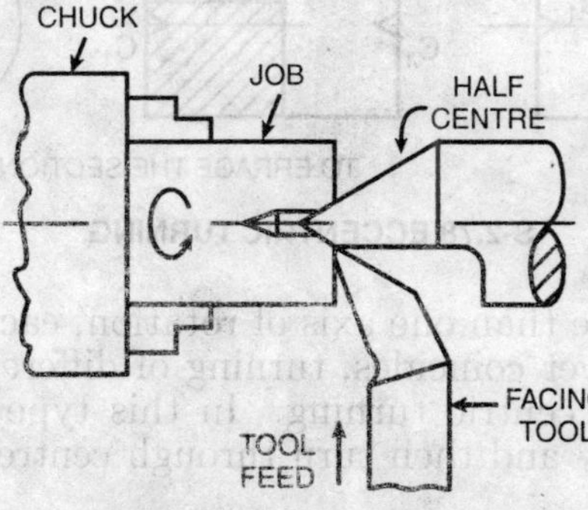

S-2.75 FACING OPERATION

A2.76 : Straight Turning

It is the operation of producing a cylindrical surface of a job by removing excess material. In this operation, the job rotates and the tool is fed longitudinally by giving the desired depth of cut.

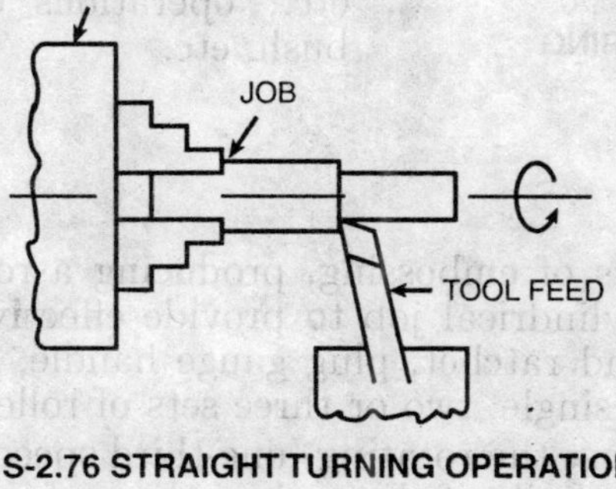

S-2.76 STRAIGHT TURNING OPERATION

A2.77 : Shoulder Turning

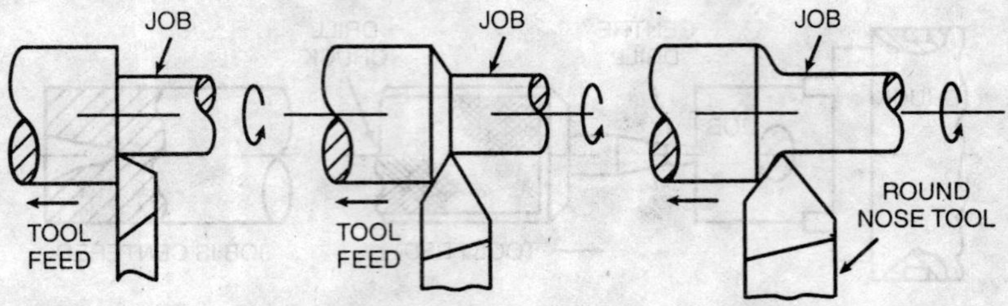

S-2.77

If a job is turned with different diameters, the step from one diameter to the other so formed, the surface is known as shoulder turning. There are several types of shoulder turning such as Square, Bevelled, Radius, etc.

A2.78 : Eccentric Turning

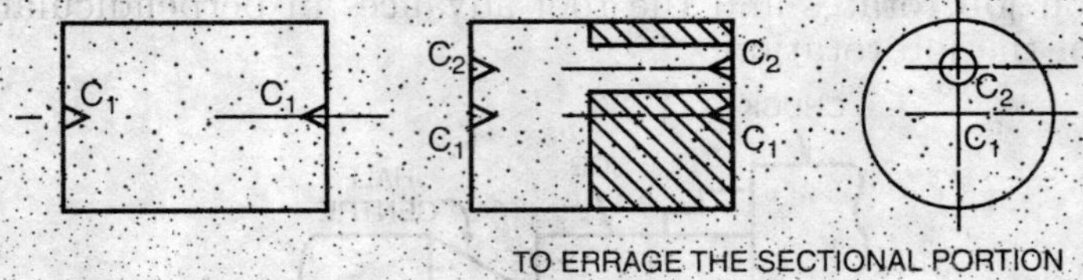

S-2.78 ECCENTRIC TURNING

When a job having more than one axis of rotation, each axis may be parallel with each other but never coincides, turning of different cylindrical surfaces of a job is known as eccentric turning. In this type of turning, first turn through centres $C_1 - C_1$ and then turn through centres $C_2 - C_2$.

A2.79 : Chamfering

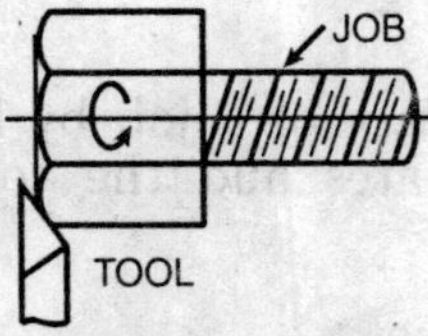

S-2.79 CHAMFERING

It is the operation of bevelling the end of a job to remove burrs, to look better, to make passage of the nut into the bolt. This operation is generally done after thread cutting, knurling, rough turning, etc., operations on ends of bolt/nut, bush, etc.

A2.710 : Knurling

Knurling is the process of embossing, producing a roughened surface on a smooth surface of a cylindrical job to provide effective gripping surface on micrometer thimble and ratchet, plug gauge handle, etc., to be rotated by hand. Knurling tools (single, two or three sets of rollers) are held rigidly on tool post, pressed against the rotating (one-third speed of turning) surface of a job, leaving exact facsimile of the tool on the surface of the job.

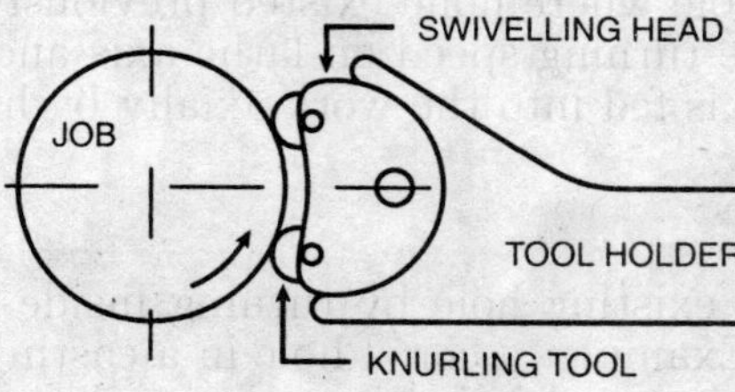

S-2.710(A) KNURLING

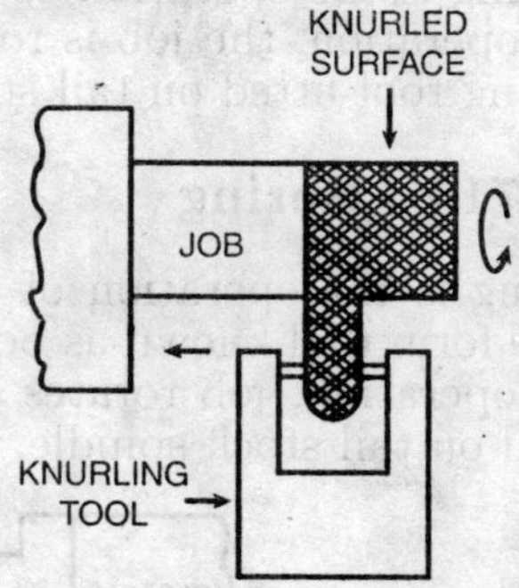

S-2.710(B)

A2.711 : Spinning

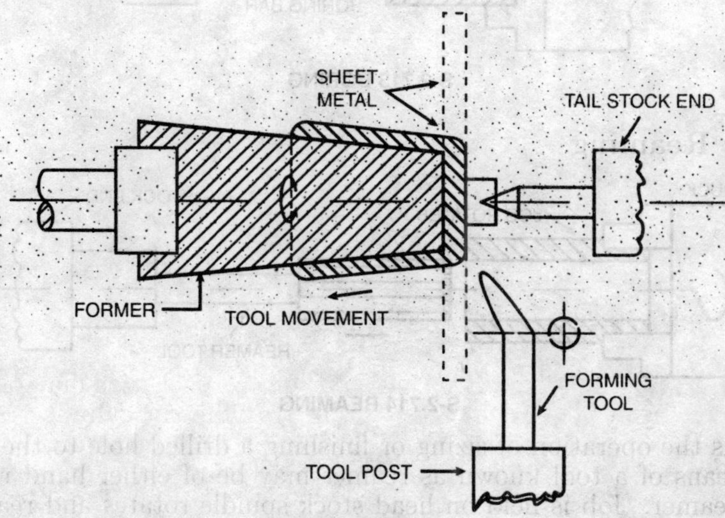

S-2.711 SPINNING

In this operation, job of this sheet metal is held in between the former and the tail stock centre rotates at a high speed with the former. The long round nose forming tool rigidly fixed on special tool post presses the job on the periphery of the former. Thus, the job is taken the exact shape of the former and the operation is known as spinning. It is a chipless machining process.

A2.712 : Drilling

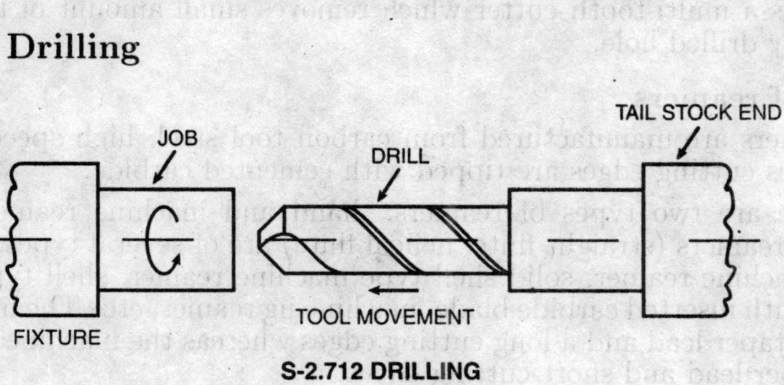

S-2.712 DRILLING

Drilling is the operation of making a hole where none existed previously. In this operation, the job is rotated at the turning speed on lathe axis and the drilling tool fitted on tail stock spindle is fed into the work axially by hand.

A2.713 : Boring

Boring is the operation of enlarging a existing hole by turning inside with some form tool known as boring tool. Example: a cored hole in a casting. In this operation, job rotates and the boring tool is fed into the work axially, fitted on tail stock spindle.

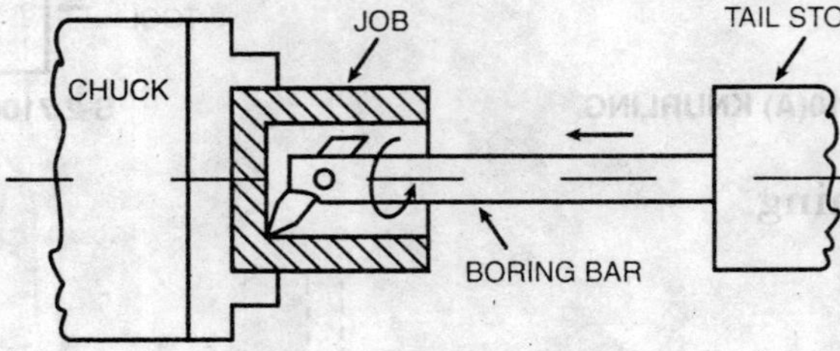

S-2.713 BORING

A2.714 : Reaming

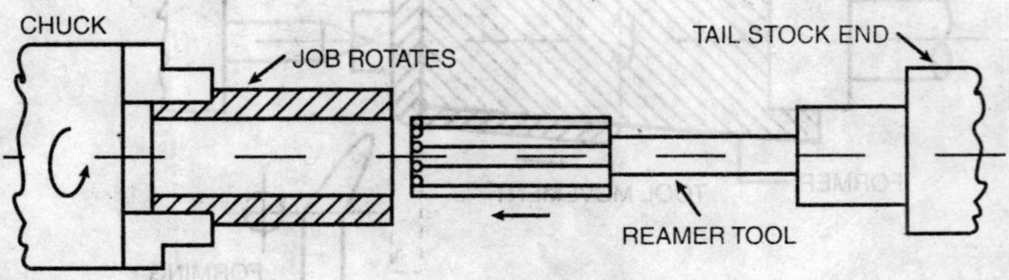

S-2.714 REAMING

Reaming is the operation of sizing or finishing a drilled hole to the required size by means of a tool known as reamer may be of either hand reamer or machine reamer. Job is held on head stock spindle rotates and reamer tool fitted on tail stock spindle is fed into the job axially.

A2.714.1 : Reaming Operation

Drilling hole is seldom accurate in size (slightly over-size). These drilling holes are quite satisfactory for holes in which bolts and rivets are placed. When high dimensional accuracy and surface finish are required, the hole is first drilled under-size and then the hole is finished by reaming operation. Reamer is a multi-tooth cutter which removes small amount of material on previously drilled hole.

Types of reamers

Reamers are manufactured from carbon tool steel, high speed steel and sometimes cutting edges are tipped with cemented carbide.

There are two types of reamers: hand and machine reamers. Again machine reamers (straight flute, helical flute) are of several types, e.g., taper shank machine reamer, solid shell type machine reamer, shell type machine reamer with inserted carbide blade, machine jig reamer, etc. The hand reamer has long taper lead and a long cutting edges whereas the machine reamer has short taper lead and short cutting edges.

Reamers have 6 to 16 flutes with usually irregular spacing along the circle for better quality of reamed holes.

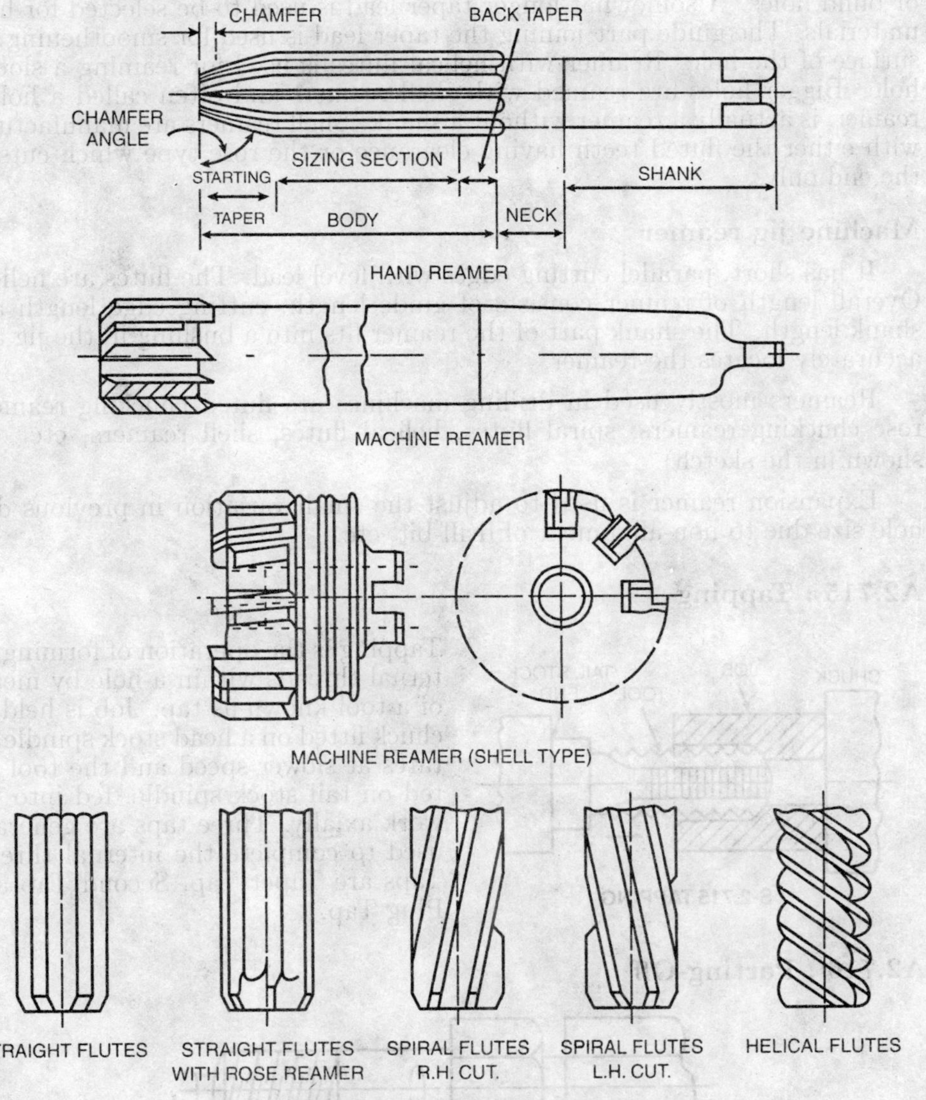

S-2.714.1 DIFFERENT TYPES OF REAMERS

Straight shank hand reamer

Elements are body, neck and shank. Body consists of chamfer, starting taper, sizing section and back taper. Starting taper comprises a cutting section with a 90 level, sizing section is to size the hole. Cutting section is the main part of the work in operation, is much longer than machine reamers. Back taper reduces friction against reamed surface and prevents the cutting to oversize holes.

Machine reamer

Machine reamers are of shorter cutting edges. A cylindrical or tapered shank is used for holding or fixing in the machine. It is used in pneumatic tool or in portable electric tool for structural and shipbuilding works. A short taper lead is to be selected on machine reamers for tough and soft material or blind holes. A somewhat longer taper lead is used to be selected for hard materials. The guide part joining the taper lead is used for smoothening the surface of the hole. Reamer with helical flutes is used for reaming a slotted hole. Bigger holes are reamed with shell reamers and often called a hollow reamer, is actually a reamer without a shank. Shell reamers are manufactured with either the fluted teeth having clearance or the rose type which cuts on the end only.

Machine jig reamer

It has short, parallel cutting edges with level lead. The flutes are helical. Overall length of reamer consists of guide length, cutting edge length and shank length. The shank part of the reamer fits into a bushing in the jig and accurately locates the reamer.

Reamers mostly used in drilling machines are fluted chucking reamers, rose chucking reamers, spiral flutes, helical flutes, shell reamers, etc. (as shown in the sketch).

Expansion reamer is used to adjust the small variation in previous drill hole size due to non-alignment of drill-bit, etc.

A2.715 : Tapping

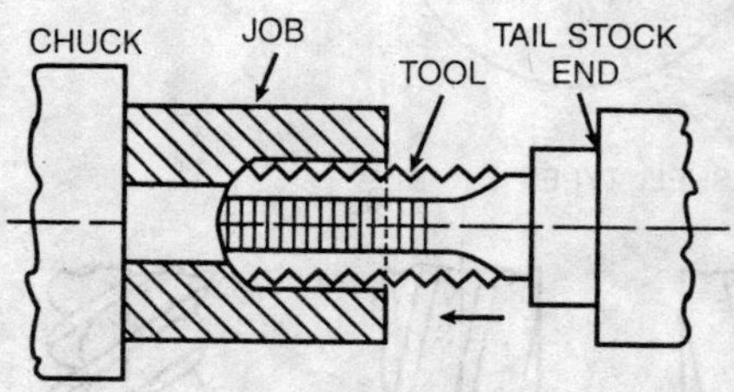

S-2.715 TAPPING

Tapping is the operation of forming internal threads within a hole by means of a tool known as tap. Job is held on chuck fitted on a head stock spindle rotates at slower speed and the tool fitted on tail stock spindle, fed into the work axially. Three taps are generally used to complete the internal thread. Taps are Taper Tap, Second Tap and Plug Tap.

A2.716 : Parting-Off

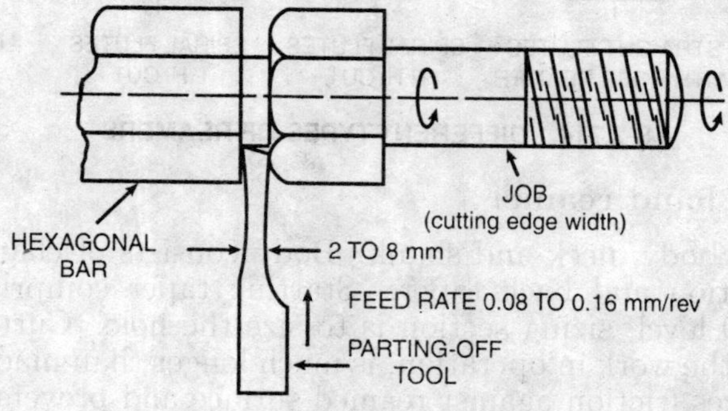

S-2.716 PARTING-OFF

It is the operation of cutting-off a bar type job after it has been machined to the required shape and size. In this operation, a bar type job is held on collect chuck, rotates to the turning speed and the parting-off tool is fed into the job very slowly until the tool reaches to the centre of the job.

A2.717 : Standard Tapers With Their Comparative Study

Standard tapers	Taper numbers	Conicity or amount of taper (*K*), $K = \frac{D-d}{L}$	Application in different machine tools.	Remark
Metric	Seven in number 4, 6, 80, 100, 120, 160, and 200	1 : 20	Modern Machine Tools.	Taper number stands for the larger diameter in mm. Advantage of having uniform angle of taper.
Morse	Seven in number 0, 1, 2, 3, 4, 5, and 6	1 : 19.002 to 1 : 20.047 amount of taper varies from size to size.	Drilling Machine and tail stock end of an engine lathe.	'0' is the smallest and '6' is the largest in size. Disadvantage of having non-uniformity of angle of taper.
Jarno	Twenty in number 1 to 20	0.6 inch per foot run.	Lathe machines.	
Brown and Sharp	Eighteen in number 1 to 18	0.5 inch per foot, except taper no. 10, which has a taper of 0.5161 inch/foot run.	Milling machine.	

A2.718 : Taper Turning In Lathe Machines

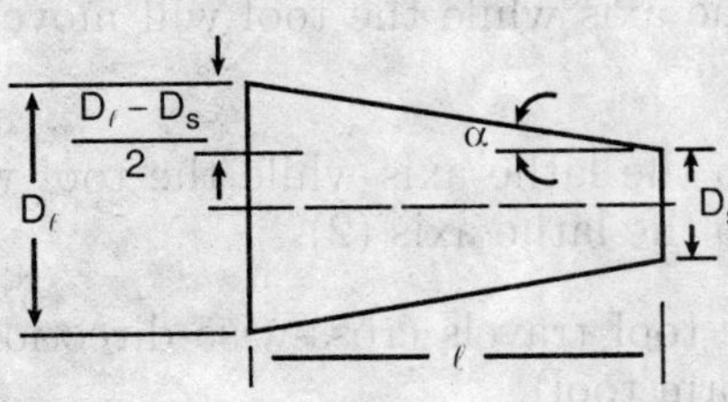

$$\tan\alpha = \frac{\text{perp.}}{\text{base}} = \frac{(D_l - D_s)/2}{\ell}$$

$$\text{or,}\quad \frac{(D_l - D_s)/2}{\ell} \times \frac{1}{2} = \frac{K}{2}$$

$$\text{or,}\quad K = 2\tan\alpha$$

S-2.718

Taper

Taper is a gradual increase or decrease in diameter of a cylindrical workpiece along its length.

Elements of taper

D_l = Large diameter in mm

D_s = Small diameter in mm

l = length of taper in mm

α = half taper angle

2α = full taper angle

Conicity, K is the ratio of the difference in diameters of large and small to its length of taper

$$\text{or,}\quad K = \frac{D_l - D_s}{l}$$

$$\tan\alpha = \frac{\text{perpendicular}}{\text{base}} = \frac{D_l - D_s}{2} \Big/ l = \frac{D_l - D_s}{l} \times \frac{1}{2} = \frac{K}{2}$$

$$\therefore\quad K = 2\tan\alpha$$

Method of expressing taper

(1) Slope : The rise in height in a certain horizontal distance, i.e., 1 in 40.

(2) Taper angle: Angle of taper, α in degrees.

(3) Conicity, $K = \frac{D_l - D_s}{l}$ (gradual increase or decrease in diameter of job per unit length).

A2.719 : Taper Turning Method In Lathe Machine

The following are the methods by which taper turning can be done in lathe machines :

1. By swivelling the compound rest.
2. By off-setting the tail stock.
3. By taper turning attachment.
4. By a form tool.
5. By combination of feeds by carriage and cross-slide.

A2.720 : Principles Of Taper Turning In Lathe Machine

Taper turning in lathe may be done either of the ways:

(a) The job will rotate about the lathe axis while the tool will move at an angle to it (1, 3 and 5).

(b) The job will rotate at an angle to the lathe axis while the tool will move in a straight path parallel to the lathe axis (2).

(c) Job rotates on lathe axis and the tool travels cross-wise directions, perpendicular to the lathe axis (form tool).

A2.721 : Taper Turning By Swivelling The Compound Rest

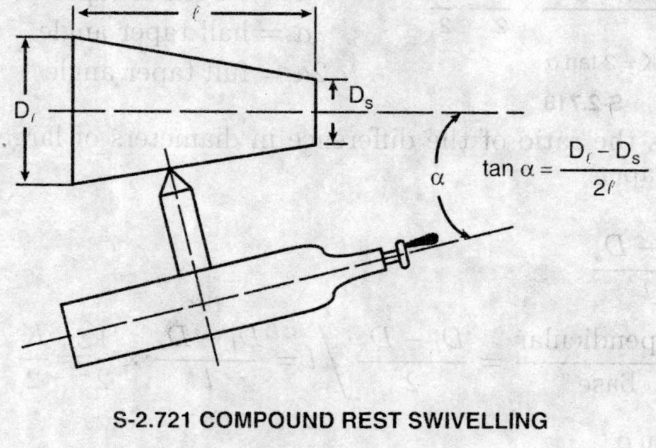

S-2.721 COMPOUND REST SWIVELLING

This method is suitable for jobs of shorter length, steeper angle, jobbing work or batch work by swivelling the compound rest holding the tool at an angle to the axis of rotation of the job. The compound rest may swivelled at any angle up to 45° on either side of the lathe axis.

$\tan\alpha = \frac{D_l - D_s}{2l}$, the compound rest is to be swivelled to an angle α.

A2.722 : Taper Turning By Off-Setting The Tailstock

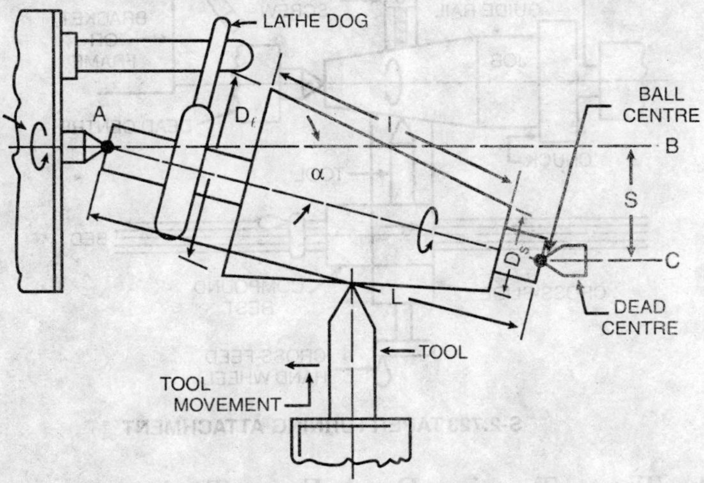

S-2.722 TAIL STOCK OFF-SETTING

This method is suitable for larger job, small taper angle of jobbing or batch work by changing or shifting the axis of the tail stock centre and hence shifted the axis of rotation of the workpiece to the required half taper angle.

The amount α of off-set to be calculated as per following method:

Tail stock set-over or off-set $= S = BC = L\sin\alpha$.

$(\because\ \sin\alpha = \frac{S}{L}$. Let $\sin\alpha \approx \tan\alpha)$

The tool is fed longitudinally either by automatically or by hand.

$$S = L\tan\alpha = L \times \frac{D_l - D_s}{2l}$$

$$S = \frac{D_l - D_s}{2l} \text{ for tapering of entire length}$$

$\therefore$ Tail stock set-over or off-set, S

$$S = \frac{\text{Entire length of work} \times \text{Conicity}}{2} = \frac{L \times \frac{D_l - D_s}{s}}{2} = \frac{L \times K}{2}.$$

A2.723 : Taper Turning By Taper Turning Attachment

This method is suitable for longer jobs and very steeper angle, accurately in mass production work. There is an attachment attached to the rear end of the lathe bed, set at an angle equal to half taper angle, engage the rear end of the compound slide with the attachment by releasing the binder screw with the cross-feed screw. Job rotates in its own axis and tool travels longitudinally in angular path with respect to the angle set in the attachment guide block. This method is limited to 10° to 12° angle of taper.

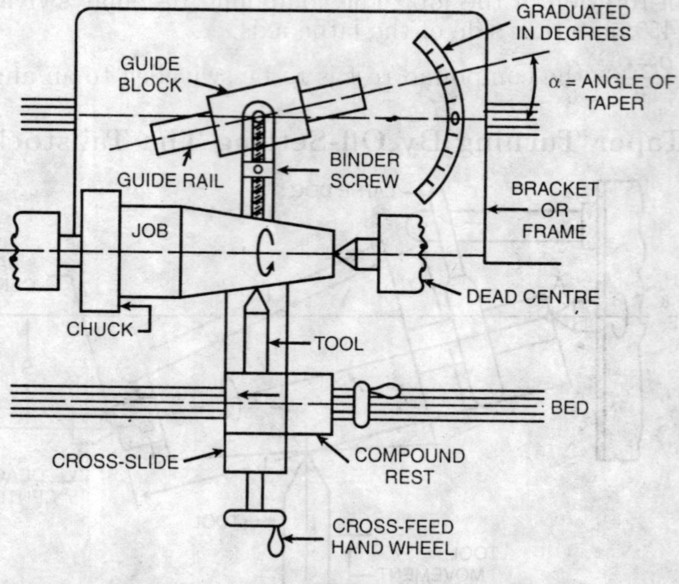

S-2.723 TAPER TURNING ATTACHMENT

A2.724 : Taper Turning By A Form Tool

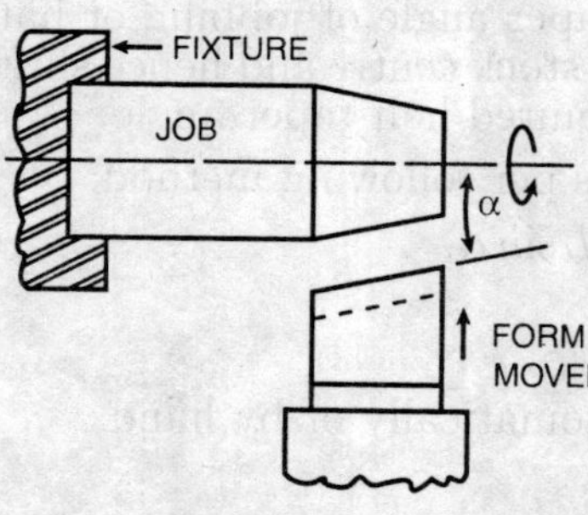

S-2.724 USING FORM TOOL

This method is mostly suitable for mass production for chamferring on bolts, nuts, bushes, etc., where length of taper is very short, by using a tool of definite shape of the taper. The tool having the same angle of taper as required for the job. Taper can be obtained simply by pressing or making crosswise movement of the tool against the job.

A2.725 : Taper Turning By Combination Feeds

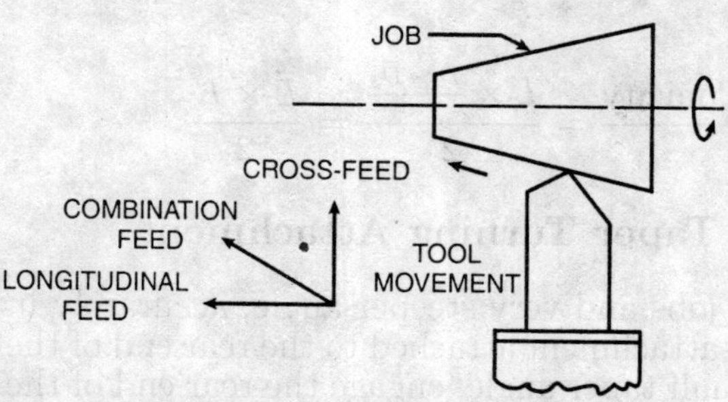

S-2.725 COMBINATION FEED

If both longitudinal and cross-feeds are given simultaneously, the tool will follow a taper path, i.e., in resultant direction as shown above. Some machines are provided with both the feeds automatically, can only be used for this type of taper turning.

A2.726 : Comparative Study Of Different Methods Of Taper In Lathe Machine

Method of taper	Principles of taper turning	Taper angles and length of taper (or limitations)	Formula used	Application	Accuracy	Remark
By swivelling the compound rest	Job rotates on lathe axis and tool moves in angular path.	Any angle 0 to 90°, short length of taper up to 150 mm approx.	$\tan\alpha = \frac{D_l - D_s}{2l}$ $\alpha = \tan^{-1}\frac{D_l - D_s}{2l}$	Shorter length and steeper angle, jobbing and batch work, bevel gear blank.	Moderate	
By off-setting the tail stock	Job rotates at an angle to the lathe axis and the tool travels longitudinally to the lathe axis.	Any angle 0 to 8°, long jobs of smaller diameter.	$S = L\tan\alpha = \frac{D_l - D_s}{2l}$ $= \frac{D_l - D_s}{2}$ for entire length $= \frac{L \times K}{2}$ where, $K = \frac{D_l - D_s}{l}$	Longer jobs and small tool angle of taper, jobbing and batch work. It is also used for internal taper turning.	Low	Used only when taper turning attachment is not possible or available.
By taper turning attachment	Job rotates on lathe axis and tool moves on guided angular path.	Any angle 0 to 12°, long jobs of steeper angle of taper	Set guide rail as per angle of taper	Longer jobs of steep angle in mass production work (mandrel)	Most	Used only for external taper

Method of taper	Principles of taper turning	Taper angles and length of taper (or limitations)	Formula used	Application	Accuracy	Remark
By a form tool	Job rotates on lathe axis and tool moves cross-wise direction, perpendicular to the lathe axis.	Very small length of taper and any angle 0 to 90°	Tool itself designed as per requirements	Mass-production for chamfering on bolts, nuts, bushes, etc.	Moderate	
By combination feeds	Job rotates on lathe axis and tool travels on resultant path.	Any length and any angle	Taper angle to be determined by trial and error method	By hand feeds for making the ball of a hammer, by power feeds for production work	Hand feed—low. Power feed—Moderate	Separate power feeding arrangement is necessary

A2.727 : Examples Of Taper Turning

Example 1.

Determine the value of conicity of taper job having the length of taper 200 mm, larger diameter = 100 mm and smaller diameter = 80 mm.

Solution: We know the formula of Conicity, $K = \dfrac{D_l - D_s}{l}$

where, length of taper, $l = 200$ mm

larger diameter, $D_l = 100$ mm

smaller diameter, $D_s = 80$ mm

$$= \frac{100-80}{200}$$

$$= \frac{20}{200} = \frac{1}{10} \quad \textbf{Ans.}$$

Example 2.

Find the larger diameter of a taper workpiece. The conicity is 1/20, length of taper 300 mm, the smaller diameter 85 mm.

Solution: We know the relation, Conicity, $K = \dfrac{D_l - D_s}{l}$

where, length of taper, $l = 300$ mm

conicity, $K = \dfrac{1}{20}$

smaller diameter, $D_s = 85$ mm

Let, larger diameter, $D_l = ?$ in mm

or, $D_l = K_l + D_s$

$$= \frac{1}{20} \times 300 + 85$$

$$= 100 \text{ mm} \quad \textbf{Ans.}$$

Example 3.

Find the angle at which the compound rest is to be swivelled or set with the following data :

Tapered portion of the work = 100 mm

smaller diameter = 50 mm

larger diameter = 75 mm

Solution: We know the relation, $\tan\alpha = \dfrac{D_l - D_s}{2l}$

where, larger diameter, $D_l = 75$ mm

smaller diameter, $D_s = 50$ mm

length of taper, $l = 100$ mm

$$= \frac{75-50}{2 \times 100} = \frac{1}{8} = 0.125$$

$$\alpha = \tan^{-1} 0.125 = 7°7' \quad \textbf{Ans.}$$

Let, α be the angle through which the compound rest to be set.

Example 4.

A shaft, 1.2 metre long, has a taper of 5 mm per metre for 600 mm of its length from one end. The maximum diameter of the shaft is 75 mm. Find the smaller diameter of the shaft.

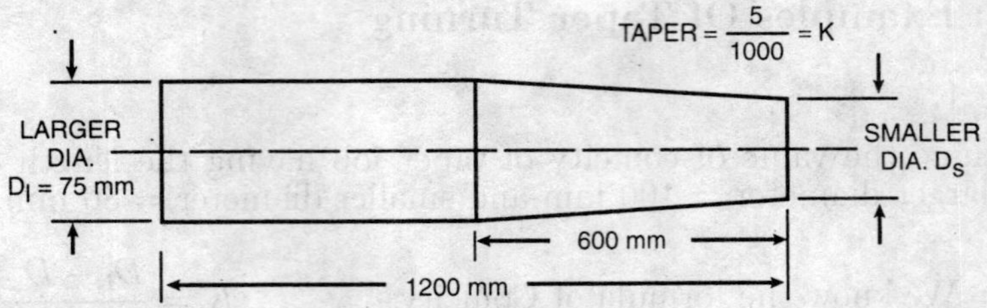

EXAMPLE NO. 4
S-2.727(4)

Solution: We know the formula,

$$\text{Conicity,} \quad K = \frac{D_l - D_s}{l}$$

$$\frac{5}{1000} = \frac{75 - D_s}{600}$$

$$\text{or,} \quad D_s = 75 - \frac{5 \times 600}{1000} = 72 \text{ mm } \textbf{Ans.}$$

Example 5.

A taper shaft of 900 mm long is to be turned for a length of 225 mm. The amount of taper is 1 : 100. Determine the tail stock off-set.

Solution: We know the relation,

$$\text{Tail-stock off-set, } S = \frac{\text{Conicity} \times \text{ Entire length of the shaft}}{2}$$

where, Conicity, $K = \frac{1}{100}$
Entire length of the shaft, $L = 900$ mm

$$S = \frac{1}{100} \times \frac{900}{2}$$

$$= 4.5 \text{ mm } \textbf{Ans.}$$

Example 6.

(a) Find out the amount of set-over of tail stock for turning a taper on a job for 2/3 of its length from one end. Larger diameter is 100 mm and smaller diameter is 80 mm. Total length of job is 300 mm.

(b) What should be the set-over to turn a taper of 1 in 20 on a job of 160 mm long such that it will have its larger diameter of 50 mm?

Solution:

Data given

Total length of the job, $L = 300$ mm

Length of taper, $l = 300 \times \frac{2}{3} = 200$ mm

Larger diameter, $D_l = 100$ mm

Smaller diameter, $D_s = 80$ mm

(*a*) We know the formula,

$$S = L \times \frac{D_l - D_s}{2l}$$

$$= 300 \times \frac{100 - 80}{2 \times 200}$$

$$= 15 \text{ mm } \textbf{Ans.}$$

Data given

Length of taper, $L = 160$ mm

Larger diameter, $D_s = 50$ mm

Conicity, $K = \frac{1}{20}$

(b) We know the relation,

Tail-stock set-over,

$$S = L \times \frac{K}{2}$$

$$= 160 \times \frac{1}{20} \times \frac{1}{2}$$

$= 4$mm **Ans.**

Example 7.

A taper pin of length 200 mm has a taper length 125 mm, material of pin is C-20. The smaller diameter of taper is 25 mm and larger diameter is 30 mm Determine (i) taper in mm/metre and in degrees, (ii) the angle to be set on compound slide for turning the taper on lathe, (iii) set-over.

Solution:

Data given

Total length of pin, $L = 200$ mm

Taper length, $l = 125$ mm

Smaller diam., $D_s = 25$ mm

Larger diam., $D_l = 30$ mm

To be determined

(i) Taper in mm/meter length of taper in degrees.

(ii) Angle to be set on compound slide, α.

(iii) Tail stock set-over, S.

We know

(i) Conicity, $K = \frac{D_l - D_s}{l}$

$$= \frac{30 - 25}{125} = \frac{1}{25}$$

i.e., For a length of 25 mm taper is 1 mm.

Taper for 1000 mm (or 1 mtr.)

$$= \frac{1 \times 1000}{25} = 40 \text{ mm}$$

$\therefore$ Taper is 40 mm/metre length **Ans.**

$$\tan\alpha = \frac{D_l - D_s}{2l} = \frac{30 - 25}{2 \times 125} = \frac{1}{50}$$

or, $\alpha = \tan^{-1}\frac{1}{50} = 1°9'$

Full taper angle, $2\alpha = 2 \times 1°9' = 2°18'$

Ans.

(ii) Angle to be set on compound slide,

$\alpha =$ half taper angle $= 1°9'$ **Ans.**

(iii) Tail-stock set-over,

$$S = L\frac{D_l - D_s}{2l} = 200 \times \frac{30 - 25}{2 \times 125}$$

$= 4$ mm **Ans.**

A2.728 : Problems On Taper Turning

Prob. 1.

Find the conicity of a taper pin having length of taper = 100 mm, larger diameter = 30 mm and smaller diameter = 25 mm [**Ans** : 1/20]

Prob. 2.

Determine the smaller diameter of a taper shaft, whose length is 300 mm, larger diameter 80 mm and conicity 1/30. [**Ans.** 70 mm]

Prob. 3.

A tapered portion of a job is 75 mm long; minimum diameter is 25 mm and maximum diameter is 50 mm. Determine the angle at which the compound rest is to be set. [**Ans.** 9°27′45″]

Prob. 4.

A shaft of 1500 mm long has a taper of 1:100 for a length of 900 mm Smaller diameter of the shaft is 80 mm. Find the larger diameter of the shaft and the amount of tail stock set-over. [**Ans.** : $D_l = 89$ mm and 7.5 mm]

Prob. 5.

A shaft of length 400 mm has a taper length of 200 mm, material is mild steel containing 0.2% carbon. The maximum diameter of taper is 50 mm and minimum diameter is 40 mm Find (i) taper in mm/metre length and in degrees (ii) angle of swivelling the compound rest for turning the taper on lathe machine and (iii) Tail stock set-over.

[**Ans** : (i) 50 mm/metre, 2°52′, (ii) 1°26′ and (iii) 10 mm]

A2.81 : Thread Cutting In Lathe

Elements of a Thread

Thread

The helical groove on cylindrical bar is known as thread.

Pitch

Pitch is the distance between two similar points of two consecutive threads.

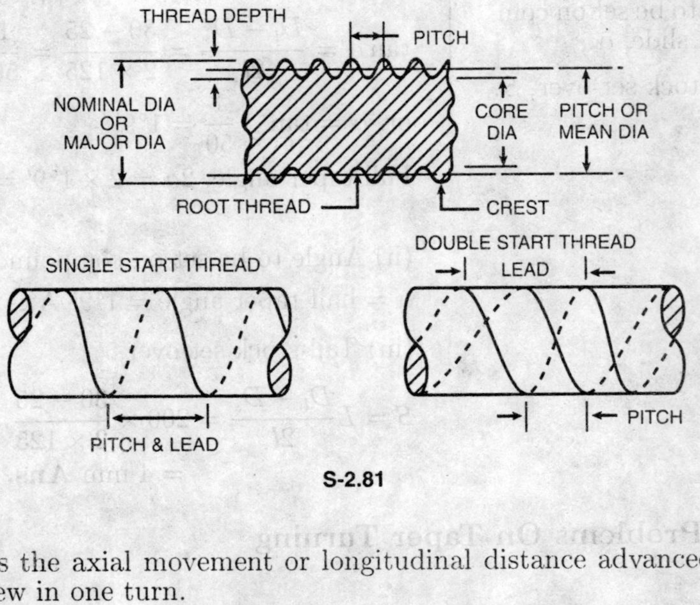

S-2.81

Lead

Lead is the axial movement or longitudinal distance advanced by a nut over a screw in one turn.

Nominal diameter (or Major diameter or Outside diameter, D_o)

It is the largest diameter of a thread.

Pitch or Mean diam. $\rightarrow (D_o + D_c)/2$ or, $\dfrac{\text{Outside diam.} + \text{Core diam}}{2}$

Core diameter (Minor diameter, D_c)

It is the smallest diameter of a thread.

Thread depth

The distance between crest and root in one side of a thread.

A2.82 : Right-Handed Thread

It is the advancement of a nut over a screw in a clockwise direction.

Left-handed thread

It is the advancement of a nut over a screw in a counter-clockwise direction.

A2.83 : Single Start Thread

When a screw has only one thread on it, the thread is known as single start thread. Lead = Pitch.

Multi-start thread

When a screw has two or more threads on it, the thread is known as multi-start thread.

$$\text{Lead} = \text{Pitch} \times \text{No. of start}$$

A2.84 : Principles Of Thread Cutting

It is to produce helical grooves on rotating cylindrical surface along lathe axis and feeding the thread-cutting tool longitudinally.

A2.85 : Thread-Cutting Operation

There is a definite relation of speed between the job and tool. This relation of speed is obtained by gear ratio depends on the pitch of the job, pitch of lead screw, number of start required. Every machine is supplied with a spur gear box (set of 23 gears) having teeth from 20 to 120 with an interval of 5 and a special gear or transfer gear is 127 no. of teeth gear for cutting metric thread. Two 20 teeth spur gears are available. Lead screw has single start threads.

Steps to be followed in thread-cutting operation on Engine Lathe

The following operational steps to be followed in thread cutting :

(1) Holding the job on machine spindle and turning to the required major diameter.

(2) Choosing suitable thread-cutting tool.

(3) Selecting slower speed (about one-third or one-fourth of turning speed) of the machine spindle.

(4) Calculating the change gear ratio with respect to the pitch of the job required, pitch of lead screw and number of start required.

(5) Fixing the calculated change gears to the Head stock spindle, Intermediate shaft and Lead screw shaft.

Calculation of change gears

$$\text{Change gear ratio} = \frac{\text{Driver teeth}}{\text{Driven teeth}} = \frac{\text{Pitch of the job} \times \text{No. of start}}{\text{Pitch of the lead screw}}$$

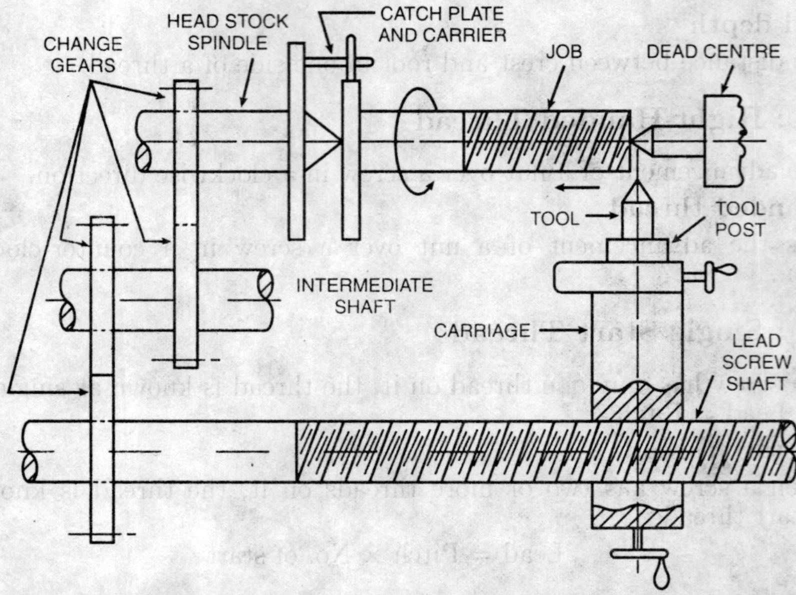

S-2.85(A) POWER TRANSMISSION FROM THE SPINDLE TO TOOL THROUGH CHANGE GEARS AND LEAD SCREW

Example :

Say, Pitch of the job = 3 mm and lead screw having 6 mm pitch for producing single start thread.

$$\therefore \text{ Change gear ratio} = \frac{\text{Pitch of the job} \times \text{ No. of start}}{\text{Pitch of the lead screw}}$$

$$= \frac{3 \times 1}{6} = \frac{1 \times 20}{2 \times 20} = \frac{20}{40} = \frac{\text{Driver gear(s)}}{\text{Driven gear(s)}}$$

[Note : Multiply by a number on Num. and Denom., such that the product will match the teeth available in a set of 23 gears in gear box.]

(6) Fixing the tool on tool post and to be aligned properly.

(7) Choosing suitable depth of cut. Three or four cuts are necessary to complete the thread.

(8) Arranging job-tool in proper position and giving desired depth of cut.

(9) Checking cutting fluid (specially cutting oil) is circulating properly.

(10) Engaging the half nut with respect to the chasing dial according to the odd/even threads.

(11) Allowing the movement of the tool up to the portions of the job necessary for thread cutting then lifting the tool from the job.

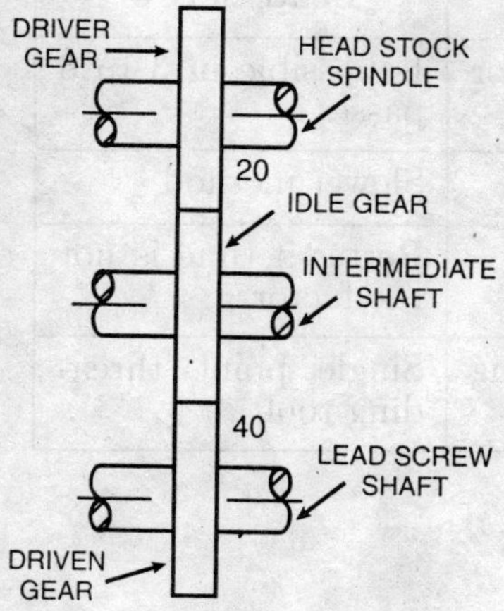

EXAMPLE S-2.85(B)

(11) Dis-engaging the half nut, move the carriage to the right side up to the position from where second cut will start. Allowing the second depth of cut and again engage the half nut with respect to chasing dial.

(12) Continuing the process until the thread-cutting operation is completed.

(13) Checking the thread by Thread Pitch Gauge.

[Note : Use any spur gear available for idle gear.]

A2.86 : Even Thread & Odd Thread

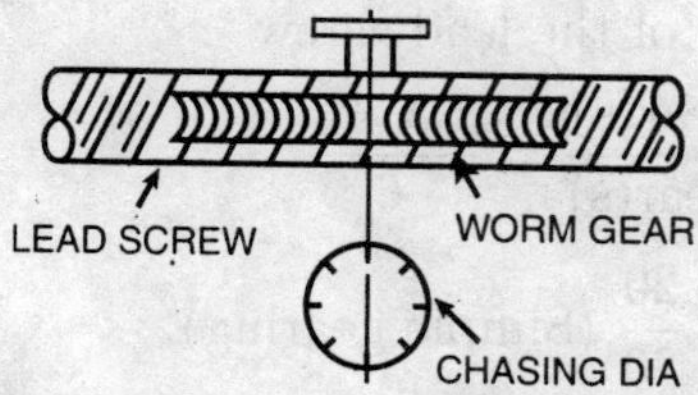

S-2.86 EVEN & ODD THREAD

Even thread are those in which the product of the pitch of the lead screw in mm and the number of teeth of the worm gear must be in exact multiple of the pitch of the job to be cut, if otherwise the thread is called odd thread.

A2.86C : Compare Chaser & Lead Screw

Points	Chaser	Lead Screw
Length of threaded portion and place of suitability	Short lead screw used mainly in production work for shorter length threading.	Long lead screw mostly used for jobbing work and can also be used for long threading in production.
Situated at	It is a tool and held on tool post when required.	It is a part of a machine and is built in the machine.
Forming and machining of threads	Cutting tools are in-built and can be sharpened easily without loosing the accuracy.	Separate tools are required for threading and can be sharpened with loosing certain accuracy.

Points	Chaser	Lead Screw
Full form of thread	Is possible in one or two passes only.	Is possible in 3 to 6 passes.
Speed	Faster method	Slower method
Accuracy	Moderate	Better if time is not the factor.
Tool	Multi-point threading tool.	Single point threading tool.

A2.87 : Examples On Thread Cutting

Example 1.

Calculate the change gears to cut a single start thread of 1.5 mm pitch in a Lathe machine having a lead screw of 6 mm pitch.

Solution:

We know the change gear ratio $= \dfrac{\text{Pitch of the job} \times \text{No. of start}}{\text{Pitch of the lead screw}}$

$$= \frac{\text{Driver gear(s)}}{\text{Driven gear(s)}} = \frac{1.5 \times 1}{6}$$

$$= \frac{1 \times 20}{4 \times 20} = \frac{20}{80} \text{ (Simple gearing).}$$

Example 2.

Calculate the change gears in an engine lathe to cut 12 threads per inch and the lead screw has 4 mm pitch.

Solution:

$$\text{Change gear ratio} = \frac{\text{Driver gear(s)}}{\text{Driven gear(s)}} = \frac{\text{Pitch of the job in mm}}{\text{Pitch of the Lead screw in mm}}$$

$$= \frac{\frac{1}{12} \times 25.4 \text{ mm}}{4} = \frac{254}{12 \times 4 \times 10} = \frac{127 \times 20}{60 \times 4 \times 20}$$

$$\left[\because \quad \text{Pitch} = \frac{1}{\text{No. of threads/inch}} \right]$$

$$= \frac{127}{80} \times \frac{20}{60} \text{ (Compound gearing).}$$

Note : Assume single start thread if not mentioned the no. of start.

Example 3.

Calculate the change gears for cutting 12 threads per inch on an engine lathe having a lead screw of 4 T.P.I.

Solution:

$$\text{Change gear ratio} = \frac{\text{Driver gear(s)}}{\text{Driven gear(s)}} = \frac{\text{Pitch of the job}}{\text{Pitch of the L.S.}}$$

$$= \frac{\frac{1}{12}\text{ inch}}{\frac{1}{4}\text{ inch}} = \frac{4}{12} = \frac{1 \times 20}{3 \times 20}$$

$$= \frac{20}{60} \text{ (Simple gearing).}$$

Example 4.

Sketch the change gear set-up for cutting thread of 3 mm pitch and the lead screw having 8 T.P.I.

Solution:

$$\text{Change gear ratio} = \frac{\text{Driver gear(s)}}{\text{Driven gear(s)}} = \frac{\text{Pitch of the job}}{\text{Pitch of the lead screw}}$$

$$= \frac{3\text{ mm}}{\frac{1}{8} \times 25.4\text{ mm}} = \frac{3 \times 8 \times 10}{254}$$

$$= \frac{120}{127} \text{ (Simple gearing).}$$

Note : Assume single start thread if not mentioned the no. of start.

In simple gearing thread is preferred than compound gearing due to less frictional power loss and quick setting of gears.

Example 5.

Calculate the change gear ratio to cut a screw having lead of 4.5 mm for a single start thread and the lead screw having 8 T.P.I.

Solution:

$$\text{Change gear ratio} = \frac{\text{Driver gear(s)}}{\text{Driven gear(s)}} = \frac{\text{Pitch of the job} \times \text{No. of start}}{\text{Pitch of the lead screw}}$$

$$= \frac{4.5\text{ mm}}{\frac{1}{8} \times 25.4\text{ mm}}$$ (Since pitch and lead are equal for single start thread)

$$= \frac{45 \times 8}{254} = \frac{45 \times 2 \times 2 \times 20}{127 \times 20}$$

$$= \frac{90 \times 40}{127 \times 20} \text{ (Compound gearing)}$$

Example 6.

Driver wheels are 50 & 20 teeth and driven wheels are 40 & 120 teeth and the lead screw of 6 mm pitch. How much will be pitch of the job.

Solution:

$$\text{Change gear ratio} = \frac{\text{Driver gear(s)}}{\text{Driven gear(s)}} = \frac{\text{Pitch of the job}}{\text{Pitch of the L.S.}}$$

$$= \frac{50 \times 20}{40 \times 120} = \frac{\text{Pitch of the job}}{6 \text{ mm}}$$

or, $\quad \text{pitch of the job} = \frac{5}{4} = 1.25 \text{ mm}$ **Ans.**

Example 7.

Find the change gear ratio for cutting a 4-start thread having a pitch of 1.25 mm and the lead screw has a pitch of 4 mm.

Solution:

$$\text{Change gear ratio} = \frac{\text{Driver gear(s)}}{\text{Driven gear(s)}} = \frac{\text{Pitch of the job} \times \text{No. of start}}{\text{Pitch of the lead screw}}$$

$$= \frac{1.25 \times 4}{4} = \frac{1.25 \times 40}{40} = \frac{50}{40} \text{ (Simple gearing)}$$

So, the driver gear having 50 teeth and driven gear of 40 teeth.

A2.88 : Problems On Thread Cutting

Prob : 1.

Pitch of the thread to be cut is 1.3 mm and the pitch of the lead screw is 6 mm. Sketch and calculate the change gears.

[**Ans.** (65 × 20)/(60 × 100)]

Prob : 2.

1 mm pitch screw thread (Internal) is to be cut in a centre lathe having a 6 T.P.I. lead screw. Calculate gearing arrangements for driver and driven assuming usual quadrant restrictions and change gears availability.

[**Ans.** 30/127]

Prob : 3.

Sketch and calculate the change gears for cutting a thread of 24 T.P.I. on English lathe of 8 T.P.I.

[**Ans.** 20/60]

Prob : 4.

Sketch and calculate the pitch of the thread having 2 start and a lead screw having 4 T.P.I. The driver gears are 80 and 40 teeth and the driven gears are 127 and 20 teeth.

[**Ans.** 4 mm]

Prob : 5.

Calculate the change gears ratio for cutting a 3-start screw having a pitch of 2.5 mm and the lead screw has a pitch of 6 mm.

[**Ans.** 25/20]

A2.91 : Cutting Speed, Feed & Depth Of Cut

Cutting speed is the continuous length of the chip removing by the cutting tool from the circumference of the job in one minute. It is expressed in metre/min. and can be evaluated from the formula

$$\text{Cutting speed, } V = \frac{\pi DN}{1000}$$

where, D is the original diameter of the job in mm, N is the r.p.m.

Cutting speed depends on the tool material, the material to be cut and the type of operation. For rough cut, greater depth of cut, normal speed and feed are used and for finish cut, small depth of cut, fine feed and higher speed are used.

For Instance : By using a high speed steel tool

Job Material	Cutting speed in metre/minute	
	Turning, drilling, boring, milling operations, etc.	Thread cutting, Knurling operations, etc.
Mild Steel, Copper	25 to 30	8
Aluminium	80 to 100	25
Cast Iron	15 to 25	9
Brass	50 to 70	20
Stainless Steel & High Carbon Steel	15	6

Example 1.

A workpiece of 100 mm in diameter is to be turned in a Lathe Machine. What will be the r.p.m. of head stock spindle to attend a cutting speed of 25 m/min.?

Solution:

$$\text{Cutting speed, } V = \frac{\pi DN}{1000},$$

where, cutting speed, $V = 25$ m/min., Job diameter, $D = 100$ mm. Let r.p.m., $N = ?$

$$25 = \frac{\pi \times 100\text{ N}}{1000}$$

$$N = \frac{25 \times 1000}{\pi \times 100} = 79.6 \text{ r.p.m.}$$

$$= 80 \text{ r.p.m. (available speed)}$$

Note : Select nearest r.p.m. available in the machine chart.

Feed

Feed in lathe work is the linear distance advances by the cutting tool in one revolution of the work piece. It is expressed in mm/revolution. Its value in Lathe work varies from 0.2 mm/rev to 1 mm/rev.

Depth of cut

It is the amount of the tool penetrates into the work. It is equal to half the difference between the diameters before and after machined.

$$\text{Depth of cut, } t = \frac{D-d}{2} \text{ mm}$$

where, D = diameter before machining in mm and d = diameter after machining, in mm.

A2.92 : Greater Depth Of Cut To Remove Excess Material

Cross-section of chip is the product of feed and depth of cut. If the cross-section of chip is increased, will increase consequently the cutting forces and the amount of heat generation. But the tool wear will be more rapid with the increase of cutting speed than the increase in cross-section of the chip. Thus the production capacity can be increased by increasing the cross-section of the chips, not the cutting speed. Cross-section of the chip will again be increased by increasing the depth of cut not the feed.

A2.93 : Machining Time In Lathe Operations

It is the time required to perform different operations in a lathe to finish the work.

Machining time for a single cut in turning operations (T_m)

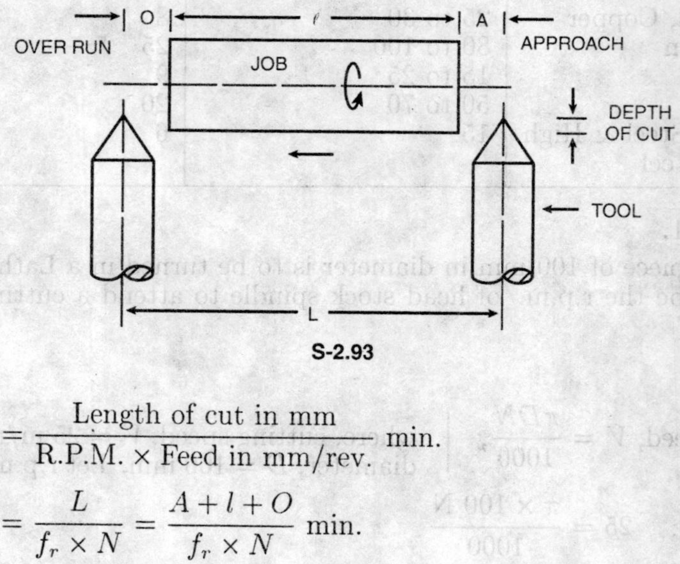

S-2.93

$$T_m = \frac{\text{Length of cut in mm}}{\text{R.P.M.} \times \text{Feed in mm/rev.}} \text{ min.}$$

$$T_m = \frac{L}{f_r \times N} = \frac{A+l+O}{f_r \times N} \text{ min.}$$

Total machining time, (for turning operation)

$$\text{Total } T_m = \frac{A+l+O}{f_r \times N} \times n \text{ min.}$$

where, total length of tool movement, L in mm
length of turning portion, l in mm
length of approach of tool, A in mm
length of over-travel of tool, O in mm
feed per revolution of the tool, f_r in mm/rev.
job speed, N in r.p.m.
no. of passes or cuts, n.

[*Note :* If not mentioned, A and O may be assumed any value in between 2 mm to 8 mm depending on job-tool size and type of operation. Neglect, over-run in some operations like step turning, shoulder turning, etc.]

A2.94 : Examples On Machining Time In Lathe Operations

Example 1. On Turning Operation

Calculate the machining time required to reduce from 60 mm dia. shaft to 50 mm diameter for a length of 1500 mm with depth of cut 2 mm for rough cut and 1 mm for finish cut. Cutting speed and feed may be assumed as 30 m/min and 0.5 mm/rev. respectively.

Solution:

Where, Machining time, T_m =? (in min.)

Assumed—Approach length, $A = 5$ mm
Over-run length, $O = 5$ mm
Length of turning, $l = 1500$ mm
Feed/rev., $f_r = 0.5$ mm/rev.
Job speed, N = ? r.p.m.
Cutting speed, $V = 30$ m/min.
No. of passes or cut = 3.
(2 rough cut + 1 finish cut)
Diameter of the job, $D = 60$ mm

Total machining time required to complete operation in 3 cuts,

$$\because \quad T_m = \frac{A + l + O}{f_r \times N} \times n$$

$$\text{And,} \quad V = \frac{\pi DN}{1000} = \frac{\pi \times 60 \times N}{1000} = 30 \text{ m/mm}$$

$$\text{or,} \quad N = \frac{30 \times 1000}{\pi \times 60} = \frac{500}{\pi} \text{ r.p.m.}$$

$$T_m = \frac{5 + 1500 + 5}{0.5 \times \frac{500}{\pi}} \times 3 \text{ min.} = 56.93 \text{ min.} \ \textbf{Ans.}$$

Example 2. On Thread Cutting Operation

Calculate the machining time required in a external thread cutting of 2 mm pitch thread on job of 20 mm diameter mild steel rod at a cutting speed of 8 m/min for a length of 100 mm on a shaft of 200 mm.

Machining time for thread cutting,

where, Diam. of the job, $D = 20$ mm
Length of threading portion, $l = 100$ mm
Cutting speed, $V = 8$ m/min.

Let job speed, N in r.p.m. = ?

Lead = pitch of the job, $p = 2$ mm

No. of cuts for external thread = 2.5 × pitch = 2.5 × 2 = 5

$$\because \quad T_m = \frac{l \times \text{no. of cuts}}{N \times \text{lead}}$$

And, $$N = \frac{V \times 1000}{\pi D} = \frac{8 \times 1000}{\pi 20} \text{ r.p.m.} = \frac{400}{\pi} \text{ r.p.m.}$$

$$T_m = \frac{100}{\frac{400}{\pi} \times 2} \times 5 \text{ min.} = 1.96 \text{ min.}$$

50% allowance on machining time to be added in thread-cutting operation for shifting the job, putting the depth of cut, engaging half nut, etc.

Allowance time, $T_A = 0.5 \times T_m = 0.5 \times 1.96$ min. = 0.98 min.

Overall time for thread cutting $= 1.5T_m$ min.

$= 1.5 \times 1.96$ min.

$= 2.9$ min. **Ans.**

[*Note :* No. of cut for internal threads = 3 × pitch of the job in mm, No. of cut for external threads = 2.5 × pitch of the job in mm, Lead = pitch in case of single start thread and Lead = pitch × no. of start for multi-start thread, $A + O$ is not considered, as allowance time is added.]

Example 3. On Facing operation

Find the machining time for a facing operation on a lathe machining for a 100 mm dia. shaft with 400 r.p.m. and feed may be taken as 0.2 mm/rev. Assume any data if required.

Solution:

Assuming single cut

Approach, $A = 5$ mm

Over-run, $O = 3$ mm

Facing length, $l = \frac{D}{2} = \frac{100}{2} = 50$ mm

Job speed, $N = 400$ r.p.m.

Feed/rev., $f_r = 0.2$ mm/rev.

No. of cut = 1

Diameter of the job, $D = 100$ mm

$$\text{Machining time, } T_m = \frac{(A + l + O)}{f_r \times N} \times \text{no. of cut}$$

$$T_m = \frac{(5 + 50 + 3)}{0.2 \times 400} \times 1$$

$$T_m = \frac{29}{40} \text{ min.}$$

$$= 0.725 \text{ min. } \textbf{Ans.}$$

Example 4. On Boring Operation

Determine the machining time, for boring operation in a lathe machine, to a job having casting internal diameter of 40 mm is to be finished to an internal diameter of 45 mm to a length of 100 mm. Job speed may be assumed as 200 r.p.m. Assume any suitable data if required.

Solution:

Assuming Feed/rev, $f_r = 0.3$ mm/rev.

Rough depth of cut = 2 mm

Finish depth of cut = 0.5 mm

Therefore, No. of cut = 2

(One rough cut and one finish cut)

Approach length, $A = 5$ mm

Boring length, $l = 100$ mm

Over-run length, $O = 5$ mm

Job speed, $N = 200$ r.p.m.

Machining time,

$$T_m = \frac{A + l + O}{f_r \times N}$$

$$T_m = \frac{(5 + 100 + 5)}{0.3 \times 200} \times 2 \text{ min.}$$

$$= 3.67 \text{ min.} \quad \textbf{Ans.}$$

Example 5. On Taper Turning Operation

A taper of 1 in 25 has to be turned on a C-20 steel job with H.S.S. tool. The larger diameter is 40 mm and length of taper is 50 mm. Determine the machining time. Assume any relevant data if required.

Data given :

Taper = 1 in 25

Job material = $C - 20$ steel

Tool material = H.S.S.

Large diameter $D_l = 40$ mm

Smaller diameter, $D_s = 40 - 2 = 38$ mm

Horizontal length of taper = 50 mm

Tapered length, $l = ?$ in mm

$l = \sqrt{p^2 + b^2} = \sqrt{50^2 + 2^2} = 50.0 \simeq 50$ mm

Assuming feed, $f_r = 0.6$ mm/rev.

Cutting speed, $V = 25$ mm/min.

Depth of cut = 1 mm will reduce 2 mm

No. of pass or cut, $n = 1$ (one)

Assuming approach length, $A = 5$ mm

Over-run length, $O = 5$ mm

$$V = \frac{\pi D N}{1000}$$

$$N = \frac{25 \times 1000}{\pi \times 40} = \frac{625}{\pi} \text{ r.p.m.}$$

$$T_m = \frac{(A + l + O)}{f_r \times N} \times n$$

$$= \frac{\pi(5 + 50 + 5)}{0.6 \times 625} \times 1 \text{ min.}$$

$$= 0.50265 \text{ min.} \quad \textbf{Ans.}$$

A2.95 : Problems On Machining Time In Lathe Operations

Problem No. 1.

A mild steel job 80 mm external diameter and 50 mm internal diameter has a length of 400 mm is to be machined to 70 mm external diameter and 55 mm internal diameter. One end of the job is to be faced for 5 mm length. Determine the machining time in (i) turning, (ii) facing, (iii) boring operations using H.S.S. tool. Assume any relevant data if required.

[**Ans.** (i) 24.9 min., (ii) 2.375 min., (iii) 13.8 min.]

[*Hints:* V = 25 m/min., f_r = 0.5 mm/rev. for turning, f_r = 0.2 mm/rev. for facing, f_r = 0.15 mm/rev. for boring, taking 10 mm as approach and 5 mm as over-run on turning and boring and 5 mm as approach and 3 mm as over-run for facing operation.]

Problem No. 2.

Calculate the machining time for a taper of 1 in 20, larger diameter 50 mm, length of taper = 200 mm with a job material M.S. and tool material H.S.S. (Assume spindle speed as 250 r.p.m. and feed/rev. as 0.5 mm/rev., rough depth of cut = 2 mm and finish depth of cut = 1 mm.)

[**Ans.** 5.04 min.]

Problem No. 3.

1.5 mm pitch threads are to be cut by a lathe machine on 30 mm diameter M.S. rod at a cutting speed of 9 m/min. to a length 200 mm. Calculate the thread-cutting time. (Assume no. of cuts = 2.5× pitch = $2.5 \times 1.5 = 3.75 \simeq 4$)

[**Ans.** 8.377 min.]

A2.96 : Questions

1. State the term Machine Shop. (A 2.11)

2. Write the relative motion of job and tool of important machining operations in different standard Machine Tools. (A 2.12)

3. What do you mean by Lathe Machine? (A 2.21)

4. Write the types of Lathe in tabulated form. (A 2.22)

5. Classify different types of Lathe. (A 2.23)

6. Specify a Centre Lathe and show in a block sketch of a Lathe. (A 2.31)

7. Represent different specifications of any particular lathe machine as supplied by the manufacturer. (A 2.32)

8. Name the different Lathe parts. (A 2.41)

9. Write the functions of Lathe parts. (A 2.42)

10. How many ways the different speeds of a head stock spindle can be attained. (A 2.51)

11. Sketch, label and explain the working of stepped pulley drive. (A 2.52)

12. State, how the number of speeds of stepped pulley drive can be increased and sketch the same, and explain its working. (A 2.53)

13. Sketch and explain how all geared head drive can supply different speeds to head stock spindle. (A 2.54)

14. Explain the feed reversing gear with sketch. (A 2.55)

15. Sketch, label its components and explain how power is transmitted from the motor to the tool end. (A 2.56)

16. Compare all geared head drive and cone pulley drive. (A 2.57)

17. Explain the term full-proof arrangement (self-locking mechanism). (A 2.511)

18. Sketch and explain how tool can move either longitudinally or transverse direction automatically (apron mechanism). (A 2.512)

19. State with example the term parts, accessories and attachment. (A 2.61)

20. State the function of a Lathe centre, sketch and show the elements of a centre. (A 2.62)

21. Make a comparative study of different types of centres with their use. (A 2.63)

22. Sketch and label the utility of Driving plate and Lathe dog for a long job mounting in between centres. (A 2.64)

23. State the utility of different types of chucks. (A 2.65)

24. Write short notes on face plate, mandrel, steady rest and follower rest. (A 2.66, 2.67, and 2.69)

25. Classify with sketches the use of different types of Mandrels. (A 2.68)

26. Compare Steady rest and Follower rest. (A 2.610)

27. Distinguish between Driving plate and Mandrel. (A 2.611)

28. Make a comparative study between Driving plate and Face plate. (A 2.612)

29. Compare Universal chuck and Independent jaw chuck. (A 2.613)

30. State the principles on which a job can be mounted on H.S. spindle. (A 2.71)

31. Write the principles of operation in Lathe Machine. (A 2.72)

32. Name at least 15 main operations can be done on Centre Lathe. (A 2.73)

33. State with sketches and show the job-tool motion of the following Lathe operations—(a) Centering, (b) Facing, (c) Straight turning, (d) Shoulder turning, (e) Eccentric turning, (f) Chamferring, (g) Knurling, (h) Spinning, (i) Drilling, (j) Reaming, (k) tapping, (l) Parting-off. (A 2.74 to A 2.716)

34. Tabulate the standard tapers with their comparative study. (A 2.717)

35. Show the taper turning elements. (A 2.718)

36. How many ways taper turning is possible in a Centre Lathe? (A 2.719)

37. State the principles of taper turning. (A 2.72)

38. Sketch and explain the following taper-turning methods :

(a) Taper turning by Swivelling the compound rest. (A 2.721)

(b) Taper turning by off-setting the tail stock. (A 2.722)

(c) Taper turning by taper turning attachment. (A 2.723)

(d) Taper turning by a form tool. (A 2.724)

(e) Taper turning by combination feeds. (A 2.725)

39. Make a comparative study of different methods of taper turning. (A 2.726)

40. Sketch and show the elements of a thread. (A 2.81)

41. What do you mean by Right-handed thread and Left-handed thread. (A 2.82)

42. State about Single-start and Multi-start thread. (A 2.83)

43. Write the principles of thread cutting. (A 2.84)

44. Write the steps to be followed in external thread-cutting operation in a Centre Lathe. (A 2.85)

45. State the term cutting speed, feed and depth of cut used in Lathe operations. (A 2.91)

46. State the reasons why greater depth of cut is used instead of cutting speed or feed for bulk removal of metals. (A 2.92)

47. Deduce the machining time calculations in Lathe operations. (A 2.93)

48. Sketch and state different types of Reamer. (A2.714.1)

49. Compare Chaser and Lead Screw. (A2.86c)

3

CAPSTAN & TURRET LATHE MACHINES

A3.1 : Comparison Among Engine, Capstan & Turret Lathe

Points 1	Engine Lathe 2	Capstan Lathe 3	Turret Lathe 4
1. Direction of spindle rotation	Mostly anti-clockwise direction of rotation.	Both the direction of rotation.	
2. Motor power	Requires less power as the machine is to do only one operation at a time.	Require about five times the power of Engine Lathe as two or more operations are generally done at a time.	
3. Spindle speeds	Less number of spindle speeds. Average and slower speeds available.	More number of spindle speeds and greater range of speeds available.	
4. Machining time	Setting and machining time is higher as the machine works on slower speeds.	In a mass production work, setting time for each job is negligible and machining time is also very less as it works more operations at a time at higher speeds.	
5. Type of labour requirement	Skilled labour is essential in tool room work for setting and machining different types of work. Labour cost is higher.	Skilled worker is necessary for setting and tooling lay-out. But semi-skilled worker can perform the machining operations. So labour cost is less.	
6. Arrangement of lead screw for cutting	There is long lead screw for thread-cutting operations.	There is no such type of lead screw but threads for shorter length may be cut by using short lead screw or chaser.	
7. Arrangement of taper turning	Any type of taper turning is possible.	Taper turning for shorter length is possible by using a formed tool.	

Points 1	Engine Lathe 2	Capstan Lathe 3	Turret Lathe 4
8. Arrangement of rear tool post	No such arrangement.	There is an arrangement of rear tool post for parting-off operations.	
9. Head stock size and rigidity	smaller size and less rigid.	Larger size and more rigid as number of operations at greater depth of cut and at higher speed are done at a time.	
10. According to the work of application	Any type of work can be done or known as versatile machine, suitable for tool room work and repair work.	Modification of Engine Lathe, maximum use in production work for higher accuracy and small jobs of bar type.	Modification of Engine Lathe maximum use in production where higher accuracy is not the factor and mainly for chucking work of big and heavier jobs.
11. Number of tools can work and hold at a time	Only one tool can work and hold at a time.	Combination cut is possible, 4 tools can be arranged at square tool post, one tool post on rear side of cross-slide and 6 or more tools can be arranged on a hexagonal tool post.	Same as capstan lathe except the rear tool post, not required for chucking work.
12. Arrangement of tail stock	There is a tail stock.	There is no such tail stock instead of which provided with hexagonal turret on sub-bed.	There is no such tail stock instead of which provided with a hexagonal turret on lathe bed.
13. Arrangement of automatic feed stop	No such arrangement.	Automatic movement of cutting tool to a pre-determined amount is possible by feed stops and trips.	
14. Number of operations can be done at a time	Only one operation is possible at a time.	Two or more operations are possible to machine at a time.	

A3.2 : Comparison Between Capstan & Turret Lathe

Points 1	Capstan Lathe 2	Turret Lathe 3
1. Type	Ram type Lathe.	Saddle type Lathe.
2. Accuracy	Greater accuracy for smaller jobs.	Comparatively less accuracy for larger jobs.
3. Job size	Light production work for bar type of works, limited to 60 mm in diameter.	Mass production work for chucking type of work. It can turn up to 200 mm in diameter.
4. Turret rest on	The hexagonal turret rests on a short slide or ram, which slides on saddle. The saddle can move on lathe bedways.	The hexagonal turret rests directly on a saddle, which can move both the ways, i.e., along bedways and traverse directions.
5. Arrangement of extra tool post or rear tool post	There is an extra tool post provided on the rear side of the tool post for parting-off, facing, etc., operations.	There is no such type of arrangement but may be provided if necessary.
6. Operators' skillness and attention require	More attention is required for higher accuracy of the job and hence necessity of skilled labour.	Mechanism is simple, less attention is required and semi-skilled worker can do it.
7. Special arrangement	In between Head stock and Ram. There is a cross-slide which is mounted on a carriage moves on bedways.	Power chucks or air operated chucks are generally provided in heavy duty turret for holding larger sizes of work quickly. Side hung type carriage may be equipped with this type.
8. Time required for job setting	Less than that for turret.	More than that for capstan.
9. Hexagonal turret may be operated	Manually or by power feed	Mostly by power feed
10. Turret travel	Moves on a short distance within the ram length.	Moves on a longer distance within the bed length.

Points 1	Capstan Lathe 2	Turret Lathe 3
11. Operator's fatigue	Less fatigue due to short travel of turret and smaller size of jobs.	More fatigue due to longer and heavier jobs having longer travel of turret.
12. Metal removal rate	Less, as lesser amount of speed, feed and depth of cut generally used.	More, due to greater amount of speed, feed and depth of cut are mostly used.

A3.3 : Capstan & Turret Lathes

These machines are the modification of an Engine Lathe. There is a hexagonal turret in place of a tail stock of an Engine Lathe. And also there is no long lead screw in these machines as having in an Engine Lathe. But short threads may be cut in these machines by using a short chaser or a form tool. These machines are very much used for batch or mass production work. Eleven or more tools may be mounted on the hexagonal Turret (6 tools), square Turret (4 tools) and rear tool post to match different operations of a particular production job. Though the initial setting time is higher but the machining time and tool changing time is much less than the Engine Lathe, will drastically reduce the overall machining time and hence the cost of production will be much less. Combination cuts may also be given in these machines. In these machines, stops are used for controlling the length of travel of the turret so that the tool may be moved to the required distance.

In case of a Capstan Lathe, the hexagonal Turret is mounted on a short slide or ram which again fitted on saddle. The saddle can move along the bedways and can be fixed to the bed in the required working position. Capstan Lathe is mostly used for bar type of work.

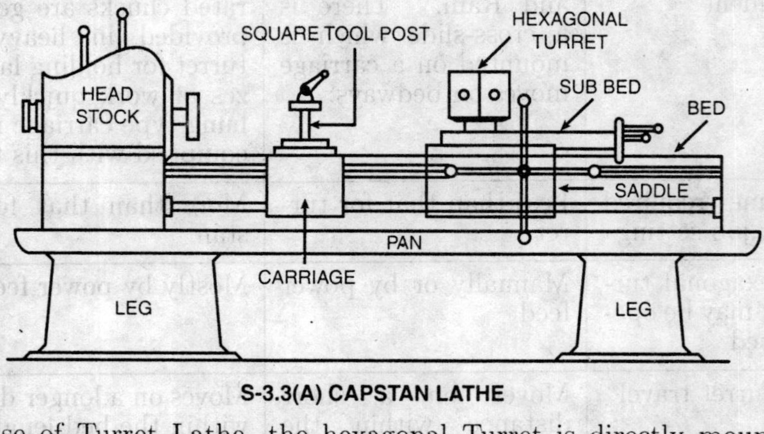

S-3.3(A) CAPSTAN LATHE

In case of Turret Lathe, the hexagonal Turret is directly mounted on the saddle. The saddle may slide directly on the bedways and having a longitudinal movement along the bedways which is limited by the length of the bed. Turret Lathe mostly suitable for chucking type of work.

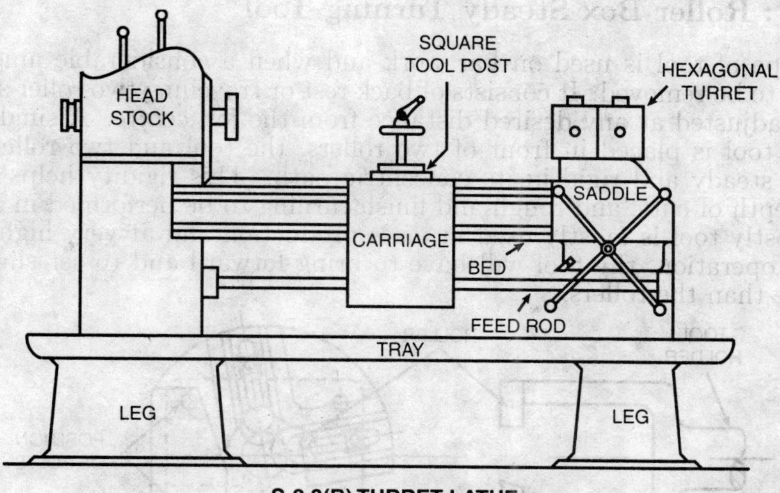

S-3.3(B) TURRET LATHE

A3.4 : Bar Feeding Arrangement In A Capstan Lathe Machine

In bar feeding arrangement, the bar is pushed forward immediately after releasing the collet without stopping the Lathe Machine. This arrangement helps to minimise the setting time of a product. The feeding bar passes through the Pedestal bushing, Bar holding chuck, Head stock spindle and to the Collet chuck. Collet chuck screwed on head stock spindle, holding the feeding bar and revolves with it with the revolution of the spindle. Bar holding chuck rotates within the sliding block with the rotation of the feeding bar. There is a rope fixed at pin on sliding block, runs over pulleys P_1 and P_2 hangs dead weight at the other free end, when the collet chuck is released by a lever (not shown in the sketch), the dead weight tends to move downward, exerts end thrust on the bar holding chuck, forces sliding block forward along with the bar holding chuck and feed the bar until it touches the workstop tool. This arrangement of feeding the bar without stopping the machine is known as bar feeding arrangement.

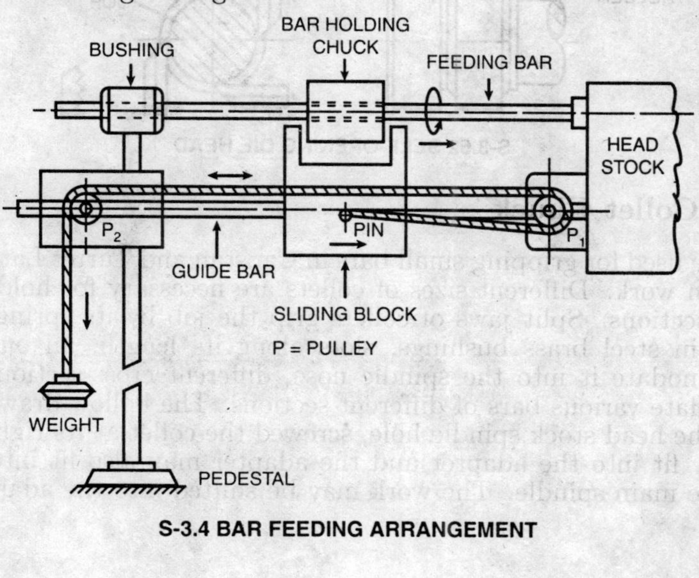

S-3.4 BAR FEEDING ARRANGEMENT

A3.51 : Roller Box Steady Turning Tool

This type of tool is used on bar work and when a considerable amount of stock is to be removed. It consists of back rest or travelling two roller steadies can be adjusted at any desired distance from the job centre. A single-point cutting tool is placed in front of two rollers, the tool and two rollers keep the job steady and rigid in its concentric path. This rigidity helps to give more depth of cuts, and rough and finish turning to be performed in a single cut. Costly tool is mostly used in mass production and at very high speed. During operation, the tool will have to bring forward and to set slightly in advance than the rollers.

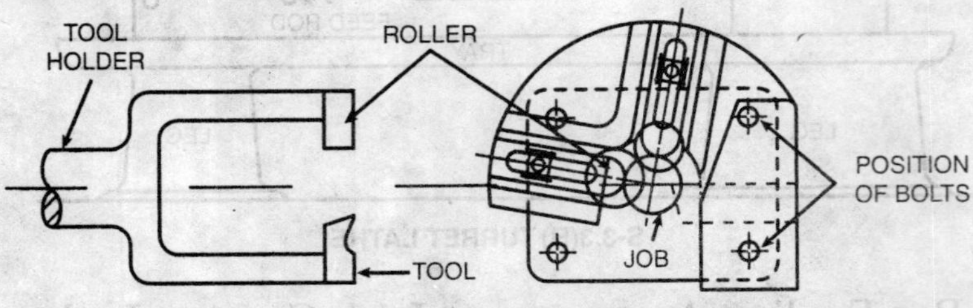

S-3.51 ROLLER BOX STEADY TURNING TOOL

A3.52 : Self-Opening Die Head

This type of tool is generally fitted with four chasers equally spaced for cutting an external thread. The pitch and size of the cutting edges of the chaser will be according to the type of the thread to be cut. Chasers may be of Radial or Tangential or Circular type. Tangential chasers are generally used on a self-opening die head due to its easy re-grinding facility and long tool life. According to the name self-opening, the die is automatically opened when the tool travel is stopped after the screw cutting operation. The die may again be closed by the handle.

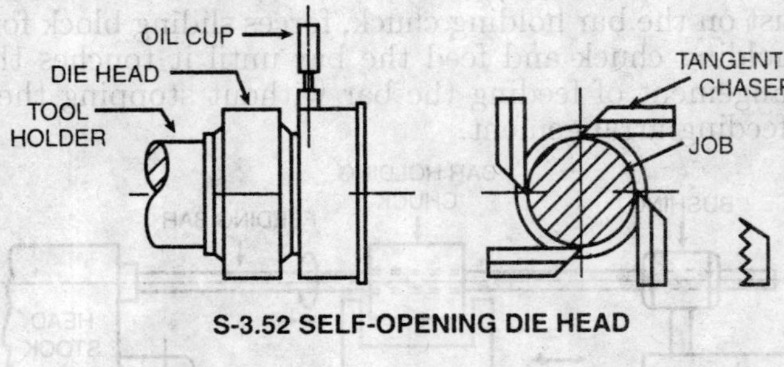

S-3.52 SELF-OPENING DIE HEAD

A3.53 : Collet Chuck

Collets are used for gripping small bars in Capstan and Turret Lathes of mass production work. Different sizes of collets are necessary for holding bars of different sections. Split jaws of collets grip the job by its springing action. It is a thin steel brass bushings. slots along its length, an outside taper to accommodate it into the spindle nose, different cross-sectional holes to accommodate various bars of different sections. The hollow draw bar passes through the head stock spindle hole, screwed the collet at its right end. The collet may fit into the adapter and the adapter may also fit into the taper hole of the main spindle. The work may be shifted into the adapter until it is gripped.

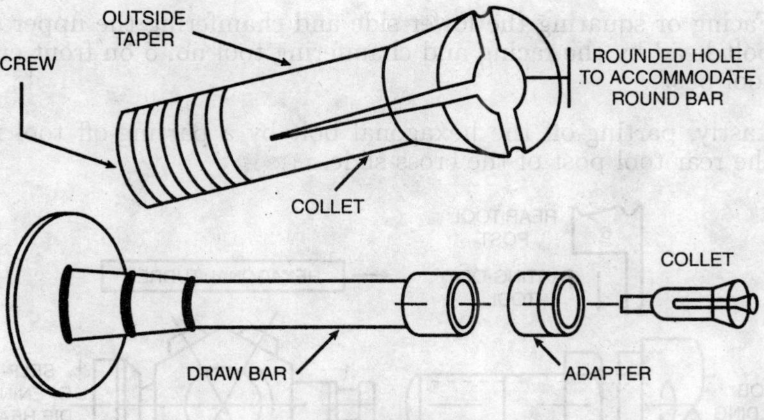

S-3.53 COLLET CHUCK

A3.61 : Production Of A Hexagonal Bolt By The Capstan Lathe

The following points are to be noticed before starting the operations:

1. Finished drawing of the product.
2. Capacity of the machine available.
3. Raw material available to the required size and composition in the market.
4. Availability of cutting tools and equipments as per operations involved.
5. Suitable speed, feed and depth of cut are to be set.
6. Use of suitable cutting fluid.
7. Checking of machines before starting towards lubrication, flow of cutting fluid, other cutting conditions, etc.
8. Setting of tools and equipments according to the sequence of operations involved.

The following sequence of operations is to be performed for the production of a hexagonal bolt :

1. Available hexagonal bar mounted on collet and rotates with it. Feeding the hexagonal bar through the collet against bar stop tool no. 1 on hexagonal turret.
2. Turn the hexagonal bar to the required diameter and length by the roller steady box turning tool no. 2 on hexagonal turret.
3. Form end of the bolt by a roller steady end forming tool no. 3 on hexagonal turret.
4. Threads are to cut by a self-opening die head with required diameter chasers tool no. 4 on hexagonal turret.

5. Facing or squaring the lower side and chamfering the upper side of a bolt head by the facing and chamfering tool no. 5 on front cross-slide tool post.

6. Lastly, parting-off the hexagonal bolt by a parting-off tool no. 6 on the rear tool post of the cross-slide.

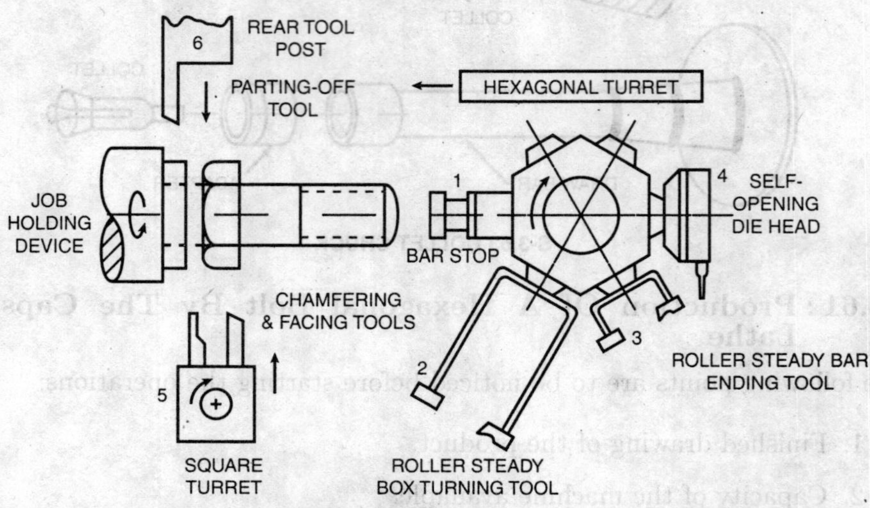

S-3.61 TOOLING LAY-OUT FOR PRODUCTION OF A HEXAGONAL BOLT

A3.62 : Production Of A Knurled And Screwed Cap As Per Drawing From A Round Bar

Before starting the operation the points stated for the production of hexagonal bolt numbers from 1 to 8 to be considered. Then, the following operational steps to be followed for the production of a cap in a Capstan Lathe machine :

1. Mounting of a available round bar on collet chuck and rotates with it. Feeding the bar through collet against bar stop tool no. 1 on 1st hexagonal turret position.

2. Drilling the required size and depth by a drill bit no. 2 on 2nd hexagonal turret position.

3. For internal thread cutting, taps (rougher, intermediate and finisher cuts) are to machine in sequence by tool nos. 4, 5 and 6 on 3rd, 4th and 5th hexagonal turret positions respectively.

4. Turning to the required diameter and length by the roller steady box turning tool no. 6 on 6th hexagonal turret position.

5. Ends chamfering and facing at a time are done by the form tool no. 7 on the 1st square turret position.

6. Knurl with adjustable knurling tool no. 8 and 9 on the 2nd and 3rd square turret positions respectively.

7. Lastly, parting-off the cap by the parting-off tool no. 10 on rear tool post position.

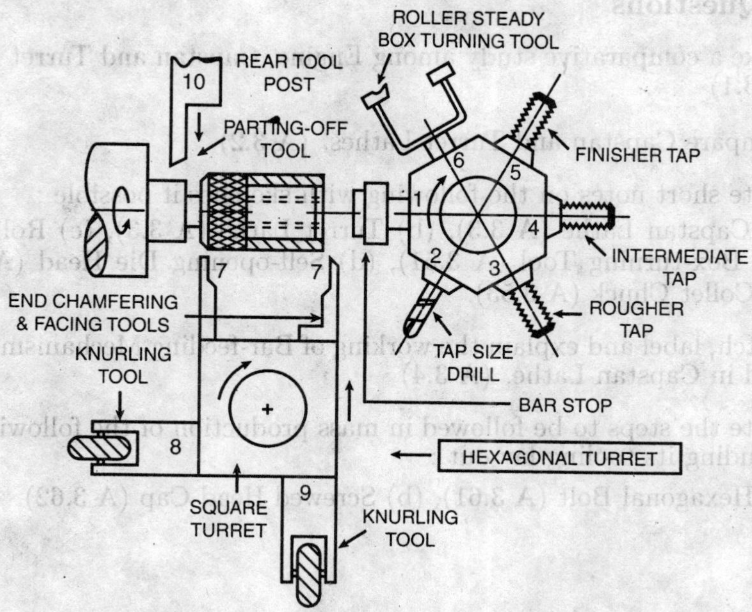

S-3.62 TOOLING LAY-OUT FOR PRODUCTION OF CAP

Operation Tool Schedule Chart for Cap Manufacturing

Machine–60 mm Capstan Lathe
Material–Round Mild Steel Bar

Sequence of operations	**Operations**	**Tools used**	**Tools position**
1.	Bar holding in Collet.	Bar stop.	1st Hex. Turret position.
2.	Drilling to the required size.	Taper shank twist drill.	2nd Hex. Turret position.
3.	Internal thread cutting by Rougher, Intermediate and Finisher taps.	Rougher tap, Intermediate tap and Finisher tap.	3rd, 4th and 5th Hex. Turret positions.
4.	Turning all.	Roller steady box turning tool.	6th Hex. Turret position.
5.	End chamfering and facing operations at a time.	Form tool.	1st square Turret position.
6.	Knurling operations.	Knurling tools 1st cut and 2nd cut.	2nd and 3rd square Turret positions.
7.	Parting-off operation.	Parting-off tool.	Rear tool post.

A3.7 : Questions

1. Make a comparative study among Engine, Capstan and Turret Lathes. (A 3.1)

2. Compare Capstan and Turret Lathes. (A 3.2)

3. Write short notes on the following with sketches if possible :

   (a) Capstan Lathe (A 3.3), (b) Turret Lathe (A 3.3), (c) Roller Steady Box-turning Tool (A 3.51), (d) Self-opening Die Head (A 3.52), (e) Collet Chuck (A 3.53).

4. Sketch, label and explain the working of Bar-feeding Mechanism mostly used in Capstan Lathe. (A 3.4)

5. Write the steps to be followed in mass production of the following jobs including its tooling lay out :

   (a) Hexagonal Bolt (A 3.61), (b) Screwed Head Cap (A 3.62).

4

DRILLING MACHINE

A4.1 : Introduction

Drilling machine is the second most important machine tool in a Machine Shop. It is mostly used for blind holes, enlarging and finishing holes in solid material. A hole is drilled quickly and at a low cost by this machine. In a drilling machine, the hole is generated by the cutting edges of rotating cutting tool known as the drill bit which exerts large force on the workpiece fixed on the table. The drilling machine sometime called "drill press" as the machine exerts vertical pressure to originate a hole. Drilling, Boring, Counter-boring, Counter-sinking, Reaming, Tapping and Spot Facing operations can be done by this machine. Drilling machines are used in machine assembly, repair shop, tool room, maintenance work, agricultural machinery, etc. The Upright drilling machine and the Radial drilling machines are widely used in the general engineering industries among drilling machines.

A4.2 : Classification Of Drilling Machines

Drilling machines are manufactured in different forms and sizes. Each machine tool designed to operate a class of work or specific job for maximum utility. Drilling machine is classified with respect to the following :

1. Type of drive used—Group drive or Individual drive.
2. Place of installation—On hand or On bench or On floor.
3. Application—Portable or Stationary.
4. Power supplied for operations—Manual or Electric.
5. No. of jobs to be machined—Single or Batch or Large number.
6. Axis position of drilling spindle—Vertical or Horizontal.
7. Wt. of the job in operation—Light or Medium or Heavy duty.
8. Number of spindle—Single or Multiple.
9. Speed of the operation—Low or Medium or High.
10. Feeding power—Manual or Automatic or Both.
11. Type of feed used—Mechanical or Hydraulic or Pneumatic or Automatic.
12. Feeding by—Table or Spindle.

13. Movement of tool head—Fixed or Sliding.
14. Work movement—Fixed or Movable.
15. Arrangement of table—Adjustable or Fixed.
16. Cross-section of column—Round or Square.
17. Skillness of the operator—Semi-skilled or Skilled.
18. Accuracy available—Moderate or High.

A4.3 : Types Of Drilling Machines

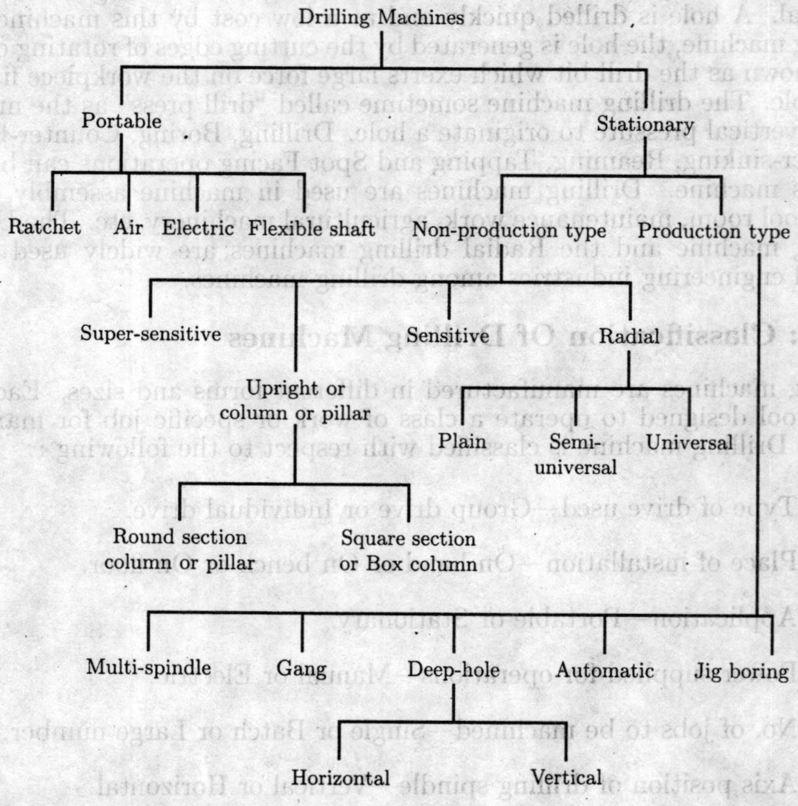

A 4.41 : Portable Drilling Machine

This type of machine is light in weight, compact in a smaller unit, easily handled or with respect to the workpiece in the workshop. Portable drilling machine is operated by hand power, pneumatic power or electric power (AC or DC). This type is mostly suitable for high speed non-productive jobs and drill size limited to 18 mm diameter maximum.

A4.42 : Ratchet Drilling Machine

This manually operated drilling machine is mostly suitable where electric power is not available or beyond the capacity of electric portable drilling machine. This machine is a very slower method and rotates the drill during the forward stroke only and the pawl actuates to slip during backward stroke.

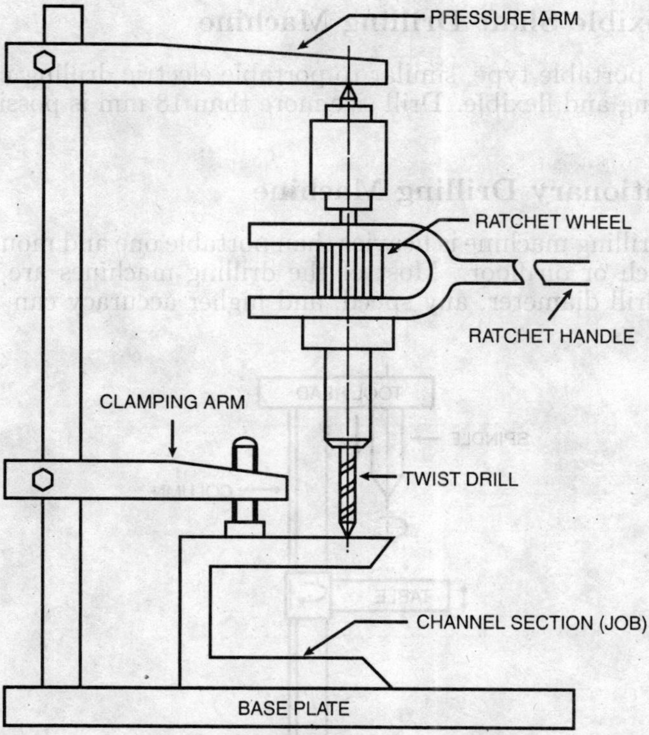

S-4.42 BLOCK DIAGRAM OF RATCHET DRILL

A4.43 : Air Drilling Machine

This is also a portable drilling machine, operated by compressed air, mostly used in mines, roads, old buildings, etc., to crack or to break.

A4.44 : Electric Drilling Machine

This is a portable drilling machine operated by electric power, maximum use among portable drilling machines, having high speed and can be moved elsewhere, but the maximum size of drilling diameter is restricted to 18 mm.

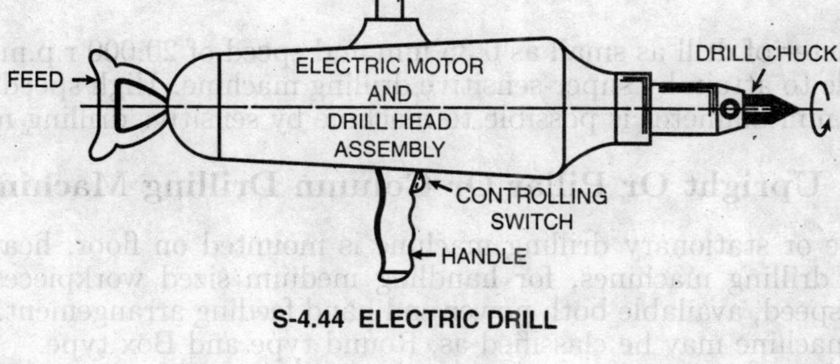

S-4.44 ELECTRIC DRILL

A4.45 : Flexible Shaft Drilling Machine

This is also a portable type, similar to portable electric drilling machine but the shaft is long and flexible. Drill size more than 18 mm is possible, by this type.

A4.46 : Stationary Drilling Machine

This type of drilling machine is heavier than portable one and mounted rigidly either on bench or on floor. Most of the drilling machines are within this group. Any drill diameter, any speed, and higher accuracy can be attained by this group.

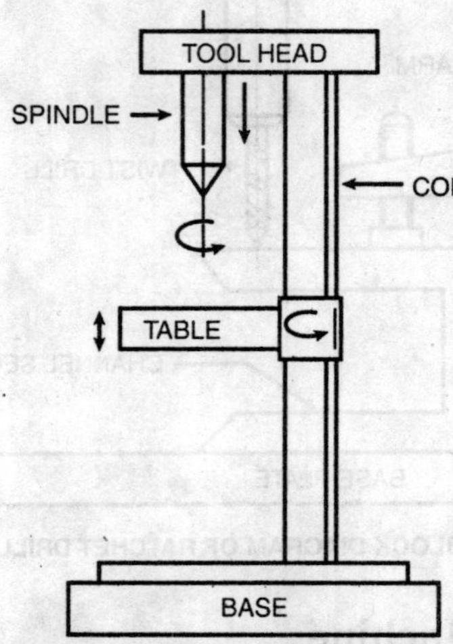

BLOCK DIAGRAM OF DRILLING MACHINE
S-4.46

A4.47 : Super-Sensitive And Sensitive Drilling Machines

Both the machines are smaller in size, generally mounted on bench, having very high speed to drill very small holes on light jobs.

There is no power feeding arrangement, feeding is purely on hand control of the operator, so that the operator can sense the feeding or can control feeding at will and accordingly named as super-sensitive or sensitive drilling machine.

Diameter of drill as small as 0.35 mm and speed of 20,000 r.p.m. or more is possible to attain by super-sensitive drilling machine. High speed and drill up to 15 mm diameter is possible to machine by sensitive drilling machine.

A4.48 : Upright Or Pillar Or Column Drilling Machine

This type or stationary drilling machine is mounted on floor, heavier than sensitive drilling machines, for handling medium sized workpieces, having medium speed, available both power and hand feeding arrangement. Upright drilling machine may be classified as, Round type and Box type.

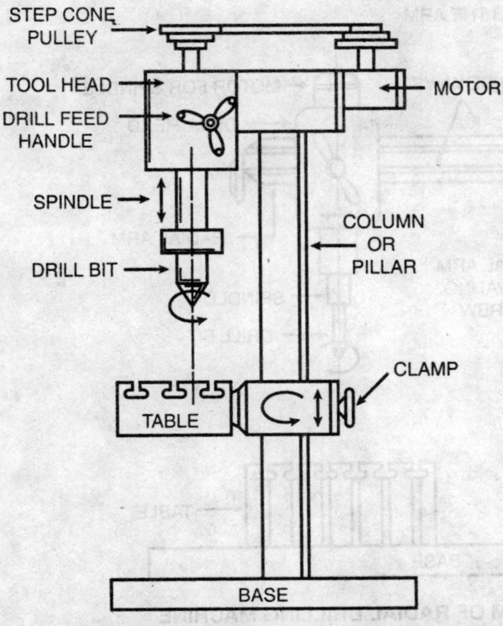

PILLAR OR UPRIGHT DRILLING MACHINE
S-4.48

Round type

In this type, column is round in section, also named as pillar drilling machine. The table holding workpiece can have the arrangement of rotating 360° about the round column for locating the workpiece under the spindle. Drill up to 50 mm diameter is possible to machine by this type.

Box type

In this type, column is square in section, heavier, more strengthy and rigid than round type. Square table can be raised or lowered by an elevating screw that gives more rigidity to the work table. Drill diameter more than 50 mm and up to 75 mm diameter is possible by this type.

A4.49 : Radial Drilling Machine

This type of machine is mounted on floor suitable for drilling medium to large and heavy workpieces are placed on the table or on the floor or in a pit. This machine having a vertical column about which a radial arm swings. The arm can also be moved up and down with respect to the column, which can be locked at any desired position as per job size. The radial arm is holding a tool post and can be moved horizontally along the arm ways, through which drilling spindle rotates and giving desired feed for tool bit.

Radial drilling machine may be classified with respect to movements of radial arm and tool head.

Plain type

1. Vertical movement of the radial arm with respect to the column, 2. Circular movement of the radial arm about column, 3. Horizontal movement of the tool along the arm ways.

Semi-universal type

In addition to the above three movements as in case of plain type, the fourth movement of the tool post can be swung about a horizontal axis perpendicular to the arm, which permits for drilling a hole inclined at any angle to the horizontal plane.

Universal type

In addition to the above fourth movement as in case of semi-universal type, the fifth movement of the radial arm is rotated on a horizontal axis. All these movements enable the universal drilling machine to drill on a job at any angle either in horizontal plane or in vertical plane or in both the planes.

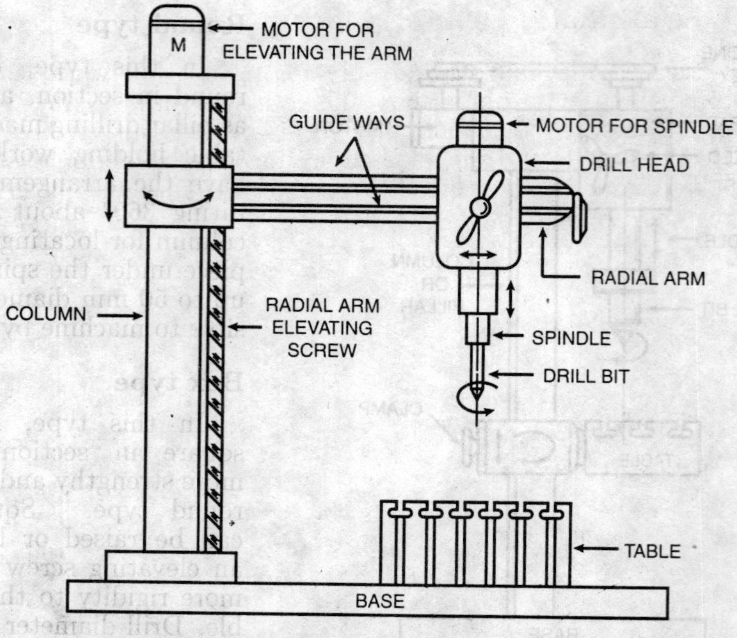

S-4.49 BLOCK DIAGRAM OF RADIAL DRILLING MACHINE

A4.410 : Multi-Spindle Drilling Machine

This machine is suitable for production of identical pieces to drill number of holes in a workpiece simultaneously by the number of spindles which are operated simultaneously from one driving shaft. In this type machine, spindles position can be adjusted as per requirements, feeding is mostly done by raising the table, while the drill head remains stationary.

A4.411 : Gang Drilling Machine

When a number (varies from four to six) of single spindle with essential speed and feed is mounted side by side on one base and has common work table, is known as the gang drilling machine. The space in between the spindles is adjusted or fixed. The speed and feed of the spindles can also be controlled independently. This is production type machine, each spindle is fixed with respective speed, feed, tool and position of the spindle as per operations required of a production job with the arrangement of job movement by using suitable jig and fixture.

A4.412 : Deep Hole Drilling Machine

This also a production type machine is suitable for drilling deep holes in rifle barrel, long shaft, etc. In this machine, job is rotated in horizontal or vertical axis at high speed and the long tool with bearing supports at required intervals, remains stationary or sometimes is rotated and using low feed for removing chips, tool is withdrawn from workpiece after penetrating to a length equal to the diameter of the tool and proper lubricant must be used to reduce frictions. According to the axis of the job-tool, the machine may be classified as horizontal or vertical type deep hole drilling machine.

A4.413 : Automatic Drilling Machine

This job-oriented machine tool is only for production work, can perform a series of drilling type operations at successive units and workpiece can transform unit to unit automatically. In a word we can say, if we feed raw material from one side and the finished product can have from other side, the job movement, unit change, speed change, feed change, etc., everything will be done by the machine itself.

A4.5 : Name The Various Types Of Drilling Machine Known To You With Their Specifications, Uses And Limitations

Classification 1	Specifications 2	Uses 3	Limitations 4
Remark.	In addition to the specification mentioned below, certain other specifications, need to mention to specify it fully— (i) Wt. of the m/c. (ii) H.P. of the motor. (iii) Floor space reqd. (iv) Speed range, no. of speed available. (v) Feed range and no. of feed available, etc.	Upright and Radial drilling machines are mostly used in Jobbing workshop, Tool room and Maintenance shop. But Gang drilling, Deep hole drilling, Multiple spindle drilling, Automatic drilling are mostly suitable in production work.	Drilling diameter, spindle travel, job holding arrangement, speed and feed range, rotational arrangement of tool and table are main limitations of different drilling machines.
Portable drilling m/c.	Maximum dia. of the drill that the spindle can hold.	1. To drill smaller holes and at any angle, 2. heavy weight or greater size of jobs which difficult to transport or lack of space.	1. High speed 1500 to 2000 r.p.m., 2. Limited to drill holes 18 mm diameter.
Retchet drilling m/c.	1. Maximum size of drilling dia., 2. Length of lever, 3. Maximum feeding range available.	1. Suitable for jobbing work, 2. hand operated, 3. drilling used only when electrical power is not available and where greater dia. drill than portable drilling is necessary, 4. in awkward situation, 5. for slower method of machining.	1.Hand operated, 2. Lever length is limited, range 300 to 600 mm rod, 3. Length of feed range 60–85 mm.

Contd...

Classification 1	Specifications 2	Uses 3	Limitations 4
Super-sensitive and sensitive drilling machine.	1. Maximum dia. of the job can be accommodated under the spindle, 2. Maximum spindle travel is available.	1. For very light work, 2. Very high speed hand feed only.	1. Very high speed (20,000 and above), 2. Hand feed only, 3. Super-sensitive drilling even to 0.35 mm dia. 4. Sensitive drilling dia. up to 15 mm.
Upright drilling m/c. (a) Round column or pillar, (b) Box column.	1. Maximum dia. of the job can be accommodated, 2. Table can be rorated 360° about its centre in case of pillar drill.	1. For medium size work, 2. Having both hand and power feeding arrangements.	The machine can work within the following parameters: 1. For Round column holes up to 50 mm dia., 2. For Box column holes up to 75 mm dia., 3. Dia. range 3 to 75 mm, 4. Speed range 75 to 3500 r.p.m., 5. Power range 0.27–10 kW, 6. Weight range 40-3500 kg.
Radial drilling m/c.—plain, semi-universal and universal.	1. Dia. of the column. 2. Lengh of arm, 3. Maximum and minimum drilling radius, 4. More range of spindle speed and feed, 5. Swinging arrangements of drill head. 6. Rotary arrangement of drill head on a horizontal axis.	1. For medium to large and heavy work-pieces, 2. To drill work-piece at any angle, 3. Most versatile, 4. Mostly used in tool room and maintenance room.	1. Drilling dia. range 35-100 mm, 2. Maximum drilling radius 1250 to 3150 mm, 3. Speed range 10 to 2,500 r.p.m., 4. Power range 2.2 to 13 kw, 5. Net wt. 3000 to 18000 kg.

Contd...

Classification 1	Specifications 2	Uses 3	Limitations 4
Gang drilling machine.	1. Arrangements of number of single spindle on a common work table, 2. Maximum space adjustable in between column and spindle.	1. Series of operation are performed by minimum movement of the workpiece on common work table, 2. Mostly suitable for production work.	Four to six spindles in a common work table.
Multiple spindle drilling machine.	1. Maximum drilling area, 2. The size and number of holes a machine can drill, 3. Maximum amount of table raising and tool lowering for feeding.	1. Number of identical holes in a workpiece can be drilled at a time. 2. Suitable for mass production work only.	Two or more spindles are driven by a common driving shaft.
Deep hole drilling machine.	Drilling dia. and maximum length of drilling hole available.	For drilling very long holes or smaller dia. such as rifle barrels, long shafts, etc.	1. Majority of deep hole drilling application in rifle hole-depth 10 to 30 times the drill diameter, 2. However, depth of diameter ratios of 100 : 1 available.
Automatic drilling machine.	Specification completely depending on the product and its operations.	1. Series of machining operations at successive units. 2. To transfer workpiece from one unit to another automatically, 3. M/c suitable for mass production work only.	Limit to a production work only within certain limit of a particular product.

A4.6 : Short Description Of Main Parts Of Drilling Machine

Definition

Drill press or drilling machine is manufactured for making holes on blind materials, tool also known as drill bit rotates with proper speed in vertical axis and penetrates into the work material giving desired axial feed, while the work material remains fixed on the table or base during the operation.

Short description of main components of Drilling Machine

Bed or Base

It is the main support of the column, the driving pulleys, the motor in case of a self-drive, etc. The top surface of the base or bed is machined and provided with tee-slot to fix any fixture or big jobs on the table.

Column

It is a vertical member must be sufficiently rigid to support the tool post including driving mechanism and the table. The cross-section of column is either round or square (also named as box type); square section is more rigid than the round one.

Table

The table is mounted round the column having three types of adjustment, radial adjustment about the column, vertical adjustment and circular adjustment about its own axis, after required adjustment the table and the arm may be fixed in position during operation. The table may be either round or rectangular surface and provided with tee-slots for fixing vice or workpiece on the table.

Transmission system or Drive

The spindle is rotated in vertical axis by the top shaft in horizontal axis through two bevel gears. Top shaft rotates from the fast and loose pulley shaft in horizontal axis through two sets of step cone pulleys, which gives different speeds of the spindle shafts. Fast and loose pulley shaft (also known as counter-shaft) gets power either from motor directly or through line shaft.

Spindle

The spindle has two motions, one is circular or rotary gets power from the top shaft through transmission system and the other is vertical motion for feeding gets power from the top shaft through spur gears, worm, sets of pinions and rack fixed on hollow cylinder, mounted on spindle guided by bearings. The spindles feeding may be hand or power operated. The lower end of the spindle has a Morse taper socket to fit drill bits or drill chuck for parallel shank drill bit. There is a rectangular slot to use drifts for removing drill bits or drill chuck from spindle nose.

A4.61 : Drill Spindle (Head) Assembly

A vertical shaft known as drill spindle, holds drill bit through a Morse taper hole at its lower end. Spindle gets power from the top shaft through two bevel

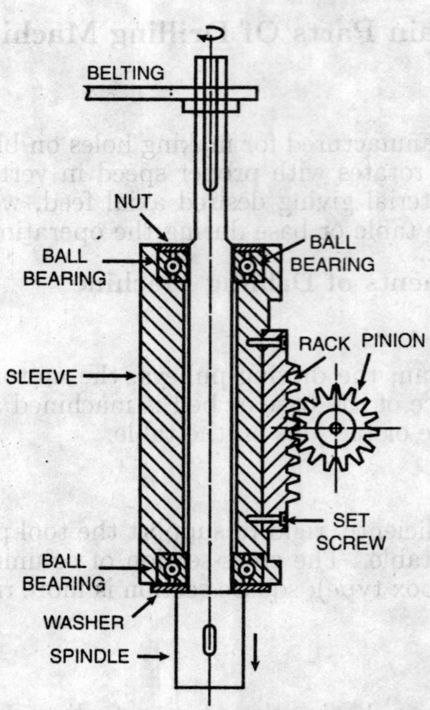

S-4.61 DRILL HEAD ASSEMBLY

gears. The vertical shaft is splined or feather-keyed so that the spindle can rotate as well as can move vertically up and down for feeding through the bevel gears, which remains and rotates, in a particular position. The spindle is rotated within a non-rotating hollow shaft placed two ball bearings in between the hollow shaft and the vertical solid shaft. The hollow shaft remains in position along the solid shaft length with the help of washer at the lower end and a nut at the top of the hollow shaft. A rack is mounted on the right side of the hollow shaft. The hollow shaft can move up and down through a pinion meshed with the rack. The pinion gets power from the top shaft through two spur gears, worm shaft, worm. There is rectangular slot on the top of the Morse taper hole to remove the drill bit after operation from the spindle by the drift.

Radial Arm

The radial arm is mounted on the column, moves radially along the column through 360 degree manually and can also be moved up and down by a separate motor. A drill head is mounted on the radial arm, can move on radial arm horizontally to its guide ways manually before operation, all the above movements, will be locked to maintain rigidity of the machine.

A4.7 : Specifications Or Size Of A Drilling Machine

The size of a drilling machine is expressed or specified by the following items :

1. Maximum size of the drill in mm that the machine can operate.
2. Table size of maximum dimensions of a job can mount on a table in square metre.
3. Maximum spindle travel in mm.
4. Number of spindle speeds and range of spindle speeds in r.p.m.
5. Number of automatic spindle feeds or feed range available in mm/revolutio
6. Morse taper no. of the drill spindle nose.
7. Power input of the machine in H.P.
8. Floor space required in square metre.
9. Net weight of the machine in tonne.

A4.8 : Specifications Or Size Of A Radial Drilling Machine

The size of a radial drilling machine is expressed or specified by the following items :

1. Diameter of the column in mm.
2. Length of the arm in mm.
3. Maximum drilling radius in mm.
4. Minimum drilling radius in mm.
5. Maximum size of the drill that the machine can operate in mm.
6. Table size or maximum dimensions of a job can mount on a table in square metre.
7. Number of spindle speeds or range of spindle speeds in r.p.m.
8. Number of automatic spindle feeds, or feed range available in mm /revolution.
9. Morse taper no. of the drill spindle nose.
10. Power input of the machine in kw.
11. Floor-space required in square metres.
12. Net weight of the machine in tonnes.

A4.9 : Tool Holding Devices

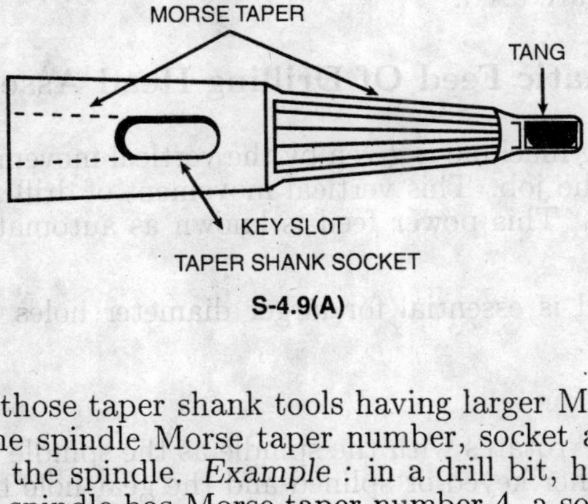

S-4.9(A)

Socket

In the case of those taper shank tools having larger Morse taper number with respect to the spindle Morse taper number, socket acts as a media to hold the tool into the spindle. *Example :* in a drill bit, having Morse taper number 6 but the spindle has Morse taper number 4, a socket of 6 to 4 may be used to connect.

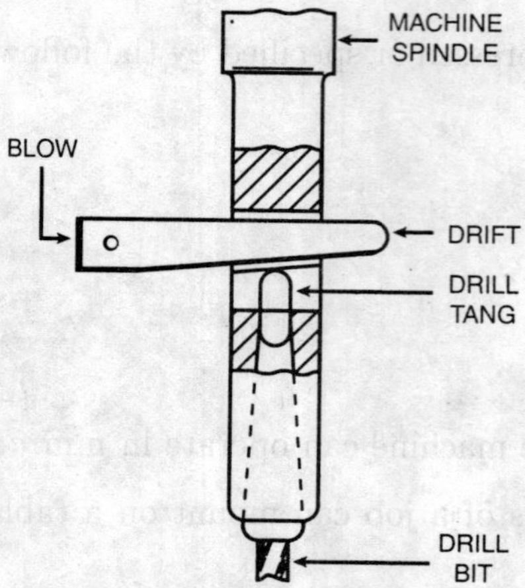

EXTRACTING OF DRILL BIT BY USING DRIFT
S-4.9(B)

Sleeve

In case of those taper shank tools having smaller Morse taper number with respect to the spindle Morse taper number, one or more sleeve(s) may be used as a media to hold the tool into the spindle. *Example* : Three sleeves are required to match with the spindle Morse taper no. of 5 from the drill Morse taper no. of 2, i.e., in the 1st sleeve 2–3, 2nd sleeve 3–4 and 3rd sleeve 4–5 Morse taper nos.

Drift

It is used for taking out the socket or sleeve or twist drill or drill chuck from the spindle by driving it into the keyslot. The above sketch itself is self explanatory.

Drill chuck

This type of chuck is suitable to hold any smaller size drills, straight shank drills and other tools not having Morse taper at the holding end.

Floating holder

The type of holder is either straight shank or taper shank for taps, counter bores, reamers, etc., are used to compensate for non-alignment of drill spindle with work, and permits self alignment of the tool. Such holder is less important when jigs are used.

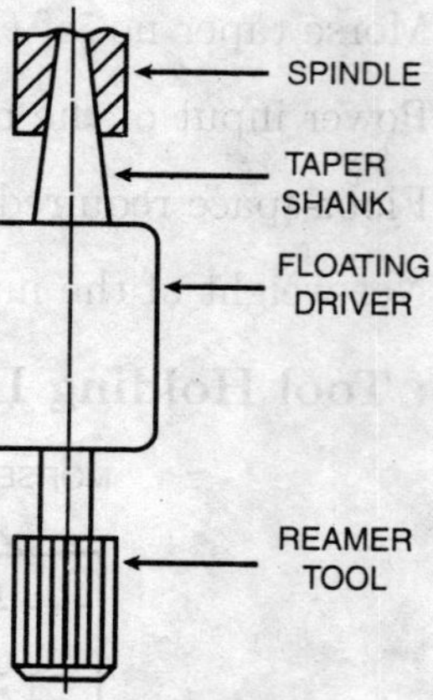

FLOATING HOLDER
S-4.9(C)

A4.10 : Automatic Feed Of Drilling Head Assembly

The feed in drilling machine is given by the vertical movement of the drill bit in a spindle into the job. This vertical movement of drill may be controlled by hand or power. This power feed is known as automatic feed motion in drilling machine.

Automatic feed is essential for larger diameter holes and more cutting pressure.

Power feed

The gear 'A' rotates with the spindle as the spindle passes through it. The spindle is feather keyed or splined and the gear hole has similar slot on it, is similarly splined so that the spindle rotates with gear and at the same time the spindle moves axially for feeding.

Gear 'B' is connected with gear 'A' and so 'B' also rotates the shaft 'S'. At a suitable point under the shaft, there is a worm which drives a pinion or a set of pinions. The last pinion is connected with the rack on the hollow shaft fitted over the spindle, the spindle with the help of the collars this hollow shaft moves axially along the spindle of the machine. Thus the automatic feed is given to the drill. In costly and precision machine, a set of gears is provided between the worm and the rack and rate of feed is adjusted accordingly by changing gears through suitable levers.

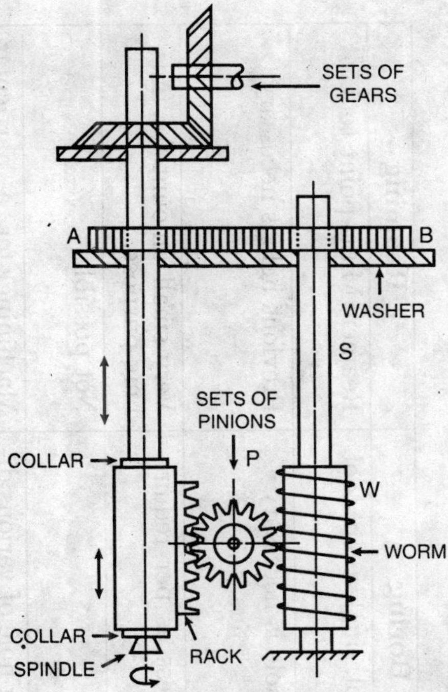

S-4.10 FEED MECHANISM OF DRILLING

Hand feed

For hand feeding a lever is attached with the shaft of last pinion meshed with 'P' and by rotating this lever hand feed is given.

A4.11 : Comparison Among Boring, Reaming And Drilling

Points	Drilling	Boring	Reaming
1. Name and type of the tool used.	Drill bit, double cutting edges.	Boring tool, single-point tool.	Reamer, Multi-point tool.
2. Whether previous hole is necessary.	Not always essential (except in case of trepanning).	previous hole is necessary.	Previous hole is necessary.
3. Amount of metal removal.	More.	More or less as per requirement.	Very small amount for finishing purpose.
4. Enlargement of hole.	Possible by using higher diameter drill.	Possible.	Not possible.
5. One tool can be used for.	Machining hole for a particular diameter of the hole.	Machining hole of various diameter of the holes.	Machining for a particular drilling hole.
6. Whether correct hole location is possible.	Not possible due to non-rigidity and non-alignment of the drill spindle and the drill.	Possible as the tool is more rigid and trace its own path.	Not possible as the tool follows the previously drilled hole.
7. Accuracy.	Less accuracy, generated hole is always slightly over-sized than the drill due to vibration.	Better surface finish and higher accuracy are possible.	Better surface finish and accuracy are possible.
8. Speed of operation.	Greater than boring and reaming.	Less than 50% speed of drilling operation.	50% speed of drilling operation.
9. Suitable for.	Faster method, general purpose.	Slower method of Machining, Tool Room Work.	Faster method of Machining, Production Work.

A4.12 : Comparison Between Boring And Reaming

Both the operations are done after drilling and generally used for better surface finishing and higher accuracy.

Points	Boring	Reaming
1. Number of Cutting point(s)	1. Single-point cutting tool is used in this operation and known as boring tool.	1. Multi-point cutting tool is used in this operation and known as reamer.
2. Speed	2. It is a slower method.	2. It is a faster method.
3. Metal removal rate	3. More metal removal is possible.	3. Metal removal is less.
4. Tool for different diameter(s)	4. One tool can be used for machining hole of various diameters.	4. One tool is fixed for a particular diameter of the hole of the job.
5. Alignment	5. In this operation tool will not follow the previous drilled hole and will trace its own path and can give correct hole location.	5. In this operation tool will follow the previous drilled hole and will not trace its own path and cannot make correct hole location.
6. Purpose	6. This operation is being done both for surface finish and also for enlargement of the hole.	6. This operation is being done only for better surface finish, not for enlargement of the hole.
7. Type of work	7. Suitable for jobbing work.	7. Used generally in mass production.
8. Cross-section of chip(s)	8. Chips are larger in size.	8. The chips are very small in size in terms of ground chips.

A4.13 : Designation Or Specification Of A Drill Bit

1. Type of shank: Shank may be of parallel or Morse taper.
2. Type of flute : Flute may of flat or straight or twist.
3. Flute length or series : It may be of short, medium or long.
4. Diameter of drill : Actual diameter of drill bit in mm.
5. I.S. code number : It deals with the composition of the drill material and other drill parameters.
6. Drill bit material : It is of Carbon Steel (C.S.) or High Speed Steel (H.S.S.) or sometimes cemented carbide tip brazed.
7. Job material : Material is Hard (H) like High Carbon Steel, Bronze or Normal (N) like Mild Steel or Soft(S) like aluminium.

8. Point angle :	It is to be selected according to the job material, i.e., for H—150°, for N—118° and for S—80°.

Designation of drill bit

Taper shank	Twist drill	(Short)	25.00	IS 599	H.S.S.	N	118.
(1)	(2)	(3)	(4)	(5)	(6)	(7)	(8)

No. at top of each element is labelled above.

A4.14 : Classification Of Drills

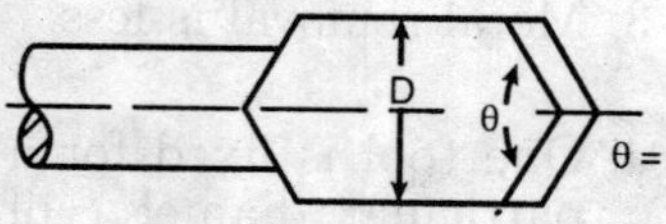

FLAT DRILL
S-4.14(1)

Drill is an end cutting tool and is used for the purpose of originating holes.

1. Flat drill : Made of tool steel and can be used where twist drills are not available. It is mainly used in wooden work.

2. D-bit drill : Made of high speed steel and generally used for deep drilling.

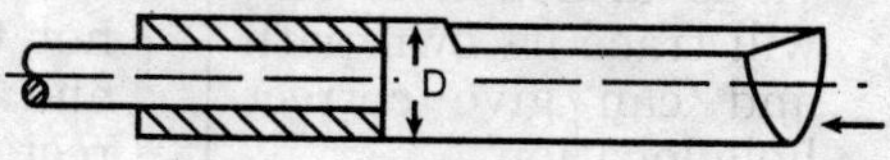

D-BIT DRILL
S-4.14(2)

3. Centre drill : It is a straight shank two-fluted twist drill used to make a centre holes on the ends surfaces of a shaft to hold the shaft in between centres on a Lathe Machine or elsewhere necessary.

4. Oil tube drill : It is used to drill a deep hole of 13 mm or more in diameter in production of steel parts. There are oil tubes (or holes) running lengthwise spirally through the body to carry oil directly to the cutting lips. Oil feeding socket is necessary with such drill to act as oil reservoir. Oil from the reservoir to the tube(s) is forced into the holes in the shank of the drill to the cutting edges to remove the chips.

This type of drill is mainly used in barrel of rifles.

5. Farmer drill : Straight fluted drill used for drilling brass and sheet metal works.

6. Twist drill : A twist drill is an end cutting tool. Its different sizes are available both in British and Metric systems.

British System Sizes

(a) Number drill : The drill sizes are available from no. 1 to 80. Diameter varies from 0.228 inch to 0.0135 inch. No. 1 is the largest size and no. 80 is the smallest size drills are made of straight shank type.

(b) Letter drill : The drill sizes are available from letter 'A' to letter 'Z'. Diameter varies from 0.234 inch to 0.413 inch. 'A' is smallest size and 'Z' is largest size.

(c) Fractional drill : The drill sizes are available from 1/64 inch to 5 inch with a division of 1/64 inch per tool up to $1\frac{3}{4}$ inch and then the steps increase gradually.

Metric System Size

This system also known as millimetre drills are available from 0.5 mm to 100 mm with an increase of 0.5 mm per tool.

A4.15 : Advantages Of Using Spiral Flutes

The following are the advantages of spiral flutes over the other flutes :

1. Diameters and spiral angles are retained to the end during re-grinding.
2. Easy removal of chips from the drill hole in the form of two curl coils.
3. Make passage for the cutting fluid to reach at the cutting edges.
4. Cutting edges are dynamically balanced.
5. Uniform torque distribution.
6. Less vibration during operation.
7. More aligned hole may be generated.
8. It increases rigidity and strength at the cutting edges.

A4.16 : Nomenclature & Angles Of A Twist Drill

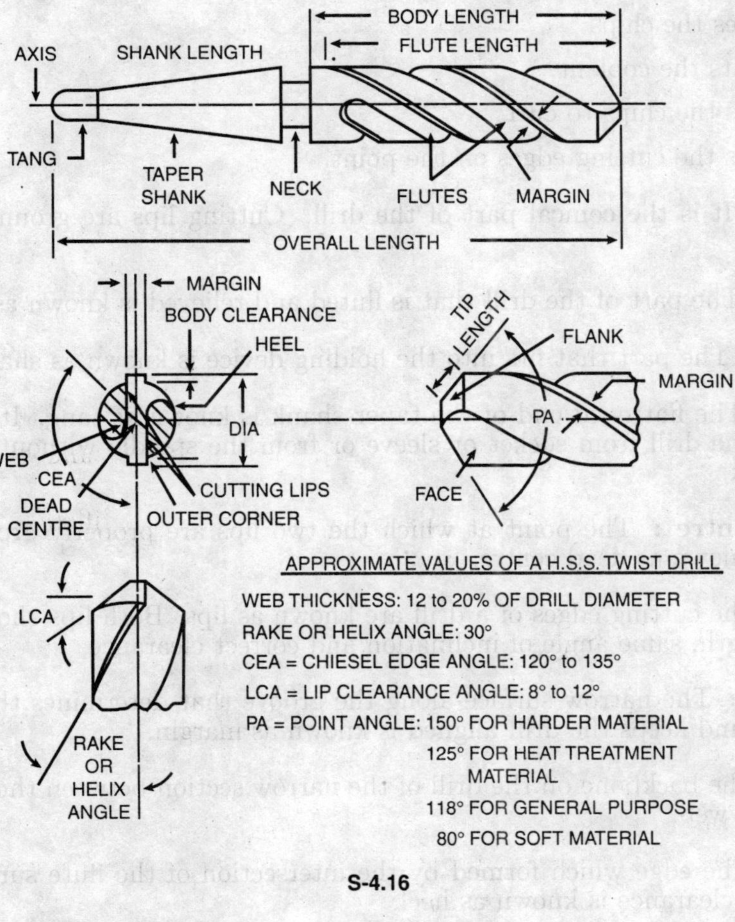

S-4.16

Twist drill is an end cutting tool. It has double end cutting edges.

Systematic naming of different elements of a twist drill and showing of different angles are known as twist drill nomenclature and angles.

A4.17 : Functional Difference Between Straight Fluted Drill And Twist Drill

Straight fluted drill is not suitable for standard practice as the automatic chip removal is not possible. It is mostly used for drilling softer materials like wooden work, brass, copper, etc. It is also not suitable for sheet metal work as it finds difficulty for lifting the tool from the sheet metal.

Twist drill may be used else where irrespective of materials soft or hard, thin or thick plates, having many advantages over the straight fluted drill, like easy chip removal, easy flow of cutting fluid at the cutting edges, uniform torque at the cutting points, more aligned, less vibration during work, etc. It is an end cutting tool.

A4.18 : Functions Of Drill Parts

Flute : The helical groove in the drill body is known as flutes. It does the following purposes:

(a) Carries the chips.

(b) Admits the coolant.

(c) Makes the chips to curl.

(d) Forms the cutting edges on the point.

Point : It is the conical part of the drill. Cutting lips are ground on the point.

Body : The part of the drill that is fluted and relieved is known as body.

Shank : The part that fits into the holding device is known as shank.

Tang : The flattened end of the taper shank is known as tang. It helps to remove the drill from socket or sleeve or from the spindle without injuring the shank.

Dead centre : The point at which the two lips are properly ground and meet is known as dead centre.

Lips : The cutting edges of a drill are known as lips. Both lips should have equal length,-same angle of inclination and correct clearance.

Margin : The narrow surface along the groove that determines the size of the drill and keeps the drill aligned is known as margin.

Web : The backbone on the drill of the narrow section between the flutes is known as web.

Heel : The edge which formed by the intersection of the flute surface and the body clearance is known as heel.

A4.19 : Drilling Operations

Drilling : The operation of producing a circular hole by removing metal by rotating the cutting edge(s) of a drill is known as drilling. Metal is removed by shearing and by extension. Drilling hole will be slightly oversized than the drill used due to the non-alignment of the drill bit and vibration of the spindle. For producing accurate hole, drill bit should be chosen slightly undersized, and subsequent reaming or boring operation is essential after drilling. For Example : 23 mm drill size may be considered to make a hole of 25 mm.

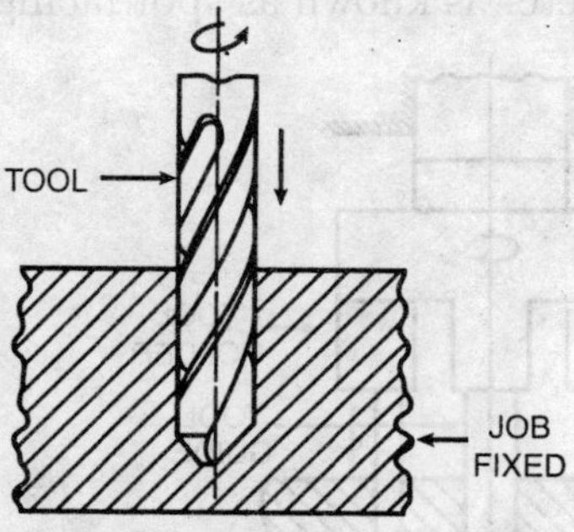

S-4.19(A) DRILLING

Reaming : The operation of finishing and sizing a previous drilled hole by using a multi-edges cutting tool named as reamer, is known as reaming. Very small amount (maximum 0.4 mm) of material removal is possible by this operation.

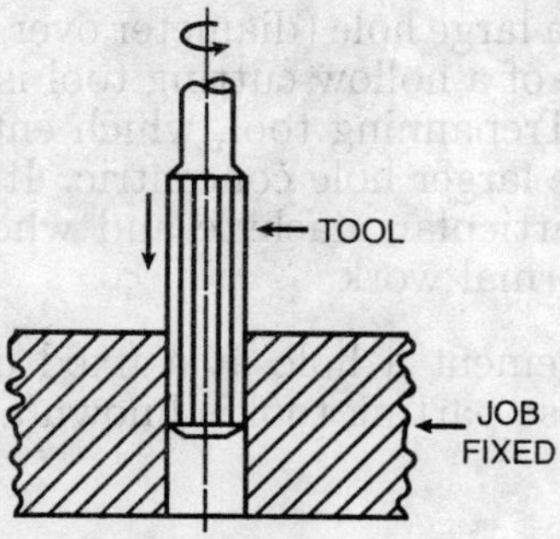

S-4.19(B) REAMING

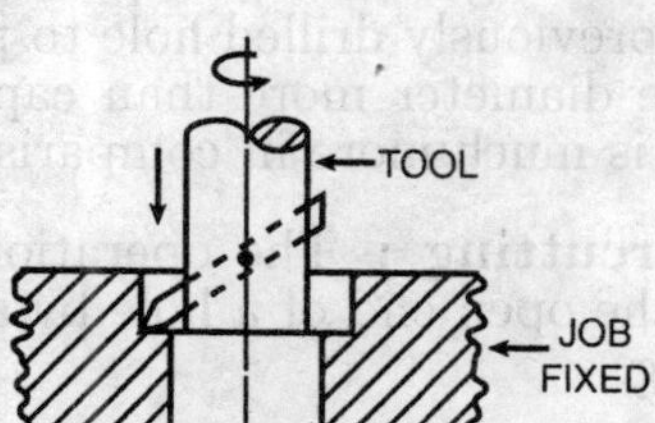

S-4.19(C) BORING

Boring : The operation of enlarging and finishing a previous casting hole throughout its length by means of an adjustable single edge cutting tool named as boring tool is known as boring.

Counter-boring : The operation of enlarging the end of a hole cylindrically, as recess for a bolt head, is known as counter-boring.

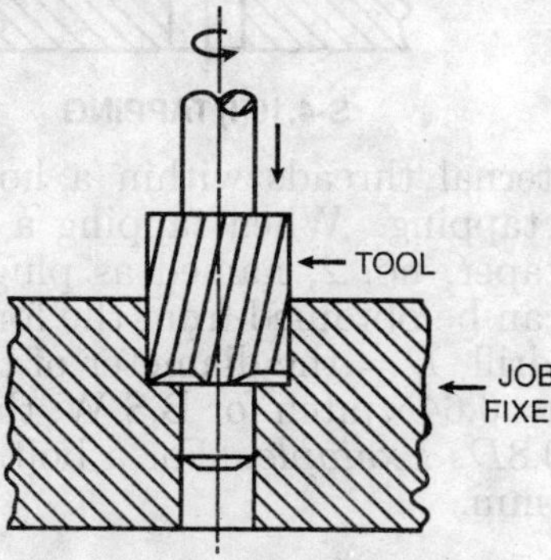

COUNTER-BORING OPERATION
S-4.19(D)

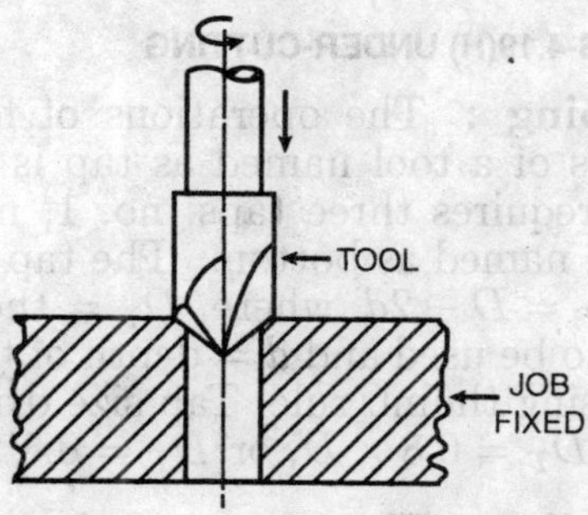

S-4.19(E) COUNTER-SINKING

Counter-sinking : The operation of making a cone-shaped enlargement of the end of a hole, as for the accommodating the screw head, is known as counter-sinking.

Spot facing : The operation of squaring and smoothing the surface around a hole, as for the seat for a nut or head of a bolt, etc., is known as spot facing.

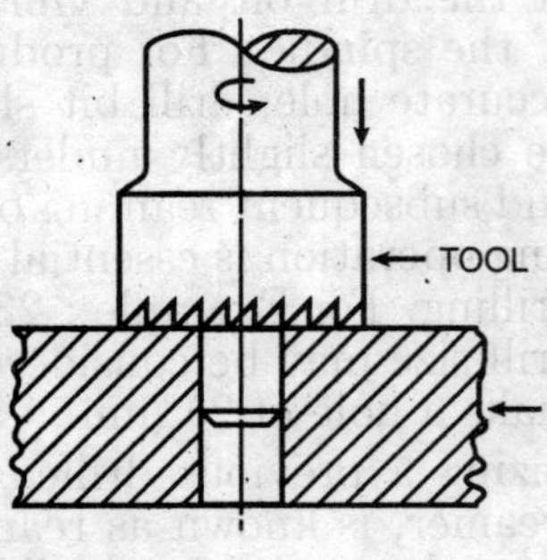

S-4.19(F) SPOT FACING

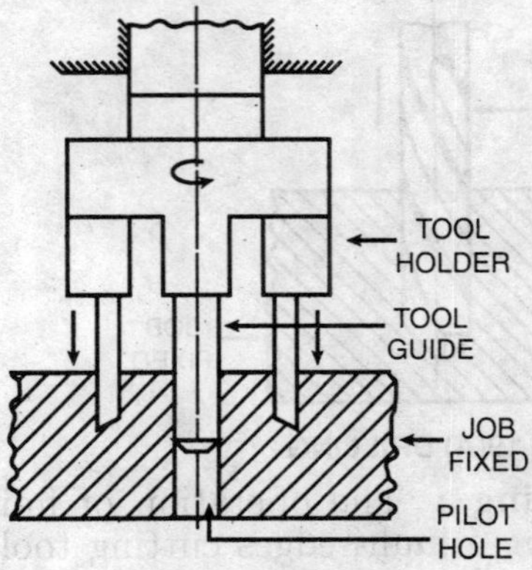

S-4.19(G) TREPANNING

Trepanning : The operation of producing a large hole (diameter over 50 mm) by removing metal along the circumference of a hollow cutting tool is known as trepanning. There is a pilot inside the Trepanning tool, which enters the small previously drilled hole to produce the larger hole concentric. It is used for the diameter more than capacity of particular machine and where hole depth is much more in comparison with normal work.

Undercutting : The operation of enlargement a hole at a fixed distance from the open end of a hole by a square nose parting tool is known as under cutting.

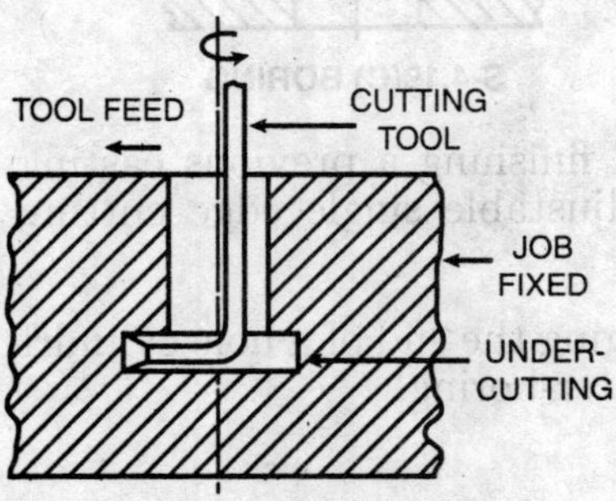

S-4.19(H) UNDER-CUTTING

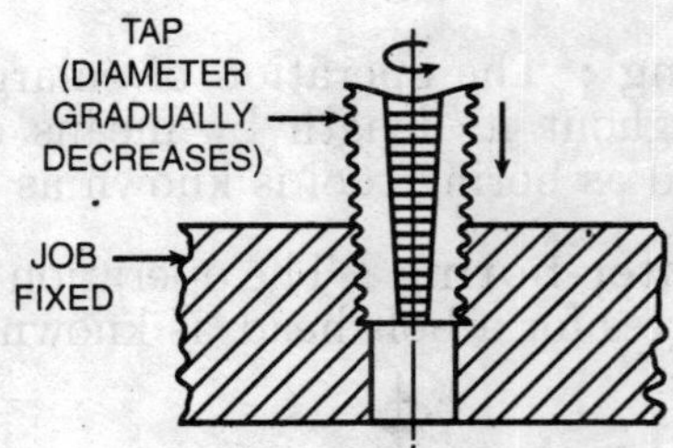

S-4.19(I) TAPPING

Tapping : The operations of forming internal threads within a hole by means of a tool named as tap is known as tapping. When tapping a blind hole requires three taps, no. 1, named as taper, no. 2, named as plug and no. 3, named as bottom. The tap size drill can be obtained from the formula as $D_T = D - 2d$, where, $D_T =$ the tap size drill, $D =$ the diameter of tap or bolt to be used and $d =$ depth of the thread $= 0.64 \times$ pitch for B.S.W. thread. By using thumb rule, Tap size drill, $D_T = 0.8D$. *Example :* For a bolt of 25 mm, $D_T = 0.8 \times D$, or $D_T = 0.8 \times 25 = 20$ mm.

Grinding : The operation of removing large amount of stock w.r.t. lapping and honing operations from hardened material and super-finishing a surface accuracy up to ±0.0025 mm by means of a tool named as grinding wheel is known as grinding. (Accuracy 0.9 to 5 Microns, 1 Micron = 0.001 mm)

Lapping : The operation of sizing hardened holes and extremely limited in stock removal is known as lapping. (Accuracy 0.08 to 0.25 Microns)

Honing : The operation of finishing relatively large holes such as automobile cylinders by means of relatively slow moving abrasives is known as honing. (Accuracy 0.12 to 1.2 Microns)

A4.20 : Cutting Speed, Feed & Depth Of Cut

Cutting speed

In drilling machine, cutting speed is the peripheral speed of a cutting point which in contact with the job.

It depends mainly on Job material and Tool material, and other dependable factors are also machine's rigidity, coolant used, type of finish required, operator's alertness, etc.

To select a spindle speed (i.e., r.p.m.) from a given cutting speed and drill diameter, we have to use the relation

$$\text{Spindle speed, } N \text{ (in r.p.m.)} = \frac{\text{Cutting speed, } V \text{ in metre/min.} \times 1000}{\pi \times \text{Diameter of drill, } D \text{ in mm}}$$

$$\text{or, } N = \frac{V \times 1000}{\pi D} \text{ r.p.m.}$$

Example 1.

Calculate the spindle speed or r.p.m. to drill a hole of 30 mm using cutting speed as 20 metres per minute.

Solution:

Data given :

Cutting speed, $V = 20$ m/min.

Diam. of drill, $D = 30$ mm

Let N be the spindle speed in r.p.m.

We know the relation,

$$N = \frac{V \times 1000}{\pi D}$$

$$= \frac{20 \times 1000}{3.142 \times 30} \text{ r.p.m.}$$

$$= 212.2 \text{ r.p.m.}$$

$$= 214 \text{ r.p.m.}$$

(nearest available speed on the machine)

Select a nearest available number on the machine chart if the calculated r.p.m. is not available.

Average cutting speed for different job materials using H.S.S. twist drill

Job Materials	Cutting speed (in m/min.)
Cast Iron	25–35
Mild Steel	20–30
Stainless Steel	12–18
Aluminium	80–100
Copper and Brass	60–70

Feed : In drilling machine, it is the distance a drill moves into the job in mm in one revolution of the spindle. It is expressed in mm/revolution.

Feed in mm/minute, $f_m =$ Feed in mm/revolution $(f_r) \times$ r.p.m. (N).

$$f_m = f_r \times N$$

Approximate feed/revolution with variation of drill size in mm

Drill size (mm)	Feed (mm/rev.)
Less than 3	0.025
3 to 6	0.075
6 to 12	0.14
12 to 25	0.25
More than 25	0.5

Example 2.

Calculate the feed/revolution to drill a hole of 20 mm in one minute to a plate thickness of 30 mm and using a spindle speed of 500 r.p.m.

Solution:

$$\text{Machining time, } T_m = \frac{\text{Total distance travelled to drill bit}(H)}{\text{Feed per revolution, }(f_r)\times \text{ Spindle speed}(N)}$$

$$\text{We know, Feed, } f_r = \frac{\text{Total distance travelled by the drill axially}(H)}{\text{Spindle speed }(N)\times T_m}$$

Total distance travelled by the drill axially,

= Plate thickness (t_p)+ Approach (A)+ Over-run (O)

$= t_p + 0.5D$ (Assuming, $A + O = 0.5D$)

where, D is the drill diameter in mm.

[NOTE : Approach is depending on drill diameter and the point angle of the drill, it is generally $0.29D$ if the point angle 118° is considered. Over-run is to be considered slightly to scrap last chip from the job end].

$$\therefore \quad f_r = \frac{t_p + 0.5D}{N \times 1} \quad [\because T_m = 1 \text{ minute}]$$

$$= \frac{30 + 0.5 \times 20}{500} = \frac{40}{500} = 0.08 \text{ mm/rev.}$$

Depth of cut

In drilling machine, it is the distance measured from the axis of the drill to the end of drill, i.e., radius of the drill, or one half of the diameter of the drill and expressed in mm.

$$\text{Depth of cut} = \frac{\text{Drill diameter}(D)}{2} \text{ mm}$$

A4.21 : Machining Time Estimation In Drilling Machine

Machining time (T_m) is the time required in minute to a total travel of drill (H) to make a hole to a depth of plate thickness (t_p)

$$T_m = \frac{H}{f_r \times N}$$

where, 'H' is the total drill travel in mm, f_r is the feed/revolution in mm/rev., N is the spindle speed in rev./min.

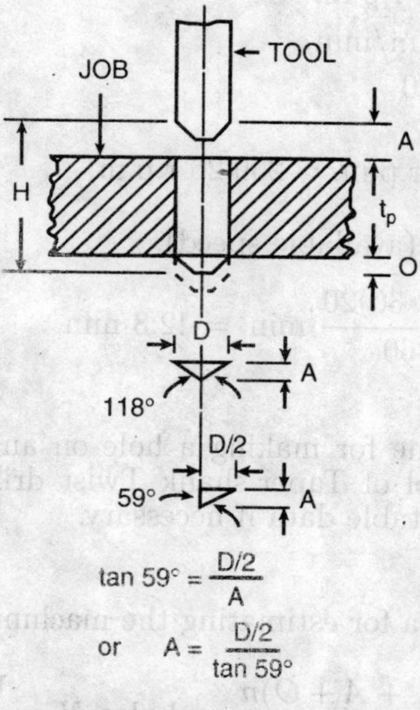

S-4.21 MACHINING TIME IN DRILLING MACHINE

But, $H = A + t_p + O = t_p + (A + O) = t_p + 0.51D$,
where, A is the approach of the drill and can be calculated with respect to the drill diameter and drill point angle. One sample calculation is shown above. 't_p' is the plate thickness in mm, O is the over-run, to be estimated approximately 2 to 10 mm, depending on drill size '$(A + O)$' can be approximately chosen as $0.5D$, where 'D' is the drill diameter in mm.

$$T_m = \frac{H}{f_r \times N} = \frac{t_p + 0.5D}{f_r \times N} \text{ min.}$$

For drilling 'n' number of holes,

$$\text{Total } T_m = \frac{(t_p + 0.5D) \times \pi n}{f_r \times N} \text{ min.}$$

Example 3.

Calculate the machining time required for making 20 holes on an M.S. plate of 40 mm thickness with the following data : drill diameter 30 mm, cutting speed = 25 m/min. and feed = 0.1 mm/revolution.

Solution:

We know from the formula of Machining time,

$$\text{Total } T_m = \frac{t_p + 0.5D) \times n}{f_r \times N} \text{ and also } N = \frac{V \times 1000}{\pi D}$$

where, plate thickness, $t_p = 40$ mm

Drill diameter, $D = 30$ mm

Feed/rev., $f_r = 0.1$ mm/rev.

Spindle speed, $N = 260$ r.p.m.

Cutting speed, $V = 25$ m/min.

Number of holes, $n = 20$

$$\therefore \quad N = \frac{25 \times 1000}{\pi \times 30} \text{ r.p.m.} = 250.26 \text{ r.p.m.}$$

$$= 260 \text{ r.p.m. (available speed).}$$

$$\text{Total } T_m = \frac{(40 + 0.5 \times 30)20}{0.1 \times 260} \text{ min.} = 42.3 \text{ min.}$$

Example 4.

Find the drilling time for making a hole on an M.S. plate of thickness 20 mm and using a tool of Taper-shank Twist drill (short)—25.00-IS-599-HSS-N-118. Assume suitable data if necessary.

Solution:

We know the formula for estimating the machining time,

$$T_m = \frac{(t_p + A + O)n}{f_r \times N} \text{ and also } N = \frac{V \times 1000}{\pi D}$$

where, Approach, $A = 0.29D$ (for point angle 118°), Over-run, $O = 0.2D$ assumed. Plate thickness, $t_p = 20$ mm Drill diameter, $D = 25$ mm, assumed, Feed/rev., $f_r = 0.15$ mm/rev. to be calculated, Spindle speed, $N =?$ r.p.m., assumed, Cutting speed for job material M.S. and Tool material H.S.S., $V =$ 30 m/min; Number of holes, $n = 1$.

$$\text{Therefore,} \quad N = \frac{30 \times 1000}{\pi \times 25} \text{ r.p.m.}$$

$$= 381.97 \text{ r.p.m.} = 380 \text{ r.p.m. (say)}$$

$$T_m = \frac{(t_p + 0.29D + 0.2D)n}{f_r \times N}$$

$$= \frac{1 \times (20 + 0.29 \times 25 + 0.2 \times 25)}{0.15 \times 380}$$

$$= 0.566 \text{ min.} = 34 \text{ seconds.}$$

A4.22 : Questions

1. Write a brief note on Drilling Machine. (A 4.1)
2. Classify the Drilling Machines. (A 4.2)
3. Write the types of Drilling Machine in tabular form. (A 4.3)
4. Give a brief note on different types of Drilling Machines. (A 4.41 to A 4.13)
5. Name the various types of drilling Machines known to you with their specifications, uses and limitations. (A 4.5)
6. Write short descriptions of main components of Drilling Machine. (A 4.6)
7. Specify a Drilling Machine. (A 4.7)
8. Write the specifications of a Radial Drilling Machine. (A 4.8)
9. Write four different types of Tool Holding Devices and give short notes with sketches where possible. (A 4.9)
10. Sketch, label and explain the automatic feed mechanism of drilling head spindle. (A 4.10)
11. Sketch, label and explain the working of Drill Head Assembly. (see within A 4.61)
12. Compare among Drilling, Boring and Reaming operations in Drilling Machine. (A 4.11)
13. Compare between Boring and Reaming operations in Drilling Machine. (A 4.12)
14. Designate a drill bit and explain its different parameters. (A 4.13)
15. Classify the drills used in Drilling Machine. (A 4.14)
16. Write the advantages of using spiral flutes (twist) drill instead of using straight flutes. (A 4.15)
17. Sketch different views of a twist drill and show its nomenclature and angles. Write also the approximate values of angles of H.S.S. Twist Drill. (A 4.16)
18. Sketch, show job-tool motion, label and short description of drilling operations. (A 4.19)
19. State the term cutting speed, feed per revolution, feed per minute and depth of cut with units used in Drilling Machine. (A 4.20)
20. Sketch and deduce the machining time calculations of drilling operation in Drilling Machine. (A 4.21)
21. Determine the spindle revolutions per minute to drill a hole of 20 mm on an M.S. plate by an H.S.S. tool.

    [*Note :* Assume cutting speed as 30 m/min]. **[Ans.** 477 r.p.m.]

22. Calculate the feed/rev. and feed/min. for drilling a hole of 35 mm on a plate of thickness of 25 mm and using a spindle speed of 300 r.p.m. Assume suitable data if required.

[*Note* : Assume $(A + 0) = 0.5D$]

[**Ans.** 0.1416 mm/rev. and 42.5 mm/min.]

23. What will be the depth of cut for drilling a hole of 30 mm?

[**Ans.** 15 mm]

24. Find the machining time required for making 50 holes on a brass plate of 20 mm thickness. The diameter of drill is 25 mm. Assume suitable data if required.

[*Note:* Cutting speed as 80 m/min. and feed per rotation as 0.2 mm / revolution]. [**Ans.** 7.976 min.]

25. Calculate the machining time for drilling a hole on an M.S. plate of thickness 30 mm and using a tool of : Straight shank—Twist drill (Medium)—12.00 IS-599-H.S.S. (N)-120. Assume suitable data if required.

(Assume feed as 0.15 mm/rev. and cutting speed as 30 m/min.)

[**Ans.** 18 seconds]

26. Explain the functional difference between a straight fluted drill and twist drill. (A 4.17)

27. Describe the purposes of drill parts. (A 4.18)

Volume II

5

MILLING MACHINES

A5.01 : Introduction

In Milling Machine, a layer of metal is removed from a workpiece to obtain the required shape, dimensions and surface finish by a rotating one (or more) multipoint cutter(s), in a particular position, fixed on extension of Machine spindle, removes metal at a very fast rate. The forward movement of the work-piece in longitudinal, cross or vertical direction is fed against the rotating multipoint cutter. The cutter is set to a certain depth of cut by raising the table. Number of cutters at a very high speed, find wide application in production work. This Machine is also suitable in Tool-room work due to its variety of operations, better surface finish and accuracy.

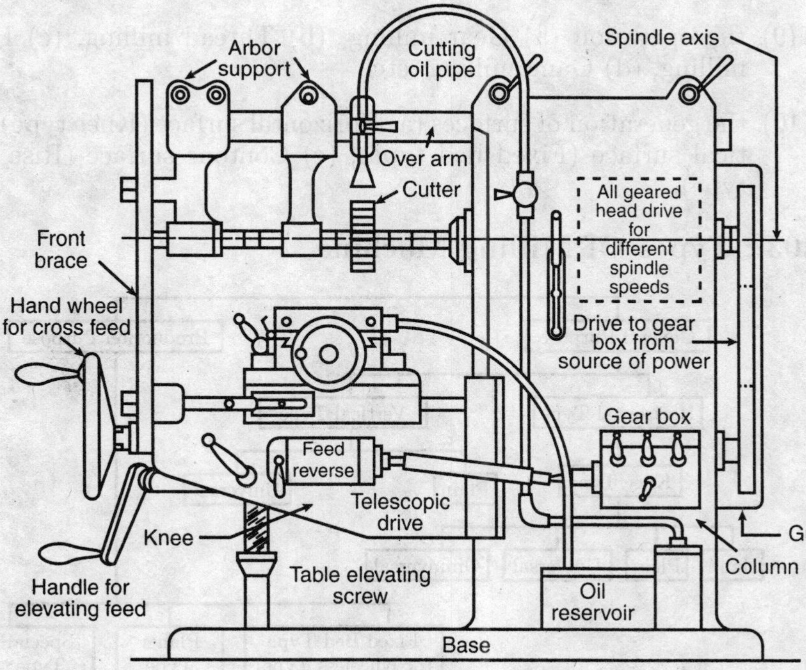

Sketch S-5.01

A5.02 : Classification

The choice of Milling Machine depends on the size of the job and the type of operations, and the number of jobs to be machined.

Numerous types of Milling Machines are available with respect to sizes, design, purposes, etc. The main types of Milling Machines may be classified with respect to—

(1) the arrangement of drive for different spindle speeds (a) Direct motor drive, (b) Cone pulley drive, (c) All geared head drive.

(2) construction of table (a) Fixed bed type or Column and Knee type.

(3) the axis of the spindle (a) Horizontal, (b) Vertical.

(4) the movement of the table (a) Plain, (b) Universal, (c) Omniversal.

(5) the motion of the table (a) Rotary table, (b) Reciprocating table (Planer type), (c) Fixed table, etc.

(6) the purpose or application (a) Maintenance or General purpose, (b) Production purpose, (c) Tool-room purpose, etc.

(7) the number of spindle heads in a machine (a) Simplex, (b) Duplex, (c) Triplex.

(8) the power for table feeding (a) Hand, (b) Gearing, (c) Hydraulic, etc.

(9) the operation (a) Gear milling, (b) Thread milling, (c) Planetary milling, (d) Gang milling, etc.

(10) the generation of surfaces (a) Horizontal surface (Knee type), (b) Vertical surface (Fixed bed type), (c) Contour surface (Rise and Fall type).

A5.03 : Types Of Milling Machine

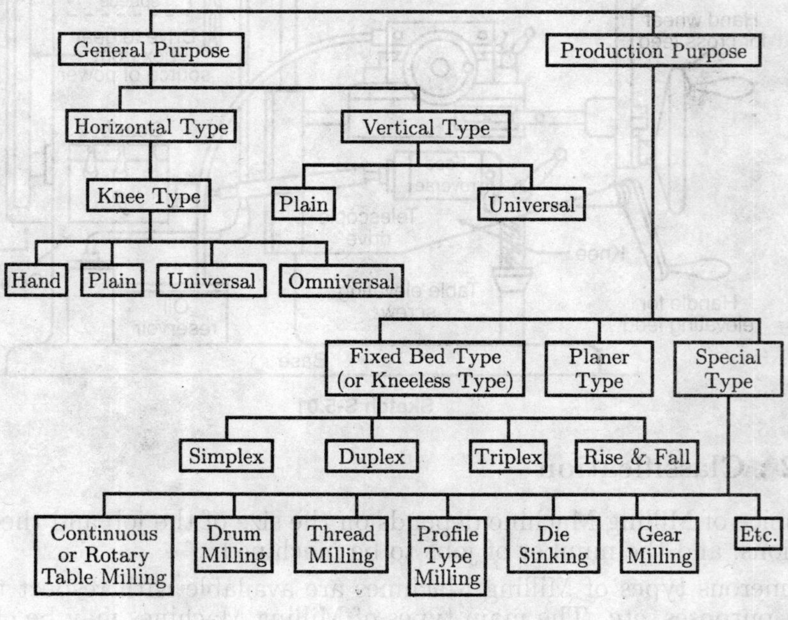

A5.04 : Short Descriptions Of Different Types Of Milling Machines

[1] Vertical Milling Machine

In this type of machine, the spindle is mounted vertically, can swivel at an angle and can adjust up and down relative to work. This machine is suitable for machining of face milling, grooves, slots, end milling, etc. This type may be of Plain or Universal type depending on the table motions.

[2] Knee Type Milling Machine

This type is also named as Column and Knee type Milling Machine. This machine is mostly suitable for General shop floor work, maintenance work, Tool-room work, etc. Knee is mounted on the vertical slides of the column and is adjustable along column axis to accommodate the work of various heights. This machine can be classified will respect to the movement of table, supply power of the table movement. The types are Hand, Plain, Universal and Omniversal Milling Machines.

[3] Hand Milling Machine

Simplest type of Milling Machine is suitable for light and simple jobs. Cutter is mounted on arbor and rotated by motor power. Hand power is used to feed the job against the rotating cutter. This machine is suitable for simple operations like keyways, slots, grooves milling, etc.

[4] Plain Milling Machine

This Machine is similar to hand milling machine but more steady and rigid in construction and having the arrangements of both hand and power feeding. This machine is suitable for general work. The machine table can be moved longitudinally, crosswise and vertically against a rotating cutter mounted on arbor.

[5] Universal Milling Machine

The machine has a horizontal spindle with table swivelling arrangement about a vertical axis. The table is mounted on a circular swivelling base graduated in degrees and can move up to 45° on either side of the normal position in addition to its normal movements longitudinal, cross and vertical. This machine is suitable for general purpose, tool-room work, for cutting helical grooves, spur, spiral, bevel gears, twist drills, reamers, milling cutters, etc.

[6] Omniversal Milling Machine

This machine has the arrangements of knee swivelling in addition to the movements of the table as in case of a universal milling machine. This machine is suitable for experimental shop-floor, tool-room work and also for taper spiral grooves in reamer. The spiral groove of varying depth as required in plunger of diesel pump can be cut by this machine tool.

[7] Fixed Bed or Kneeless Type Milling Machines

In this type, bed is fixed, suitable for machining of large, heavy and rigid production jobs. Rotary cutter is keyed on vertical arbor, can travel vertically and may adjust horizontally to provide cross adjustment. The table has the arrangement of cross feeding only. Fixed bed may be of the following types:

[7a] Simplex Type

The Machine is provided with only one vertical spindle head.

[7b] Duplex Type

The Machine is provided with two vertical spindle heads on either sides of the table.

[7c] Triplex Type

The Machine is provided with three vertical spindle heads, two spindles are arranged on either sides of the table and the third one is mounted on a cross-rail. These numbers of spindle heads facilitate the cycle of operation automatically in a regular sequence for mass production work.

[8] Rise and Fall Type Milling Machine

This is also a fixed bed type milling machine, having table longitudinal movement only and the spindle carrier carries a horizontal spindle. To machine the surfaces in different planes, curved profiles, blind or hidden surfaces can be done automatically by hydraulically or electrically controlled feeding, synchronizing the vertical movement of the spindle carrier and longitudinal movement of the table.

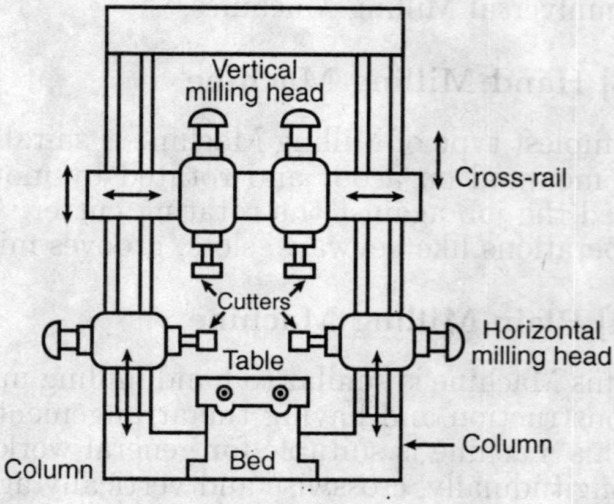

Sketch S-5.04 [9]

[9] Planer Type Milling Machine (or Plano Milling)

This is also named as Plano Milling. At least one spindle head is mounted on vertical direction and at least, two spindle heads are mounted on traverse directions.in either sides of the table. Number of workpieces and number of surfaces can be machined by the number of cutters simultaneously when the table reciprocates to provide feed. This machine is mostly suitable for mass production of heavier jobs.

[10] Continuous or Rotary Table Type Milling Machines

This is a modification of vertical milling machine. Two or more face milling cutters are mounted on vertical spindles. Number of jobs are clamped on round table, rotates on vertical axis and providing feed by the cutters. Loading and unloading of the jobs are carried out by the operator during operations and facilitates continuous production.

[11] Drum Milling Machine

Jobs are clamped on both faces of the drum rotates about horizontal axis. Two or more face milling cutters rotate also about horizontal axis and provide feed. Jobs are machined completely after one complete turn of the drum. This is also a continuous process suitable for production work.

[12] Thread Milling Machine

This machine of various designs is available for milling of hob threads, worms, etc. This is also mainly for production work.

[13] Profile Type Milling Machine

This is also a production type of machine, will mill various profiles or contours by controlling the cutter. The cutter is regulated by means of finger in contact with a same profile or contour of a template. Power feeding is mostly used.

[14] Die Sinking Milling Machine

This is most versatile type of machine, in which the cutter can be moved in three co-ordinate directions and thus any surface inclined in both the planes can be machined. This machine is suitable for production of dies, steel moulds engraving, etc.

[15] Gear Cutting Milling Machine

Gears may be cut by knee type machines like plain or universal or omniversal milling machine. Gear milling operation is similar to gear hobbing. Hobbing tools or Hobs are in the shape of inserted teeth milling cutters, revolving and feeding like milling cutters.

A5.05 : Specifications Of A Horizontal Milling Machine

1. Manufacturer type and code number, i.e. Knee type or Fixed bed type, etc.
2. Size of the machine table, i.e. Length × Breadth in mm^2.
3. Table movement, i.e. longitudinal, cross, vertical in mm; rotational, inclined, etc. in degrees.
4. Knee movement in degrees.
5. Diameter of the arbor in mm and spindle nose Morse taper number.
6. Range and number of spindle speed (speed in r.p.m.).
7. Range and number of table feeds, i.e. longitudinal, cross and vertical in mm/minute.
8. Available power of the machine in kw.
9. Net weight of the machine in tonne.
10. Floor space required in square metre.

A5.06 : Comparison Between Lathe Machine And Milling Machine

Points	Lathe Machine	Milling Machine
(1) Cutting tool and its cost.	Single-point tool and cost of tool is much less.	Multipoint tool and cost of tool is much more.
(2) Movement of job and tool.	Job rotates and the tool having translatory motion for feeding.	Tool rotates and the job having translatory motion for feeding.
(3) Type of job suitable for machining.	Generally suitable for tool room and repair type of jobs.	Mostly suitable for mass production works.
(4) Number of tools and number of operations can be at a time.	Only one tool and one operation can be performed at a time.	One or more tools and many operations can be machined at a time.
(5) Cost of subsequent tools for subsequent operations.	Less, as one tool can be utilized for different operations by tool grinding.	High, as separate tools are needed for various types and size of operations.

A5.07 : Comparison Between Plain Milling Machine And Universal Milling Machine

Points	Plain Milling Machine	Universal Milling Machine
1. Table movement	The table has three movements, i.e., (1) longitudinal, (2) cross and (3) vertical.	The table has four movements, i.e., (1) longitudinal (2) cross, (3) vertical and (4) the table can also be swivelled to the right or left in the horizontal plane and the maximum angle in one way is 45°.
2. Type of gear cutting	Spur gear or rack can be cut.	Any type of gears, i.e., spur, helical, bevel, etc., can be cut.
3. Cutting of helical gear	The helical gear cutting is not possible as the table cannot be swivelled.	The helical gear cutting is possible by swivelling the table.

Points	Plain Milling Machine	Universal Milling Machine
4. Field of applications	It is used mainly for production jobs.	It is mainly used for jobbing work and specially in the tool-room, maintenance works.
5. Location of the machine tool	It is located in shop floor, where machining of production jobs is done.	It is located in the tool-room and special machining operations.
6. Rigidity of machine	More rigid and heavier as the movement of the table is restricted and can perform heavier milling operations.	Less rigid; due to the movement of the table, it is more versatile.
7. Possibility of using variety of attachment	Difficult	Possible (arrangement of using dividing head equipment,vertical milling attachment, rotary table, etc.)

A5.08 : Different Parts Of A Horizontal Milling Machine (Knee Type)

1. Base :
2. Column : Cutting Oil Reservoir
   : Cutting Oil Pump
   : Drive to gear box from source power
   : All geared head drive for different spindle speeds
   : Table feed gear box
   : Spindle shaft
3. Knee : Work table
   : Feed gear box
   : Telescopic drive
   : Handwheel for longitudinal feed
   : Handwheel for cross feed
   : Handle for elevating feed
4. Table Elevating Screw
5. Overarm : Arbor support
   : Additional arbor support
   : Support for cutting oil pipe
6. Front Brace :
7. Spindle :
8. Arbor : Spindle nose
   : Cutter(s)
   : Spacing collars
   : Bearing support(s) or Yoke

A5.09 : Functions Of Main Parts Of A Knee Type Milling Machine

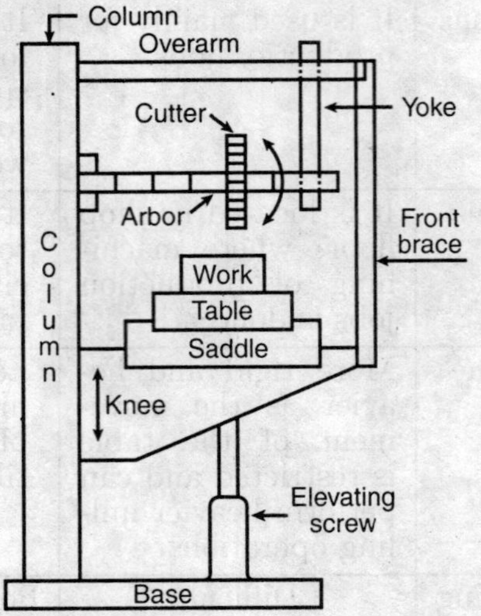

Sketch S-5.09

[1] Base

The base is the foundation member of all the machine parts. It is generally made of grey cast iron to absorb shock and vibrations. It carries column on one side and elevating screw, table, arbor and overarm on the other side.

[2] Column

The column is the main casting mounting vertically on one side of the base. This is box shaped and fitted with gearing arrangement for different spindle speeds and table feeds. The knee which supports the table is secured, and moves up and down with respect to the vertical dovetail guideways on the front vertical face of the column. A jack for elevating the knee and table is mounted to the base of the column. It carries the spindle that extends to arbor and also holds the overarm at the top.

[3] Knee

It is a rigid grey iron casting that supports and provides adjustment for the height of the table, operated by the table elevating screw and slides up and down on the vertical dovetail guideways of the column. It carries the table feed mechanism and controls to feed in longitudinal, cross, vertical and rotational, etc., either by hand power or machine power. Power feed is provided with adjustable stop to limit its travel in either directions. There is a saddle in between the knee and table which slides on a dovetail guideways set exactly at 90° to the column face. This arrangement helps to move the table horizontally for cross feeding by hand or power.

[3(a)] Table

Table is the position where the job is mounted either directly by clamps or by a fixture or in the dividing head. It rests on saddle and travels longitudinally. The table travel is limited by the adjustable stops in either directions during power feeding. In case of a universal milling machine, there is an arrangement of table to swivel horizontally around the centre of its base, required for manufacturing of twist drills, worms, etc. The table is mounted on circular base, which is graduated in degrees.

[3(b)] Feed Gear Box

The variation of power feed to all movements, is obtained through the feed gear box. The power from the feed gear box is transferred to the knee by a telescopic shaft, maintaining a flexible drive throughout the range of the knee movement. The number of feeds and the amount of feed (i.e., in mm/minute or mm/revolution of the arbor) are depending on the gearing arrangements in the feed gear box. The longitudinal, cross, vertical feeds are being attained by hand or power by operating the controls like handle, handwheel, lever, etc., placed in front of the knee.

[4] Table Elevating Screw

The lower side of the elevating screw is mounted rigidly on the base and other end is connected with the knee. Vertical adjustment of the knee can be attained by operating the elevating screw either by hand or power through the handwheel or the lever which are placed at the lower side of the knee. Differential screw jacks or hydraulic jacks may be used depending on the requirements of the machine size and rigidity.

[5] Overarm

The overhanging arm is clamped on the top of the column and carries one (or more) bearing support(s) at different position of the arbor (nearer to the cutter and at the free end of the arbor) to ensure rigidity of the arbor. The bearing supports are adjustable, sliding parallel to the arbor.

[6] Front Brace

This is also an extra support and stiffener which is fitted between the knee and overarm ensuring further rigidity to the knee and the arbor. There is a rectangular slot in the lower part of the front brace for the adjustment of the height of the knee relative to the cutter.

[7] Spindle

The spindle is mounted on the upper part of the column. It receives power from the motor through belts, gears, clutches, etc., and can be rotated at different speeds by the step-cone-pulley drive or by gearing arrangement. It should be sufficiently rigid and concentric for maintaining greater accuracy in the machining. Front end of the spindle is being extended from the column face and connected with the arbor through a tapered hole in which cutters are mounted for machining operations.

[8] Arbor

The arbor is the extension of milling machine spindle on which all the various cutters are mounted. The arbor is set in the spindle cone (Morse taper nose) and tightened by the draw bolt held in position by a bush nut and further rigidity and alignment is attained by using two keys on the key block. The arbor diameter is chosen according to the hole diameter in the cutter. Indian standard specification arbors with Morse taper shanks are available in 13, 16, 22, 27, 32, 40, 50 and 60 mm in diameter.

A key is fitted throughout the whole length of the arbor, which engages with a keyway in the cutters to provide drive. The cutters are kept in position by using adjusting rings (collars) of different widths and having the diameter slightly larger than the arbor diameter serves as a bearing and the whole set-up is kept in place by a left-hand nut of the arbor. A set of twenty adjusting rings provided with a milling machine consists of rings from 1 mm to 50 mm wide: 1.0, 1.1, 1.2, 1.3, 1.4, 1.5, 1.6, 1.7, 1.8, 1.9, 2, 3, 5, 8, 10, 15, 20, 30, 40, 50. The arbor may be supported at the free end and nearer to the cutters from the over-hanging arm to ensure more rigidity. The ejection is effected by unscrewing the draw bolt, which is due to being held captive at the end by the lock nut, causes the arbor to be forced forward.

A5.10 : Distinction Between Horizontal Milling Machine And Vertical Milling Machine

Sl. No.	Points	Horizontal Milling Machine	Vertical Milling Machine
1.	Principle of Job-Tool Motion	Multipoint cutter rotates in horizontal axis and Job gives linear feeds.	Multipoint cutter rotates in vertical axis and Job gives linear feeds.
2.	Position of Spindle	Spindle is horizontal, parallel to the work surface.	Spindle is vertical or perpendicular to the work surface.
3.	Spindle head swivelling arrangement	No such arrangement.	Spindle head connected to vertical column may be swivelled at an angle for milling angular surface.
4.	Spindle movement	No movement of spindle is possible.	Vertical spindle may be moved up and down relative to the work.
5.	Number of Spindles and Cutters can machine at a time	Number of cutters in one spindle may mill at a time.	Two or more spindles are holding only one cutter by each spindle at its bottom end to mill at a time.
6.	Types	May be hand, plain, universal and omniversal type milling machines.	May be plain and universal type milling machine.

contd...

Sl. No.	Points	Horizontal Milling Machine	Vertical Milling Machine
7.	Suitable Cutters	Plain milling cutter, Side milling cutter, Angle milling cutter, etc.	End milling cutters, Face milling cutters, T-slot milling cutters, etc.
8.	Milling Operations	Plain milling, Side milling, Angular milling, Gang milling, Saw milling, Gear cutting, Keyway milling, etc.	Face milling, Profile milling, End milling, T-slot milling.
9.	Cutting capacity and surface finish	Less	Cutting capacity is 15 to 20% higher, and better surface finish than the horizontal milling machines.

A5.11 : Fixing And Working Of A Milling Arbor

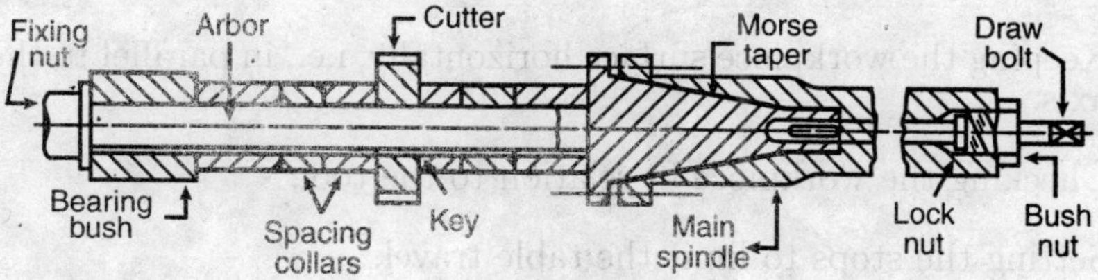

SKETCH S-5.11

The following steps are to be taken for fixing of arbor on milling spindle, fixing of cutter on arbor and working of milling operations:

1. Mounting the arbor on Milling Spindle.
2. Mounting the milling cutter and adjusting ring on the arbor.
3. Checking the cutter for true alignment by using an indicator.
4. Checking the workpiece in relation to the tool.
5. Setting the stops to limit the table travel.
6. Setting the spindle rotational speed and feed per minute.
7. Setting the depth of cut in milling.

(A) Mounting the Arbor on Milling Spindle

The following steps are to be taken:

1. Cleaning the arbor thoroughly.

2. Inserting with its conical end into the spindle hole.

3. Making the slots in the arbor flange coincident with the key blocks at the spindle end.

4. Fixing the arbor with the tightening rod (draw bolt). The conical portion of the arbor should tightly enter the spindle conical hole.

(B) Mounting the Milling Cutter on Arbor

1. Setting the cutter and adjusting rings (Collars) in position on the arbor.

2. Adjusting rings (Collars)and cutters should be cleaned thoroughly before being set.

3. Checking the cutter for true alignment by using an indicator.

4. Cutter is keyed with arbor.

(C) Mounting the Workpiece

1. Setting up a universal vice to a required angle on the machine table.

2. Mounting the workpiece in a universal vice.

3. Keeping the workpiece surface horizontally, i.e., in parallel to the cutter axis.

4. Checking the workpiece in relation to the tool.

5. Setting the stops to limit the table travel.

6. Selecting and setting the suitable spindle speed, feed and depth of cut.

7. Arranging the supply of suitable cutting fluid.

8. Setting the spindle rotational speed.

Before starting the operation, the longitudinal table with the workpiece has to be moved close towards the cutter, then the feed is given and the cutting fluid pump will be started. The workpiece can be milled accurately and smoothly in one or more cuts.

A5.12 : Mounting Of Cutter

Fixing a Cutter on Milling Machine Arbor

The following steps are to be taken for mounting and fixing a cutter on milling machine arbor:

1. Using box wrench, unscrew the locking screw and withdraw machine overarm.

2. Remove the bearing support by unscrewing it.

3. Clean the arbor thoroughly and insert it with its conical end into the spindle hole, making the slots in the arbor flange coincident with the key blocks at the spindle end and then fix the arbor with the tightening rod (Draw bolt). The conical portion of the arbor should tightly enter the spindle conical hole.

4. Set the cutter and adjusting rings in position on the arbor. Keyed the cutter with arbor. Care should be taken to the direction of the cutter spiral flutes must be opposite to the spindle hand of rotation. Adjusting rings and cutters should be thoroughly cleaned before being set.

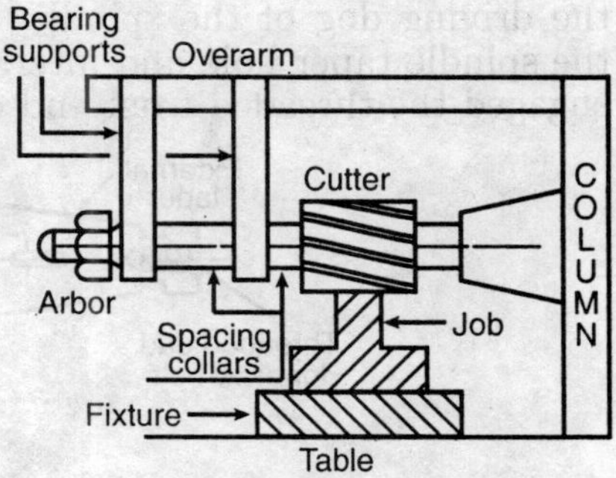

Sketch S-5.12

5. Set the bearing supports in position as shown above.

6. Clamp the cutter on arbor by tightening the nut.

7. Fix the overarm and lubricate the bearing.

8. Use indicator to check the concentricity of the arbor and cutter.

A5.13 : Cutter Holding Devices

The devices which are used for holding and rotating the milling cutters with the machine spindle during the operation are known as cutter holding devices. These are designed to accommodate different sizes and types of cutters.

The arbors of different types, adapter, collet, screw-on arbor, etc. are the main cutter holding devices.

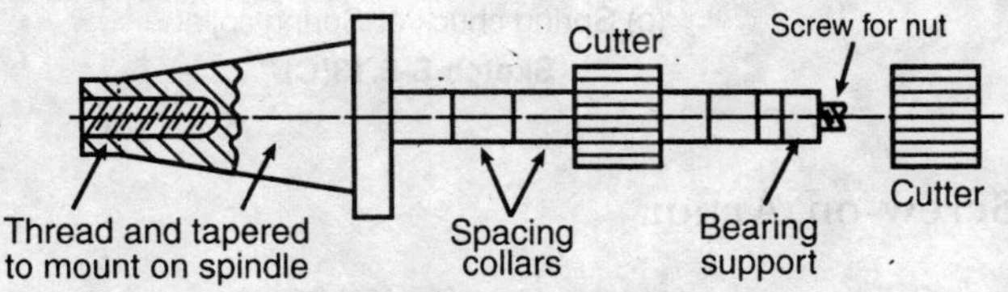

(a) Standard arbor

Sketch S-5.13(A)

[1] The Arbors

A Milling Machine Arbor is the extension of milling machine spindle, designed to hold and rotate the cutters. This type of arbors are having taper at one end and short internal screw inside the taper end to be screwed with the draw bolt of the Machine Spindle, and a short shaft on the other side on which the bored cutter is to be keyed in position, the rest void portion is to be filled by spacing collars. The end of the arbor shaft is tightened by a nut to keep in position of the cutter, spacing collars, bearing support, etc. Sketch and more details have been described in Art. No. A5.11. Arbors may be of different types like arbor for thread nose spindle, arbor for face milling cutter, arbor for shell-end mills, fly-cutter arbor, etc.

[2] The Adapter

Adapters are used for holding arbors, collets and T-slot cutters, shell-end mills, etc., having tapered shanks. If the taper size of the cutter shank is smaller than the spindle nose, the adapter(s) may be used to hold these cutters. There are two slots on the flange of the adapter to accommodate the driving dog of the spindle. External taper on adapter corresponds to the spindle taper hole and internal screw on the taper side of the adapter is engaged the threaded front end of the draw bolt.

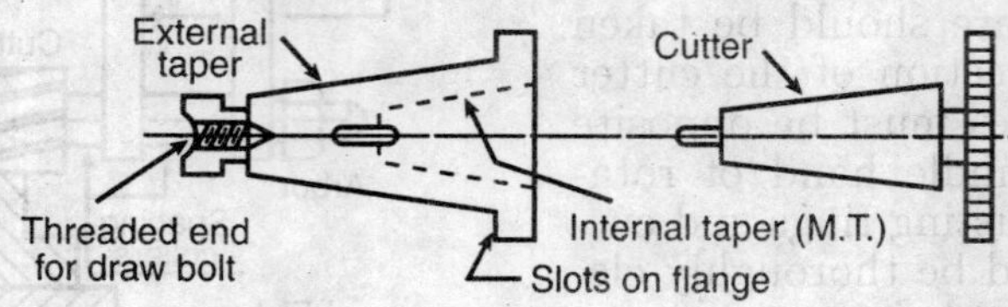

(b) Adapter

Sketch S-5.13(B)

[3] The Spring Collet or Spring Chuck

The cutters those having straight shank are held by a device known as spring collet or spring chuck. There is an external taper on spring collet to fit in an adapter and there is an internal hole to grip the straight shank cutter. Different types of spring collets are available to suit the various sizes of the cutter shanks.

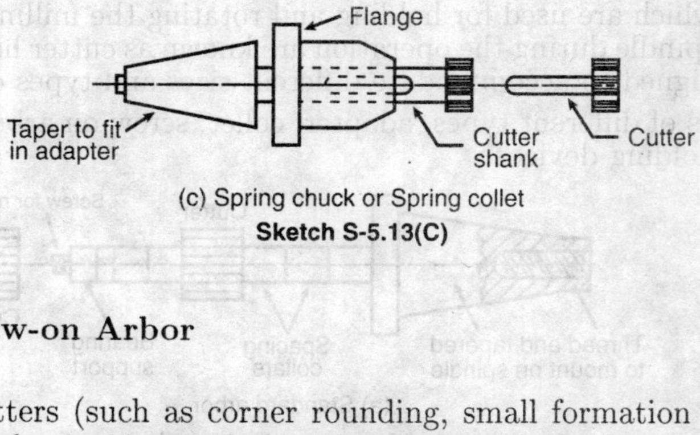

(c) Spring chuck or Spring collet

Sketch S-5.13(C)

[4] The Screw-on Arbor

The small cutters (such as corner rounding, small formation of concave or convex, etc.) having internal thread are screwed on the threaded nose of an arbor. The arbor may be mounted on the spindle in the usual way. To avoid the looseness of the cutter on the arbor during operation, left-handed threads are used for left-hand cutters and right-handed threads are used for right-hand cutters.

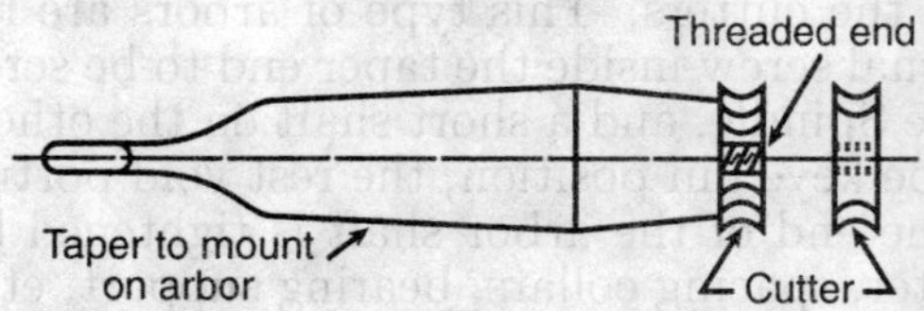

(d) Screw-on arbor

Sketch S-5.13(D)

A5.14 : Work Holding Devices

Job is to be clamped firmly and securedly on the milling machine table to machine smoothly. Job may be clamped in machine vices, or with clamps and clamping screw or by angle plates or by using special fixture for mass production job.

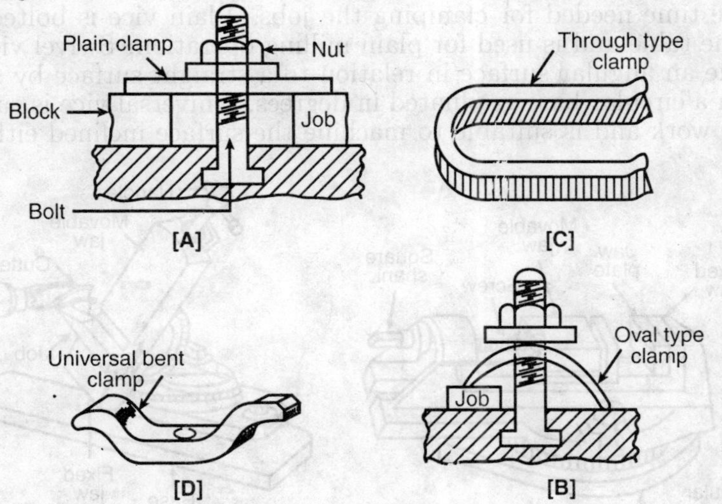

Sketch S-5.14 [1]

[1] Clamps and Bolts

These are applied for holding large or irregular shape jobs directly on the machine table. Different designs of clamps (such as Plain and off-set (A), Oval type (B), Through-type (C), Universal bent (D), etc.) may be used for different patterns of jobs.

[2] Angle Plates

These are applied for holding the jobs with two planes located at right angles. Angle plates are clamped on the machine table and the job is supported on its face by clamps and bolts. These strips are clamped with the angle plate by clamps and the base of the angle plate is provided with tongue which enters the table slot.

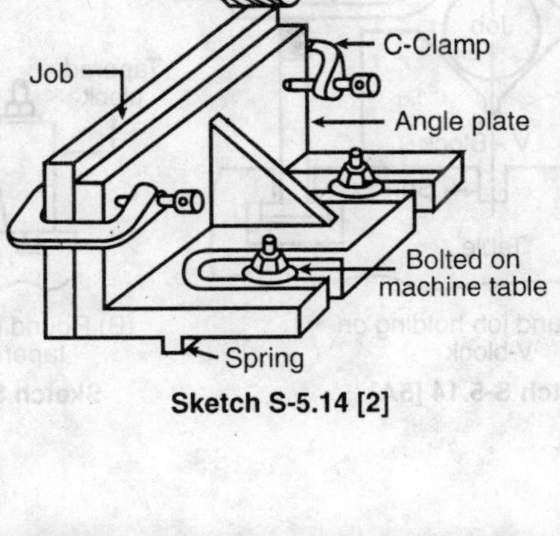

Sketch S-5.14 [2]

[3] Machine Vices

It is the most useful work holding device for quick loading and unloading arrangements. It may be plain, swivel and universal. Machine vice may be operated manually or air-operated or hydraulic drive. Special interchangeable jaws and pads may be used for holding irregular jobs, which will considerably reduce the time needed for clamping the jobs. Plain vice is bolted directly on machine table and is used for plain milling operation. Swivel vice is used to machine an angular surface in relation to a straight surface by swivelling the job on a circular base graduated in degrees. Universal vice is suitable for tool-room work and is suitable to machine the surface inclined either of the planes.

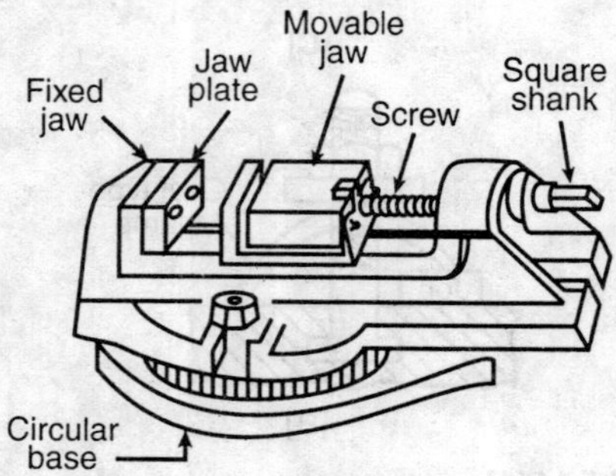

Swivel Vice

Sketch S-5.14 [3A]

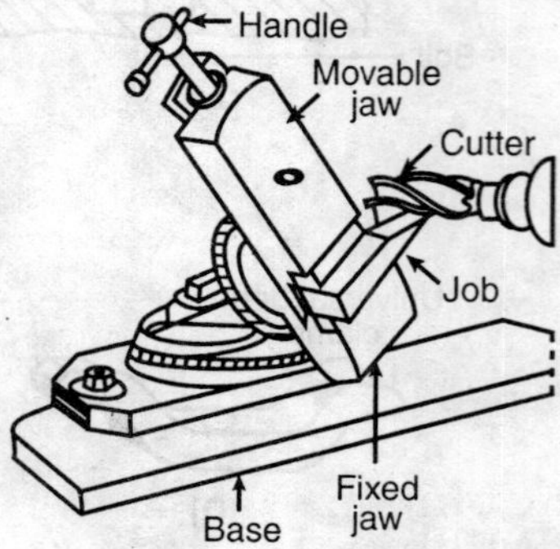

Universal Vice

Sketch S-5.14 [3B]

[4] Special Fixtures

It is a specific job-oriented holding device, specially designed to ensure higher accuracy, quick loading and unloading facility, easy clamping at minimum time. Special fixture is only suitable where large number of identical parts to be machined.

[5] V-Blocks

It is used for holding round or circular jobs directly on machine table to machine slots, key-ways, etc., which will prevent the job from any sideway movement.

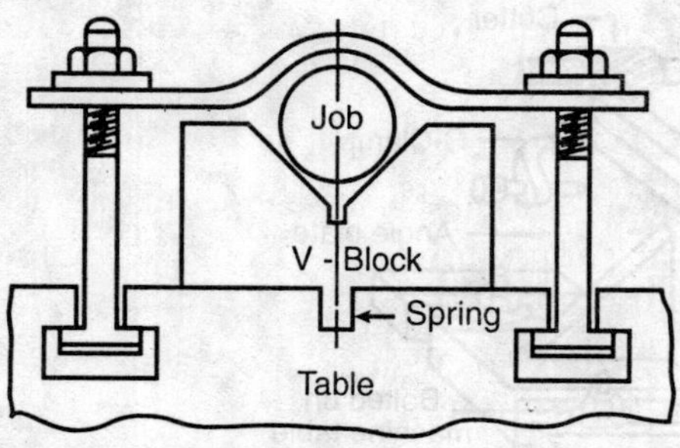

(A) Round job holding on V-block

Sketch S-5.14 [5A]

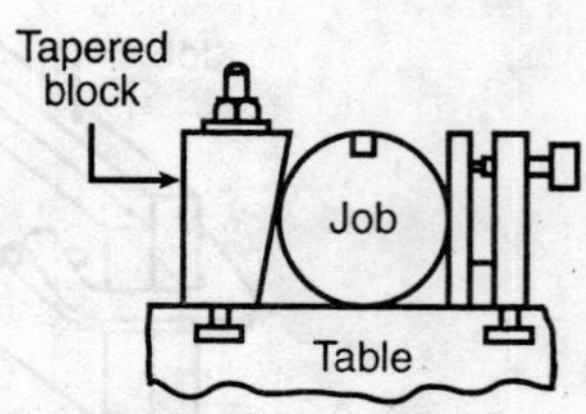

(B) Round job holding on tapered block

Sketch S-5.14 [5B]

A5.15 : Up Milling And Down Milling

Up Milling	Down Milling
In Up Milling (also known as Conventional Milling), the cutter rotates against the direction in which the work is feeding.	In Down Milling (also known as Climb Milling), the cutter rotates in the same direction as that in which the work is feeding.

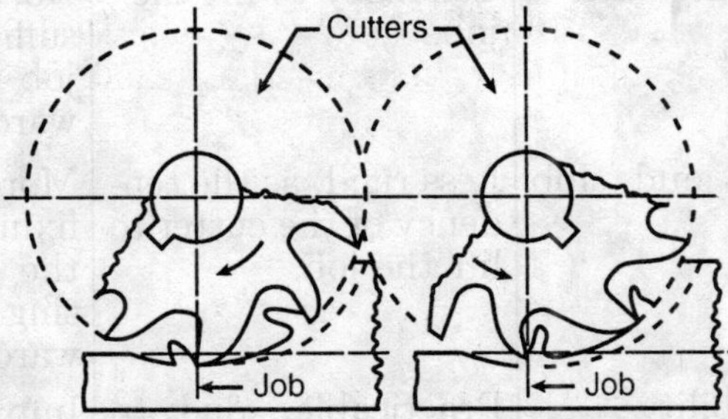

Sketch S-5.15

A5.16 : Comparative Study Between Up Milling And Down Milling

Serial No.	Points	Up Milling [A]	Down Milling [B]
1.	Sketch.	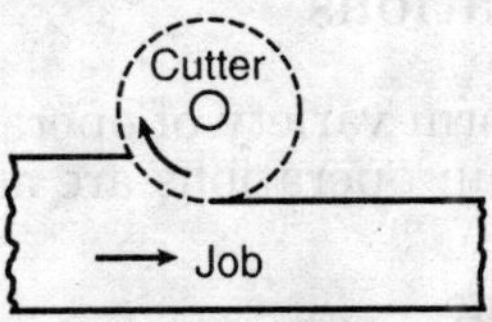 (A) Up milling **Sketch S-5.16 [A]**	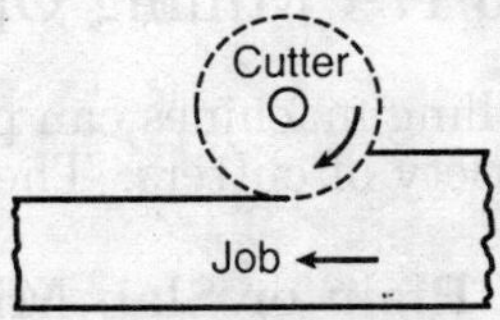 (B) Down milling **Sketch S-5.16 [B]**
2.	Alternative name	Conventional Milling	Climb Milling
3.	Definition	The cutter rotates against the direction in which the work is feeding.	The cutter rotates in the same direction as that in which the work is feeding.
4.	Job-Tool motion	Opposite direction	Same direction
5.	Chip thickness	Thickness varies from minimum to maximum.	Thickness varies from maximum to minimum.
6.	Cutting forces	Forces vary from zero to maximum.	Forces vary from maximum to zero.

Contd...

Serial No.	Points	Up Milling [A]	Down Milling (B]
7.	Use of cutting fluid	Difficult	Easy
8.	Surface finish	Better	Better, if it is free from backlash error.
9.	Job lifting tendency	Tendency to lift the job.	No, as cutter forces sufficiently on the job to press downward.
10.	Rigidity and Job fixing	Less rigid, as the tendency of the cutter to lift the job.	More rigid and job fixing easy due to the tool itself pressing the job downwards.
11.	Practicability	Practicable, due to cutting forces varying from zero to maximum and no-effect of backlash.	Impracticable, due to backlash error between the threads of the table screw and nut and also for maximum bite initially and decreases gradually.

A5.17 : Milling Operations

Milling machines can perform variety of operations on different jobs by using variety of cutters. The main operations are as follows:

[1] Plain or Slab Milling

The Plain Milling operation is the production of flat or horizontal surface parallel to the axis of the cutter.

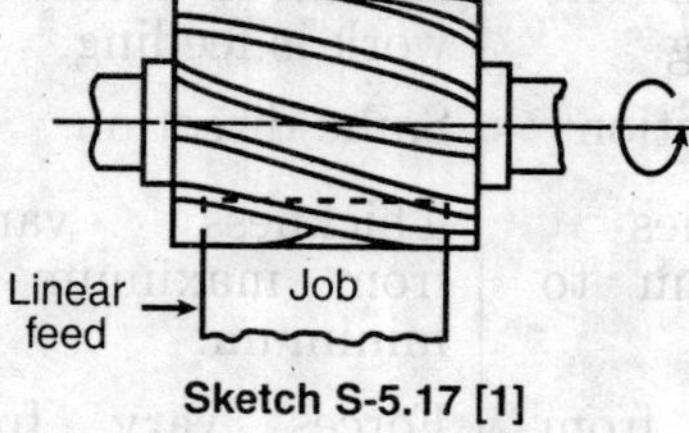

Sketch S-5.17 [1]

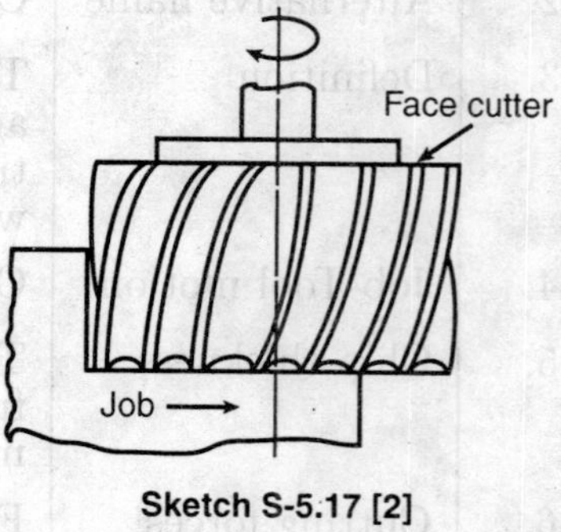

Sketch S-5.17 [2]

[2] Face Milling

The Face Milling operation is also the production of flat surface which is at right angles to the axis of rotation of the face milling cutter.

[3] Angular or Bevel Milling

The Angular Milling operation is the production of flat surface which is at an angle to the axis of the cutter.

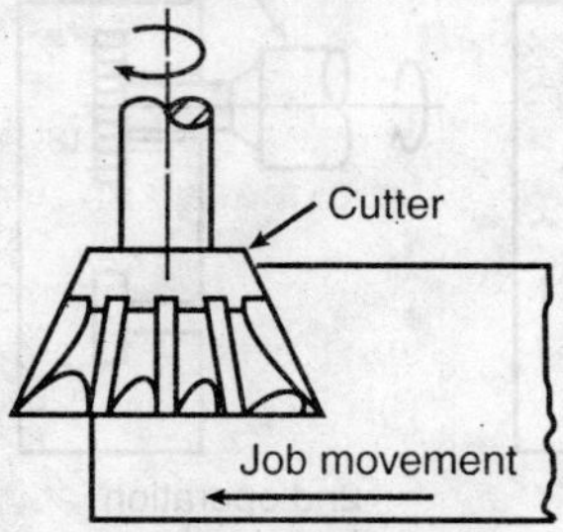

Sketch S-5.17 [3]

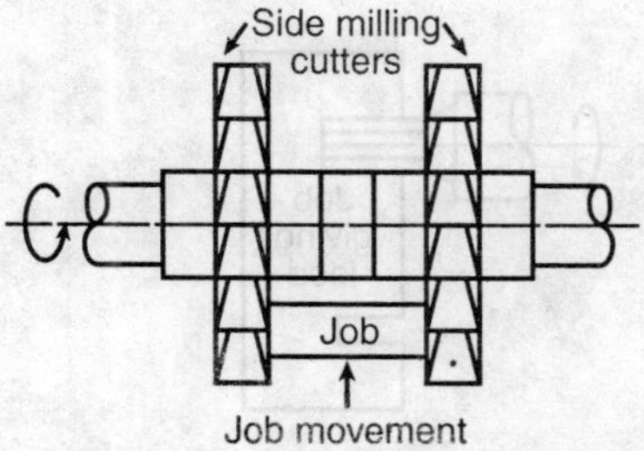

Sketch S-5.17 [4]

[4] Side Milling

The Side Milling operation is the production of a vertical flat surface on the side face of a job by using a side milling cutter.

[5] Straddle Milling

The Straddle Milling operation is the production of two vertical flat surfaces on the both sides of the job by using two side milling cutters which are separated by collars. (See same sketch as shown in Side Milling, S-5.17 [4]).

[6] Form Milling

The Form Milling operation is the production of irregular contours by using the cutters having the same profile corresponding to the surface to be generated.

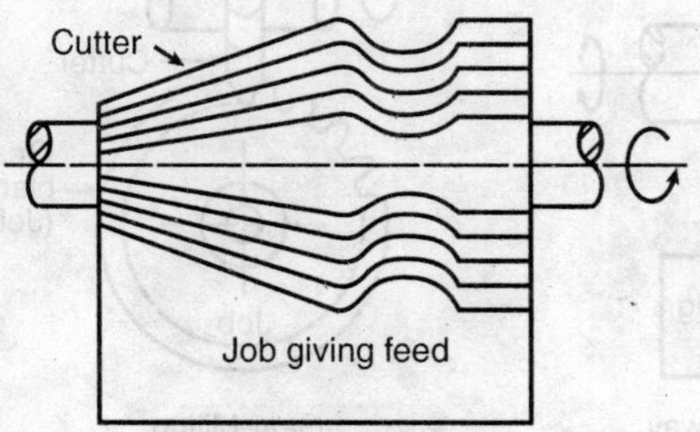

Sketch S-5.17 [6]

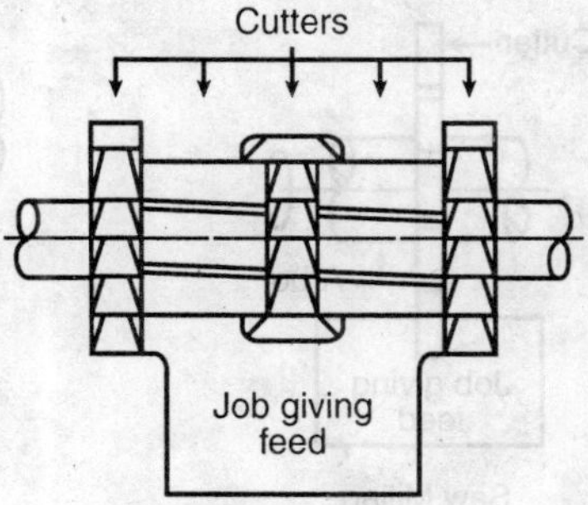

Sketch S-5.17 [7]

[7] Gang Milling

Gang Milling operation is the production of many surfaces of a job simultaneously by feeding the table against a number of required cutters. This method of operation saves machining time and hence is widely used in mass production.

[8] End Milling

The End Milling operation is the production of both peripheral and face milling operations simultaneously, generates vertical, horizontal or angular surfaces by using a end milling cutter. End Milling cutters may be used for milling of slots, grooves or key ways, etc.

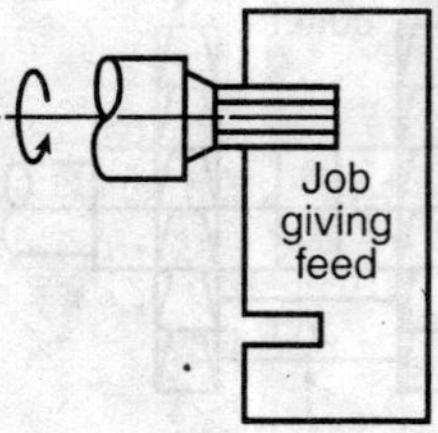

End Milling

Sketch S-5.17 [8]

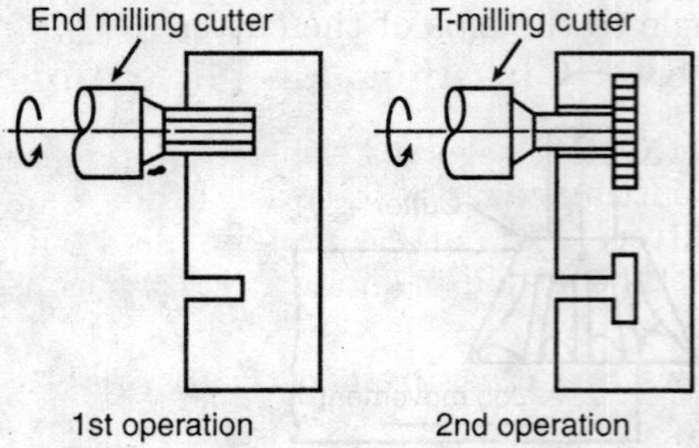

Milling Tee-Slots

Sketch S-5.17 [9]

[9] Milling Tee-slots

The operation is performed in two steps. In first operation, the end milling operation or a plain slot is made by using an end milling cutter and in second operation Tee-slot is made by using the Tee-slot cutter to enlarge the slot and to mill the bottom face of the slot.

[10] Saw Milling

Saw Milling operation is the production of narrow slots or grooves on a job by using a thin plain milling cutter or saw milling cutter. Saw milling operation is used for parting-off, grooving, screw-heads, slotted-links, etc.

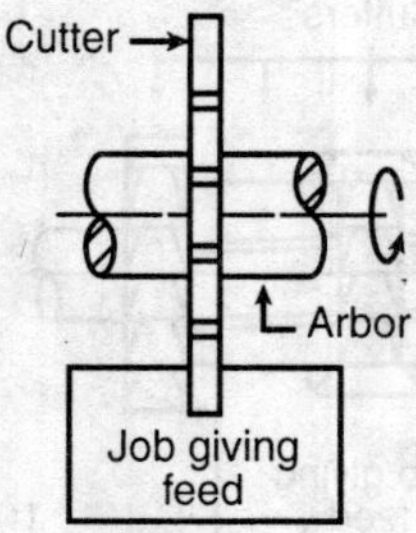

Saw Miling

Sketch S-5.17 [10]

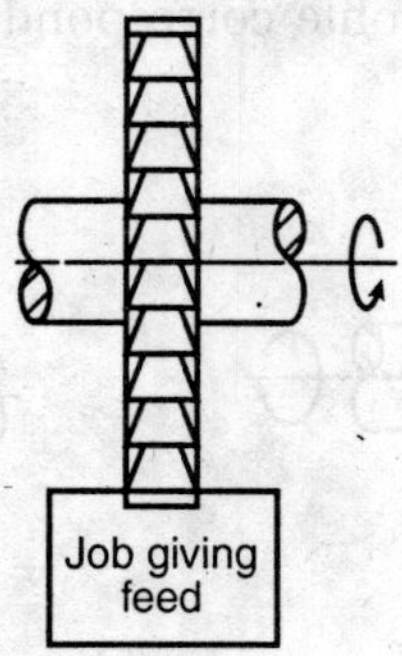

Miling Keyway

Sketch S-5.17 [11]

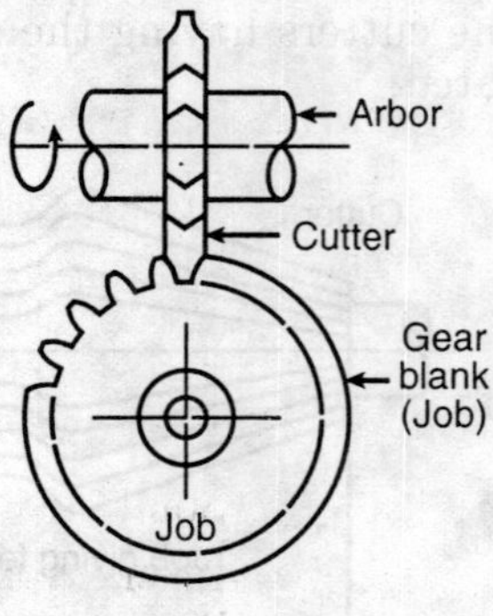

Gear Miling

Sketch S-5.17 [12]

[11] Milling Keyways, Grooves, Slots, etc.

These operations are the production of recesses in shafts of various shapes and sizes by using either a plain milling cutter or a saw milling cutter or a side milling cutter.

[12] Gear Milling

In this operation, Gear blank is held in between universal dividing head and tail stock on a machine table.

Form relieved cutter should have the profile corresponds exactly with the tooth space of the gear. Details of manufacturing gear is discussed in Article No. A5.25.

A5.18A : Classification Of Milling Cutters

Milling Cutter revolves either in horizontal or in vertical axis, having one or more cutting edges of identical forms equally spaced on the circumference of the cutter. The cutters are classified with respect to their material used in the teeth, method of mounting, direction of rotation of the cutter, purpose, etc.

Cutters are classified with respect to—

1. The material used in the teeth:

(a) High Speed Steel
(b) Cast Non-ferrous Stellite Cobalt
(c) Sintered Carbide Tips

2. Method of mounting the milling cutters:

(a) Cutters mounted on arbor
(b) Straight or tapered shank cutters mounted on a spindle adopter
(c) Face cutters mounted on the end of the short arbors.

3. Direction of rotation of the cutters:

(a) Right-hand cutters rotate in a counter-clockwise direction
(b) Left-hand cutters rotate in a clockwise direction when viewed from the end of the spindle.

4. Purpose or use of the cutter:

(a) Standard (or versatile work) milling cutter
(b) Special (or product-oriented or operation-oriented) milling cutter.

5. The constructional feature of the cutter:

(a) Solid cutter (High carbon steel or High speed steel)
(b) Inserted blade cutter (High speed steel or Cast non-ferrous stellite cobalt)
(c) Tipped cutter (Cast non-ferrous stellite or Sintered carbide).

6. The shape of the tooth:

(a) Milled tooth cutters
(b) Form relieved cutters.

7. The angle and tooth pitch for milling of various materials:

(a) Small tooth pitch for hard materials
(b) Medium tooth pitch for soft materials
(c) Larger tooth pitch for light metals.

8. The general shape and type of work:

(a) Plain milling cutters

(b) Side milling cutters

(c) End milling cutters

(d) Angular milling cutters

(e) Form milling cutters

(f) Metal slitting milling cutters, etc.

A5.18B : Nomenclature And Angles Of Plain Milling Cutter

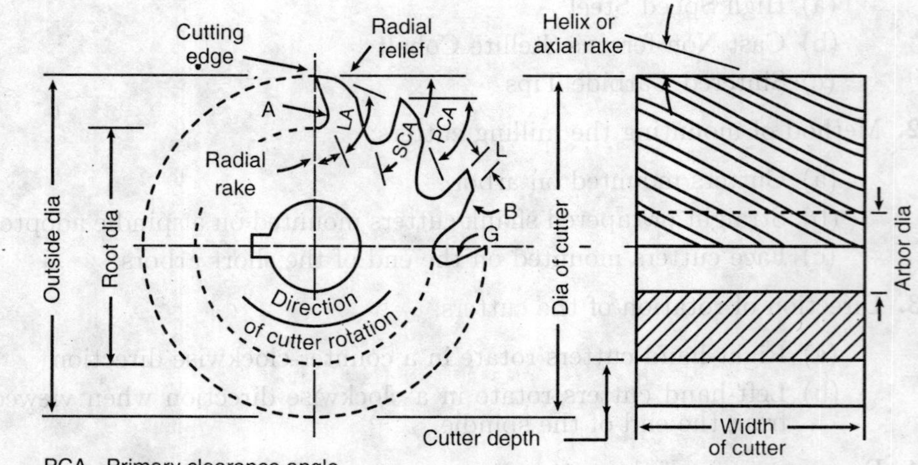

PCA - Primary clearance angle
SCA - Secondary clearance angle
LA - Lip angle, B - Back of tooth, L - Land
G - Gash hole or chip hole, A - Tooth face

Sketch S-5.18 B

Naming of different parts of a plain milling cutter and showing the different angles is known as nomenclature and angles of a plain milling cutter.

A5.18C : Elements Of Plain Milling Cutter Angles

[1] Radial Rake Angle

It is the angle between the flute face and a radial reference plane drawn from the cutter axis to the cutting edge. Positive radial rake angles for high speed cutters may vary from 10 to 15 degrees. Greater rake angle may be used for soft materials like Aluminium, etc. Positive rake is most satisfactory for most materials comprising of good shearing or cutting ability and strength. Zero axial rake for straight teeth of plain milling cutters may be used for saw type and narrow type of machining.

Negative rake angles (both radial and axial) for high speed milling with carbide-tipped cutters is used for milling hard materials, which increased cutting forces and power, keeps cutting edges more stronger and safer against impact.

[2] Lip Angle

It is the angle between the face of the tooth and the back of the cutting edge. It improves the tool life and better ability to resist shock loads. Lip angle = [90°–(Relief angle + Radial rake angle)].

[3] Relief Angle

It is the angle between the land and a tangent to the cutter from the tip of the tooth. It eliminates the tendency of the teeth to rub the work. The angle may vary from 3 to 5 degrees, with higher angles 8 to 10 degrees for smaller diameter cutters and soft work material and vise versa.

[4] Clearance Angle

Primary Clearance Angle

It is the angle between the back of tooth and a tangent to the cutter from the tip of the tooth.

Secondary Clearance Angle

It is the angle between the secondary clearance surface of the tooth and a tangent to the cutter from the tip of the tooth.

Clearance angle is to be considered as high as possible, which will increase the sharpness of the cutting edge but too much of it will tend to weaken the cutting edge. Course teeth are more efficient for removing metal than fine teeth.

[5] Helix Angle

It is the angle between the axis of a cylindrical helical cutter and with the plane of helical cutting edge. The angle depends on the type of cutters and its application.

[6] Fillet or Root Radius

The curved surface at the bottom of the gash space which joins the back of tooth to the face of the next tooth of a plain milling cutter is known as fillet.

[7] Tooth Face

The portion of gash space adjacent to the cutting edge on which the chip impinges in a plain milling cutter is known as tooth face.

[8] Gash Space or Chip Space

The space between the back of one tooth and the face of the subsequent tooth in a plain milling cutter is known as gash space or chip space.

[9] Land

The top portion of a tooth between the cutting edge and back of tooth is known as Land.

A5.19 : Milling Machine Attachments

Standard special attachments are used in Milling Machine to facilitate different kinds of operations required for a specific job. Attachment adds the versality of the machine, adaptability, efficiency, convenience and also giving greater working capability.

Most useful milling attachments are as follows:

1. Vertical Milling attachment.
2. Universal Milling attachment.
3. Rotary Milling attachment.
4. Slotting Milling attachment.
5. Rack Milling attachment.
6. Universal Spiral Milling attachment.
7. Dividing Head attachment.

[1] Vertical Milling Attachment

Vertical Milling attachment can be mounted on the face of the column of plain and universal knee type milling machine, which will act the same as in the case of a vertical milling machine. The speed of vertical milling spindle will be the same as that of the speed of the original knee type machine. There is an arrangement of spindle swivelling for machining angular surfaces.

[2] Universal Milling Attachment

Universal Milling attachment is similar to the vertical milling attachment but in addition to that, it may be set at any angle in both the planes. Higher spindle speed than the original machine speed may also be attained for operation.

[3] Rotary Milling Attachment

It is also named as Circular Milling attachment. It is a work holding device bolted to the top of the table of a plain or universal knee type milling machine. The surface of any profile such as various types of Cams, Circular Tee-slots, etc., can be machined by combining three or four movements of the table and the rotary movement of the attachment.

[4] Slotting Milling Attachment

The knee type milling machine may be altered to a slotter by mounting the slotting attachment on the face of the column and can be set at any angle between 0° and 90° on either side of the centre-line for machining angular surfaces. In slotting attachment, there is an arrangement of converting from the circular motion of the machine spindle to the reciprocating motion of the single-point tool for milling internal and external keyways, splines, etc. The length of stroke can be adjusted from 0 to 10 cm.

[5] Rack Milling Àttachment

Rack Milling attachment can be mounted at right angles to the universal milling machines spindle and becomes parallel to the table top surface. It is used for machining racks, threads, worms, etc.

[6] Universal Spiral Milling Attachment

It can be mounted on the face of the column of a plain or universal knee type milling machine for machining spiral grooves having a helix angle more than 45° (degree). The attachment is used for machining gears, worms, screw threads, twist drills, spiral milling cutters, etc.

[7] Dividing Head Attachment

Dividing Head is an attachment mostly used in knee type milling machine in making various milling cutters, reamers, multiflute drill, taps, spur-bevel-helical gears, etc. Dividing head does the following functions:

1. For setting the job at the required angle to the machine table.
2. For dividing the periphery of a job into equal number of divisions (i.e., for turning the job about its axis through a predetermined angle).
3. For continuous rotation of the job in cutting helical grooves.

A5.20 : Universal Dividing Head

Universal Dividing Head is an important job holding and accurate indexing to any fraction of a revolution of the job mostly used on a knee type milling machine. The jobs are usually mounted on centres of the dividing head and the tail stock, and the long jobs may be supported by the supporting blocks. The job may be set in vertical or horizontal or in inclined positions relative to the table surface. The job may also be rotated periodically for indexing. Continuous rotary motion of the job is done for milling helical gearing, twist drill, etc. Universal Dividing Head is used in the production of spur gears, bevel gears, splines, twist drills, helical gears, etc.

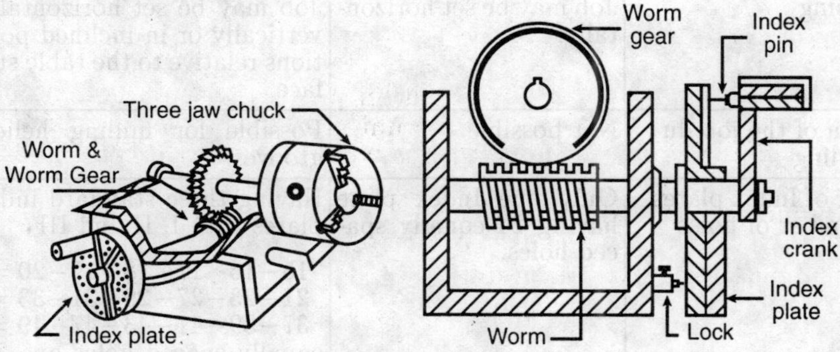

Sketch S-5.20

The main components of Universal Dividing Head are spindle with worm gear, worm shaft with index plate and sector arm, change gears for differential indexing connected to worm shaft. The worm is single threaded and worm gear has 40 teeth. Job is fixed on the spindle.

Proper index plate and change gears are mounted in proper position. For indexing by Universal Dividing head job to be fixed in position.

1. By rotating the index plate to the desired number of complete rotations and number of holes for a definite number of indexing.

2. Rotation of the index plate, transfer to the job through worm, worm wheel and spindle.

3. Rotation of the change gears may also be transferred to the job through worm shaft, worm and worm gear and then to the spindle.

4. Difference of speed of the index plate and change gears of the job gives .the actual differential indexing.

A5.21 : Comparative Study Between Plain And Universal Dividing Heads

Types of Dividing Head: The types of dividing head are as follows:

1. Plain type for direct or rapid indexing with index plate.

2. Universal type for simple and differential indexing, with index plates and set of interchangable gears

3. Optical type for precised angular indexing, etc.

Points	Plain or Simple dividing head	Universal dividing head
Position of the Index crank	Index crank is connected to the tail end of the spindle.	Index crank is connected to the worm shaft and worm wheel is mounted on the spindle.
Suitable for	Mass production and direct or rapid indexing.	Any type of Indexing.
Job setting	Job may be set horizontally.	Job may be set horizontally, vertically or in inclined positions relative to the table surface.
Rotation of the job during milling	Not possible.	Possible for milling helical grooves.
Number of Index plates with number of holes	Only one Index plate having 24 equally spaced holes.	Having three standard index plates with I, II and III 15 – 16 – 17 – 18 – 19 – 20 = I 21 – 23 – 27 – 29 – 31 – 33 = II 37 – 39 – 41 – 43 – 47 – 49 = III equally spaced holes on a hole circle.
Gearing arrangement in the dividing head	No	Yes, available gears are: 24–24–28–32–40–44–46–47–48–52–56–58–64–68–70–72–76–84–86–100

A5.22 : Types Of Dividing Heads And Indexing

Indexing is the method of dividing the periphery of a job into equal number of divisions.

[A] Direct (or Rapid) Indexing Dividing Head

Index plate having 24 holes is mounted to the tail-end of the spindle in dividing head directly. Job is mounted in between the spindle end of the dividing head and the tail stock end, both are bolted on the machine table. In a index plate having 24 holes, the divisions possible by this plate are 24, 12, 8, 6, 4, 3, 2, by rotating the plate manually for 1 hole, 2 holes, 3 holes, 4 holes, 6 holes, 8 holes, and 12 holes respectively after each cut. This type is useful for small number of divisions on the periphery and for mass production work.

[B] Dividing Head for Simple (or Plain) Indexing

In this system, indexing is effected with an index crank linked with the spindle through a worm gear. The worm wheel has 40 teeth while the worm has single start thread, both are meshed to each other. The required index plate may be mounted on worm shaft and the job is mounted with the worm wheel shaft. Therefore, for every 40 revolutions of the worm by the index crank, the job or worm wheel shaft will rotate only once.

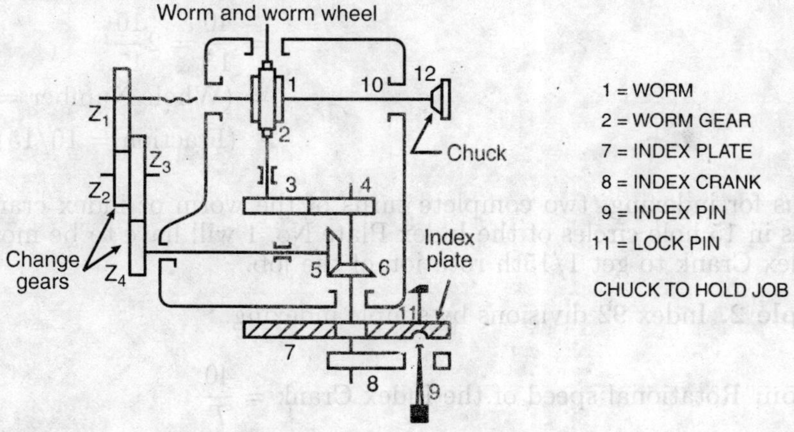

Sketch S-5.22

So the rotational speed of the Index Crank,

$$n = \frac{\text{Worm-Gear ratio (40:1)}}{\text{Number of divisions required, } T}$$

$$\text{or,} \quad n = \frac{40}{T}$$

In simple indexing, the three standard index plates having six concentric hole circles of different numbers of equally spaced holes on hole circle. The hole patterns on the index plates are as follows:

Plate Number	I –	15, 16, 17, 18, 19, 20	Numbers are equally spaced holes on a hole circle.
Plate Number	II –	21, 23, 27, 29, 31, 33	
Plate Number	III –	37, 39, 41, 43, 47, 49	

Rotational speed of the Index Crank,

$$n = \frac{40}{\text{Number of divisions required on the job}} = \frac{40}{T}$$

(Number of holes to move on the Index Plate)

$$\text{or,}\quad n = \frac{40}{T} \qquad \cdots (1)$$

$$= \text{Whole Number} \times \text{Fraction}$$

For the whole number, the worm or index crank must be rotated through the number of revolutions equal to whole number for each division required of the job.

And for the fractional portion, the numerator and denominator should be converted to a suitable factor so that the denominator is any one number of holes available in the index plate.

Example 1. To set the dividing head to get 15 divisions on a job.

Solution: Rotational speed of the Index Crank $= \dfrac{40}{T}$

$$= \frac{40}{15} = 2\frac{10}{15}$$

$= (\text{Whole Number} = 2) \times$

$(\text{Fraction} = 10/15)$

Thus for indexing, two complete turns of the worm or index crank and 10 holes in 15 hole circles of the Index Plate No. 1 will have to be moved by the Index Crank to get 1/15th rotation of the job.

Example 2. Index 92 divisions by simple indexing.

Solution: Rotational speed of the Index Crank $= \dfrac{40}{T}$

$$= \frac{40}{92} = \frac{10}{23}$$

= No whole number,

only fraction $\dfrac{10}{23}$

Therefore, the Index Crank is to move 10 holes in 23 hole circles of Index Plate No. II to get 1/92th rotation of the job.

[C] Differential Indexing

Differential indexing is employed where simple indexing not possible, i.e., the numbers 53, 61, 71, 79, 83, 97, etc. The two separate movements of the crank can be attained simultaneously by using a gear train between the tail end of the spindle and an auxiliary shaft [See Sketch S-5.22].

Diagram of universal dividing head is shown above. The (single start) worm 1 is engaged with worm gear (having 40 teeth) 2, spindle 10 is indexed by rotating crank 8 with spring loaded pin 9 relative to fixed index plate 7 which is row of circular holes equally spaced. The crank pin 9 is to set a required hole on a circular row on index plate 7. The rotary movement of the crank 8 is transmitted to the spindle 10 through bevel gear 5 and 6 with ratio 1, spur gears 3 and 4 with ratio 1 and a worm 1, worm gear 2 with ratio 1/40. The spindle 10 will make $1/T$ of a revolution for dividing the circumference into 'n' equal parts.

$\therefore$ The dynamic balance equation is $n.1.\ \dfrac{1}{40} = \dfrac{1}{T}$

$$\text{or,}\quad n = \frac{40}{T} = W + F$$

$$= W + \frac{N}{D} \times \frac{M}{M} \quad (T \text{ may be less than or greater than } 40)$$

where, T = Number of divisions required on job

W = Number of whole turns of the crank

F = Fractional turn of the crank

N = Numerator of the fraction F

D = Denominator of the fraction F

M = Common multiplier with N and D to suit the denominator equal to any number available on the index plates I, II and III

Taper shank stud 12 is inserted into the conical hole of the spindle rear end 10. The spindle is linked through a change gear train Z_1, Z_2, Z_3 and Z_4 through bevel gearing 5 and 6, Index shaft and Index plate 7. In differential indexing, by disengaging the lock pin 11, the movement of spindle is obtained by the combination of two motions (1) the turning of the index crank as per simple indexing and (2) the turning of the index plate from the spindle through the change gear trains and bevel gears.

For First Motion: Turning of the index crank as per simple indexing. Number of division required on the periphery of the job, i.e., the number to be indexed = I.

Number assumed T which it close to I, possible by simple indexing and change gear ratio or gear train is possible to calculate from the available twenty gears.

Available Gear

Change Gears	**Special Gears**
24 – 24 – 28 – 32 – 40 – 44 – 48 – 56 – 64– 72 – 86 – 100	46 – 47 – 52 – 58– 68 – 70 – 76 – 84

$\therefore$ Rotational speed of the Index Crank $= \dfrac{40}{I}$ $\cdots$ (1)

For Second Motion: Turning of the index plate from the spindle through change gear train and bevel gears with ratio 1.

$$\text{Change gear ratio} = \frac{\text{Driver gear(s)}}{\text{Driven gear(s)}} = \frac{(T - I) \times 40}{T}$$

The above two motions of the index plate may move in the same direction or opposite to each other depending on the gearing (either simple or compound) and the assumed number T chosen (either greater than or less than the number to be indexed I).

$$\text{Total rotational movement of the Crank} = \frac{40}{T} + \frac{(T - I) \times 40}{T}$$

$(T - I)$ is +ve when $T > I$, both motions will be in the same direction.

$(T - I)$ is −ve when $T < I$, both motions will be in the opposite direction.

To attain the required direction of motion, the number of idle gears to be chosen accordingly.

Use 1	idle gear for positive simple gearing
Use 2	idle gears for negative simple gearing (equal number of teeth)
Use no	idle gear for positive compound gearing
Use 1	idle gear for negative compound gearing

A5.23 : Examples On Differential Indexing

Example: Divide the periphery of circular gear blank by 93 divisions.

Solution: Rotational speed of the Index Crank $= \dfrac{40}{93}$

Number 93 by simple indexing is not possible. So, 93 is to be indexed by differential indexing.

Number to be indexed $= I = 93$

Number assumed $= T = 95$

$$\text{Total rotational speed of the Index Crank, } n = \frac{40}{95} + \frac{(95-93) \times 40}{95}$$

$$= \frac{8}{19} + \frac{2 \times 40}{95};$$

$$\left[\because \frac{2 \times 40}{95} = \frac{16}{19} = \frac{16 \times 4}{19 \times 4} = \frac{64}{76}\right] \qquad = \frac{8}{19} + \frac{64}{76}$$

For the first motion, the index crank will have to move 8 holes in 19 hole circles on Index plate number I for 92 times.

And, for the second motion, setting 64 no. teeth on taper shank stud at position Z_1, 76 teeth on horizontal bevel gear shaft at position Z_4 and according to the condition using one idle gear on the idle shaft in between Z_1 and Z_4 for positive simple gearing, i.e., 64/76.

Alternative Solution. Choosing the assumed number T less than the number to be indexed I.

Say, $T = 90$; where, $I = 93$

$$\text{Total rotational speed of the Index Crank, } n = \frac{40}{90} + \frac{(90 - 93) \times 40}{90}$$

$$\text{or,}\quad n = \frac{8}{18} + \frac{-120}{90} \left[\because \frac{-120}{90} = -\frac{4}{3} \times \frac{3}{3} = \frac{-4 \times 16}{3 \times 16} \times \frac{3 \times 8}{3 \times 8} = -\frac{64}{48} \right]$$

$$\text{or,}\quad n = \frac{8}{18} + \left[-\frac{64}{48} \right]$$

$$\text{or,}\quad n = \frac{8}{18} + (-)\frac{64}{48}.$$

For the first motion, the index will have to move 8 holes in 18 hole circles on Index plate no. *I* for 92 times.

And for the second motion, setting 64 no. teeth on taper shank stud at position Z_1, 48 no. teeth on Z_4 and 24 no. teeth on Z_2 of the idle shaft, last denominator 24 no. teeth on Z_3 and according to the condition using one idle gear either between Z_1 and Z_2, or between Z_3 and Z_4 for negative compound gearing.

For negative simple gearing, two idle gears of 24 teeth each are required in two different axes in between 64 no. teeth as drive gear and 48 teeth driver gear.

A5.24 : Problems On Differential Indexing For Practice

Prob.1 Index no. 23, 92, 156 and 245 simple Indexing.

Ans. For 23 – 1 complete turn +17 holes in 23 circles,
For 92 – 10 holes in 23 hole circles,
For 156 – 10 holes in 39 hole circles,
For 245 – 8 holes in 49 hole circles.

Prob.2. Determine the number of revolutions of the Index Crank and change gears in milling a gear with the number of following teeth—(a) 83, (b) 127, (c) 209.

Ans. For (a) $n = \frac{20}{43} + \frac{72 \times 40}{24 \times 86}$ if assumed number is 86 and no idle gear

For (b) $n = \frac{12}{39} + \frac{48}{52}$ if assumed number is 130 with one idle gear

For (c) $n = \frac{4}{21} + \frac{32 \times 24}{84 \times 48}$ if assumed number is 210 with no idle gear

A5.25 : Milling A Spur Gear In Milling Machine

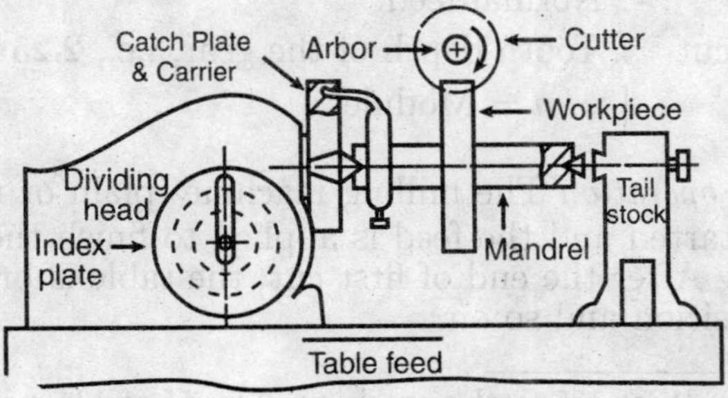

Sketch S-5.25

Problem. Write the steps to be followed in cutting a spur gear of module = 4 and teeth = 58.

Solution: In spur gear milling operations, the following steps to be followed:

1. Preparation of gear blank with respect to following particulars—

(a) Gear blank diameter = Module (No. of teeth +2)

$$= m(T+2) \quad \text{where, module, } m = 4,$$
$$= 4(58+2) \quad \text{no. of teeth, } T = 58$$
$$= 240 \text{ mm}$$

(b) Tooth depth = 2.25 × Module = 2.25 × 4 = 9 mm.

2. *Selection of cutter* with respect to the type of gear and number of teeth from the standard table for correct number of form cutter. For cutting 58 teeth spur gear operation need cutter no. 2 selected from standard table.* [see footnote]

Fixing of cutter in proper position on a milling machine arbor.

4. *Indexing of 58 teeth*

Rotational speed of the Index crank, $n = \frac{40}{T} = \frac{40}{58} = \frac{20}{29}$

The index crank is to be moved by 20 holes in 29 hole circles for 57 times by using simple indexing plate no.II.

5. *Fixing of job* on mandrel in between universal dividing head and tail-stock, mounted on milling machine table. The job, i.e., gear blank is pressed tight on the mandrel and the dog is attached to the end of the mandrel by a set screw for positive drive.

6. *Alignment of cutter* with respect to the work axis by raising the table. Depth of cut is given equal to the full depth (i.e., 9 mm) of the gear tooth by raising the table.

7. *Selection of cutting speed, feed and depth of cut.*

Speed	: It should be slightly slower than the plain milling operation.
Feed	: Normal feed.
Depth of cut	: Tooth depth of the gear, i.e., 2.25 m = 9 mm. [$\because m$ = Module]

8. *First tooth operation* The milling machine (plain or universal or omni-versal) is started and the feed is applied to finish the first tooth space of the gear. After the end of first cut, the table is brought back to the starting position and so on.

***Note:** Cutter no. 1 for 135 teeth to rack, no. 2 for 55 to 134, no. 3 for 35 to 54, no. 4 for 26 to 34, no. 5 for 21 to 25, no. 6 for 17 to 20, no. 7 for 14 to 16, no. 8 for 12 and 13.

9. *Subsequent tooth operations*

After the completion of the first tooth operation, the blank is indexed by 20 holes in 29 hole circles for the next tooth space and the operation is repeated for 57 times till the rest of the teeth are being cut.

Note: Full depth of cut may be attained gradually, generally in 3 to 6 passes depending on the amount of depth, job-tool material, machine condition and other cutting conditions.

10. *Checking of teeth.*

A5.26 : Milling A Helical Gear In Universal Or Omniversal Milling Machine

Machine required: Universal or Omniversal Milling Machine.

Cutter required: Helical Milling Cutter.

Milling a helical gearing is of producing helical grooves on the periphery of a work. The work (gear blank on mandrel) is mounted on Milling Machine Table in between universal dividing head and a tailstock end. The work is rotated by a table gear train between the work spindle of dividing head with the table feed screw. The helical milling cutter is mounted on Milling Machine Arbor. The work is fed longitudinally passed the milling cutter. The work is rotated to the calculated values against the longitudinal feeding of the table, makes helical grooves on the periphery of the work.

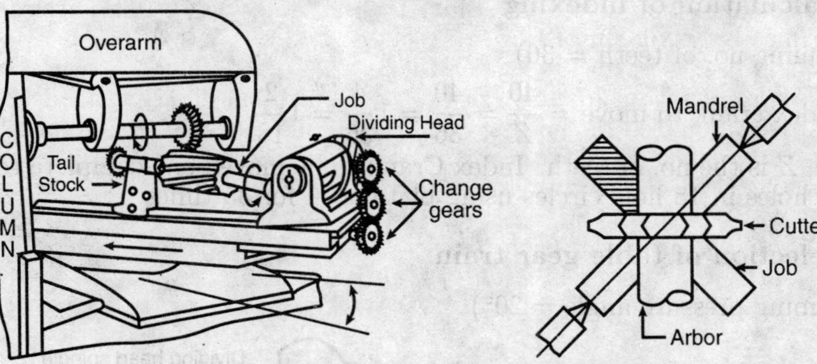

(A) Helical milling set-up (B) Helical gear operation

Sketch S-5.26 (A) & (B)

The following operational steps to be followed in milling a helical gear.

[1] Preparation of gear blank

(Assuming module = 4 mm, helix angle 45° and no. of teeth = 36)

Blank diameter or Outside diameter, $D_B = Zm + 2m_n$

where, Z = No. of teeth $= Zm + 2m\cos\beta$, in mm

m = Module in mm $= 36 \times 4 + 2 \times 4 \times \cos 45°$

m_n = Normal module $= 150$ mm

$= m\cos\beta$ and where β is the helix angle in degrees.

[2] Calculation of helix angle

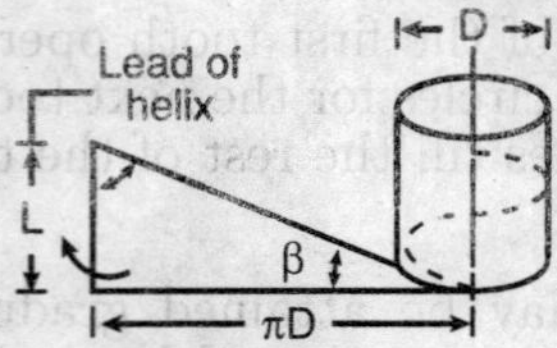

Sketch S-5.26 [2]

The helix angle is the acute angle between the tangent to a helix and the straight generator of the cylinder on which it lies.

$$\therefore \quad \tan\beta = \frac{L}{\pi D_p}$$

where, D_p is the pitch circle diameter of work in mm $= Zm$, L is the lead of helix in mm

$$\text{or} \quad L = \pi D_p \tan\beta$$
$$= \pi Zm \tan\beta$$
$$= \pi \times 36 \times 4 \times \tan 45°$$
$$= 452.4 \simeq 450 \text{ mm}$$

[3] Fixing of gear blank (work) on mandrel

The mandrel is supported in between the universal dividing head and the tail stock on milling machine table. The table is swivelled to the calculated helix angle.

[4] Calculation of indexing

(Assuming no. of teeth = 36)

$$\text{Index crank to move} = \frac{40}{Z} = \frac{40}{36} = 1\frac{4}{36} = 1\frac{2}{18}$$

where, Z is the no. of teeth. Index Crank to be moved one complete rotation and 2 holes in 18 hole circles using plate no. I for 35 times.

[5] Selection of table gear train

(Assuming pressure angle = 20°)

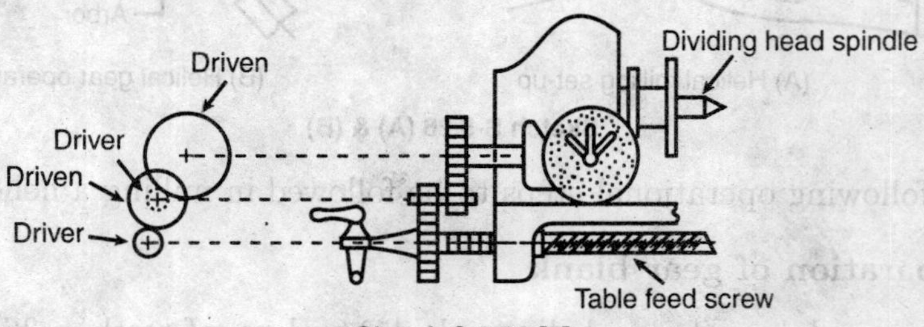

Sketch S-5.26 [5]

$$\frac{\text{Driver gears}}{\text{Driven gears}} = \frac{\text{Lead of the machine table}}{\text{Lead of the work}} = \frac{40 \times \text{Lead of the lead screw}}{\text{Lead of helix } (L)}$$

$$= \frac{40 \times 4}{450} = \frac{8}{9} \times \frac{2}{5} = \frac{8 \times 8}{9 \times 8} \times \frac{2 \times 20}{5 \times 20} = \frac{64 \times 40}{72 \times 100}$$

[6] Choice of proper cutter and mounting the same on arbor

Selection of cutter with respect to the equivalent number of teeth in case of an involute gearing.

$$Z^1 = \frac{Z}{\cos^3 \beta} = \frac{36}{\cos^3 45°} = 101.8 \simeq 102$$

where Z^1 is the equivalent no. of teeth in involute gearing.

Cutter no. 2 is selected of 102 no. of teeth assuming 36 teeth and helix angle of 45°. Cutter is mounted on arbor in such a way that the axis of the work and the cutter are perpendicular to each other on horizontal axis.

[7] Setting of gear train between the Machine table screw and Universal dividing head worm wheel spindle as per calculation shown in A5.26 [5].

[8] Selection of speed, feed and depth of cut; speed should be slightly lower than plain milling operation, feed should be normal and tooth depth equals to $2.25\ m_n = 2.25 \times 2.83 = 6.36$ mm.
where, m_n = normal module $= m \cos \beta = 4 \times \cos 45° = 4 \times \frac{1}{\sqrt{2}} = 2.83$

Cutter is in touch with the peripheral surface of the work and the required depth of cut is given by raising the table. The table must be swivelled to the required helix angle of 45°.

[9] Operation of first and subsequent teeth

For the operation, the machine is started, the first tooth space is milled in one or two cuts. At end of each cut, the table will be brought back to original position, indexing as mentioned above is performed after withdrawing the index pin from index plate. After placing the crank pin in position, the second and subsequent teeth of the helical gearing are performed.

Problem for Practice

1. A helical gear has helix of 45° and module of 4 mm. Pitch of table screw is 6 mm and no. of teeth is 8. Calculate the following:

A. Lead of helix in mm (Ans. 100 mm)

B. Blank diameter in mm (Ans. 55 mm)

C. Indexing [Ans. 5 complete turns by Index Crank]

D. Table gear train [Ans. (72 × 32)/(24 × 40)]

E. Normal module [Ans. 2.83 mm)

F. Tooth depth [Ans. 6.355 mm]

2. A helical gear, has no. of flutes = 6, lead of helix = 600 mm on a blank of 75 mm in diameter. [Assume pitch of table screw as 6 mm.]

Calculate: (A) Indexing [Ans. 6 turns +12 holes in 18 hole circles]
(B) Change gears [Ans 40 : 100]
(C) Helix angle [Ans. 21° – 25 min. approximate]

A5.27 : Gear Cutting Machines

A5.27.1 : Gear Driving Systems

The gear drive is positive as teeth and tooth spaces mesh with each other. Gears are transmitting torque and motion from one shaft to other.

Sl. No.	Gear Driving Systems	Sketches
[1]	Two spur gears with external tooth systems. Two shaft axes are parallel and the direction of rotation is opposite. The teeth are straight.	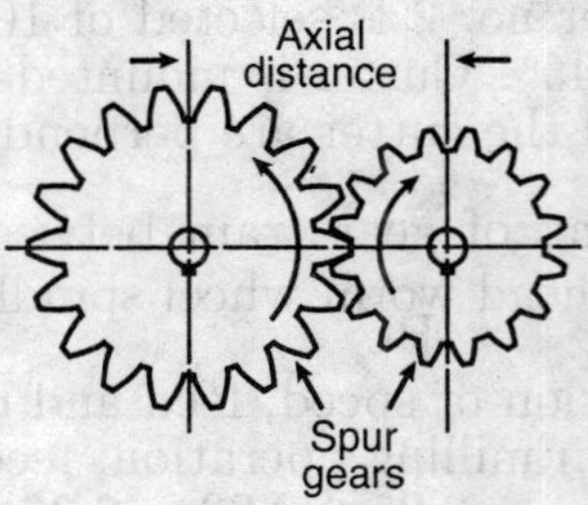 Sketch S-5.27 [1]
[2]	Two spur gears with internal tooth systems. Two shaft axes are parallel and have a parallel direction of rotation with small centre distance. The teeth are straight.	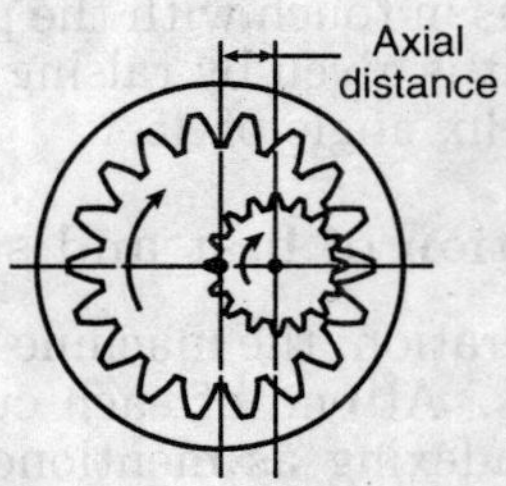 Sketch S-5.27 [2]
[3]	**Rack and Pinion system.** The rotary motion of the pinion is converted into a parallel, straight line motion of the straight tooth rack and vice versa.	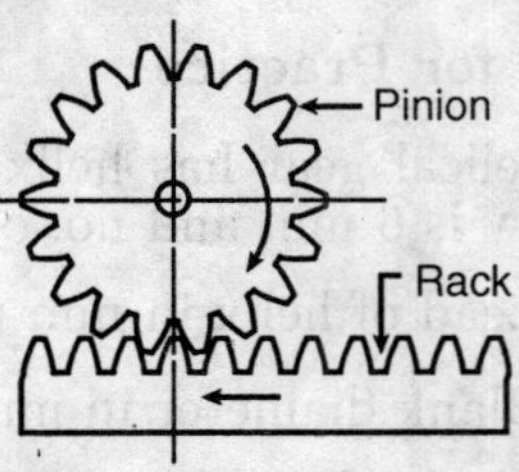 Sketch S-5.27 [3]
[4]	**Helical Gearing system**. The shaft runs parallel with helical teeth. Gradual engagement of the teeth helps to run the gears more silently. The axial pressure developed is being absorbed by the thrust bearing.	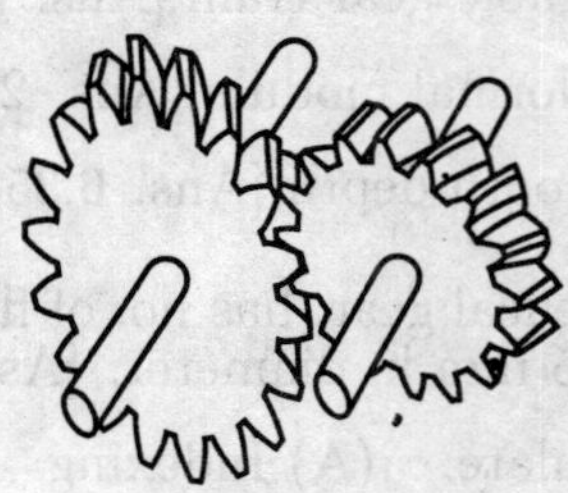 Sketch S-5.27 [4]

Contd...

Sl. No.	Gear Driving Systems	Sketches
[5]	**Herring-bone Gearing system**. The teeth are double helical. The shaft runs parallel but applied for heavier transmission, in which axial pressure is compensated.	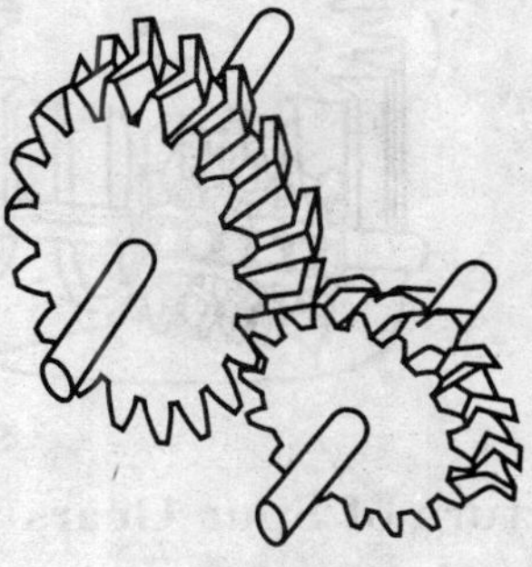 Sketch S-5.27 [5]
[6]	**Bevel Gearing system**. The teeth are bevel form with straight, helical and spiral teeth. The bevel gears have intersecting axis.	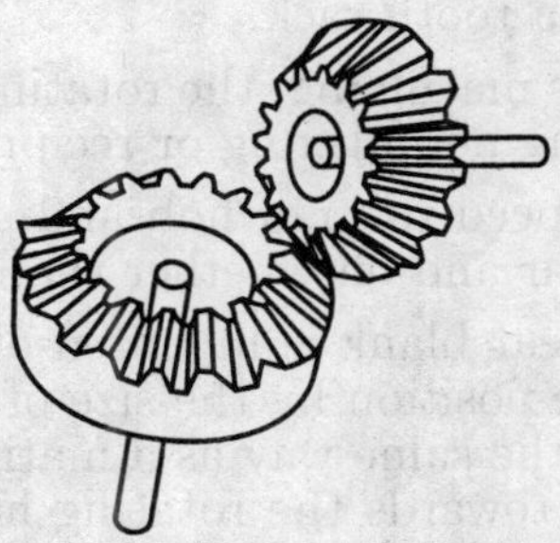 Sketch S-5.27 [6]
[7]	**Spiral Gearing system**. These are toothed gears with their axes non-interacting and in different planes.	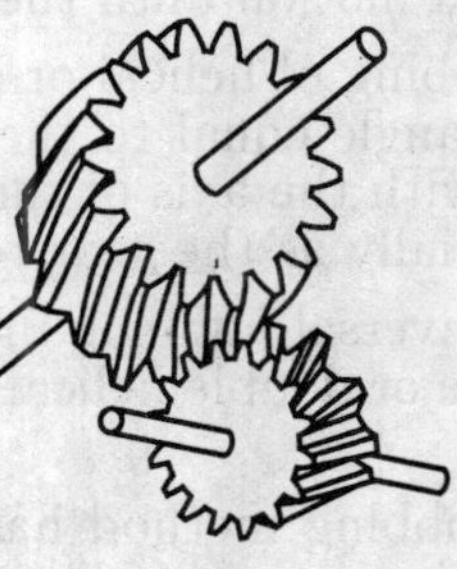 Sketch S-5.27 [7]
[8]	Worm gear pair or worm and worm wheel; worm acts as a driving wheareas worms wheel acts as driven members and is most suitable for heavy loads. The shaft axes are non-intersecting and non-parallel. Gearing arrangement is compact and runs smoothly.	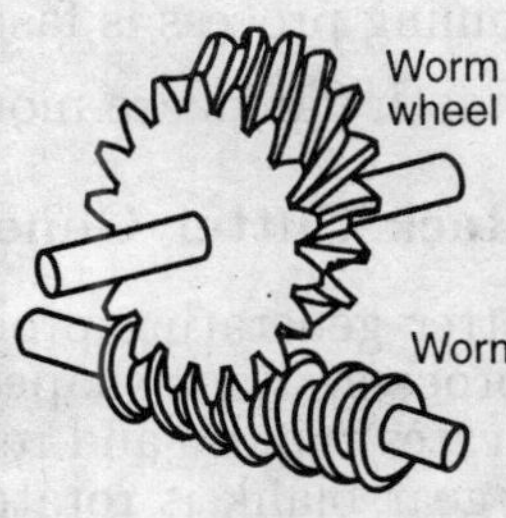 Sketch S-5.27 [8]

A5.27.2 : Gear Hobbing

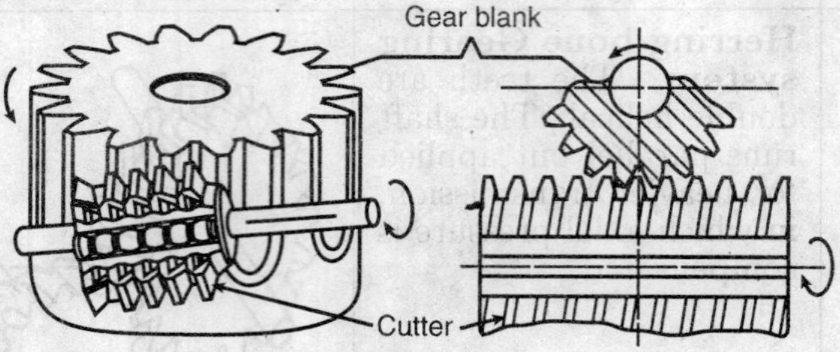

Sketch S-5.27.2

Manufacture of Spur Gears by Hobbing Machine

Hobbing is a generating system. A gear hob is a worm shaped with large number of cutting teeth. The tooth form is trapezoidal, similar to the tooth profile of a tooth rack.

In this machining, the rotating hob is in mesh with rotary gear blank and there being no indexing or reciprocating movements.

The speed ratio of hob and gear blank depends on the number of teeth on the gear and on whether the hob is single-threaded or multi-threaded.

The gear blank is mounted on the machine table and the hob is also set in oblique position by the size of its pitch. Both receive a positive drive and rotate in the same way as a mating worm and worm gear. The gear blank is moved in towards the rotating hob until the proper depth is achieved. The lead velocity of the hob is the same as the pitch line velocity of the gear. As soon as the proper depth is reached, the hob-head with the hob, performs a vertical feed motion until the teeth are complete.

For hobbing of helical or spiral gears, the hob must be moved around an additional angle equal to the helix angle of the gears. Worm gears may be machined with the axis of the hob at right angles to the gear and the hob is fed tangentially as the gear rotates.

This universal hobbing machine can be used for the manufacture of spur gears, single or double helical gears, worm gears, spiral gears, worms, splines, etc.

Gear hobbing method has the following advantages over the other type of gear cutting processes :

(1) One hob cutter can be used for all numbers of teeth of a pitch.

(2) Machining process is faster.

(3) The tooth flanks are more accurate and the pitch is more equal.

A5.27.3 : Rack Cutter Generating Process

The rack cutter generating process is also called shaping of teeth with the generating process. Rack-shaped cutter or pinion-type cutter mounted in a ram, reciprocates rapidly and removes metal only during the vertical cutting stroke. The gear blank is rotated slowly in its own axis and lateral motion parallel to the rack-shaped cutter until the tooth is completed. Then the machine table returns to its starting position; in this way the table with workpiece is shifted for shaping the second and subsequent teeth. Spur and helical gears with external tooth system can be shaped.

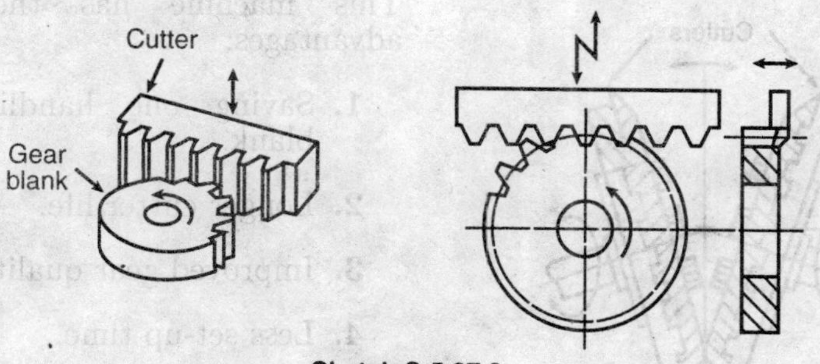

Sketch S-5.27.3

A5.27.4 : Pinion Cutter Generating Process

In this method both internal and external teeth can be shaped. The pinion-type cutter can perform the vertical cutting stroke. The generating motion is created by the rotary motion of the cutter and the workpiece. The workpiece moves off from the cutter and returns automatically to the initial working position for shaping the second and subsequent tooth. This is being done during the withdrawal of the ram, after shaping of one tooth. The process continues till all the teeth are completed.

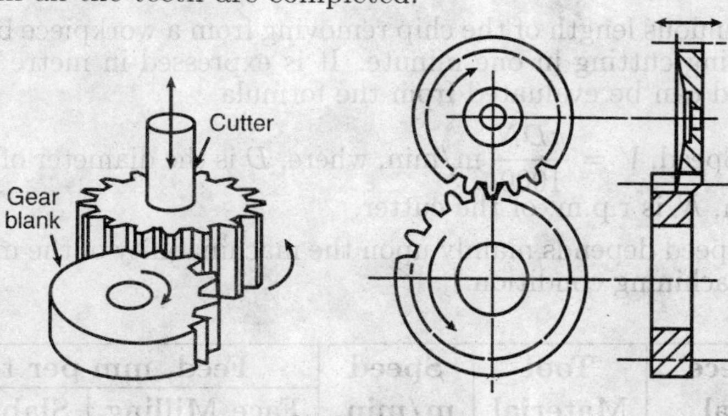

Sketch S-5.27.4

A5.27.5 : Manufacture of Worms and Worms Wheels

Worm can be machined either in Milling Machine or in Screw Cutting Engine Lathe, but the worm wheels can be manufactured by gear hobbing.

A5.27.6 : Manufacture of Bevel Gears or Bevel Gear Generators

Accurate bevel gears can be manufactured by the bevel gear generator. In generating process the two cutters simulate a tooth of a mating gear. The two cutters cut off chips alternatively. At the beginning, the cutter head is fed into the blank for rough cutting a portion of the whole depth. The cutter is mounted on the cradle and the work then rolls down to the bottom of the generating roll for rough shaping the tooth. The work is then fed to full depth for finishing the tooth space. The bevel gear blank is then indexed and the process is repeated until all the teeth are cut. The above operations of the machine are automatic.

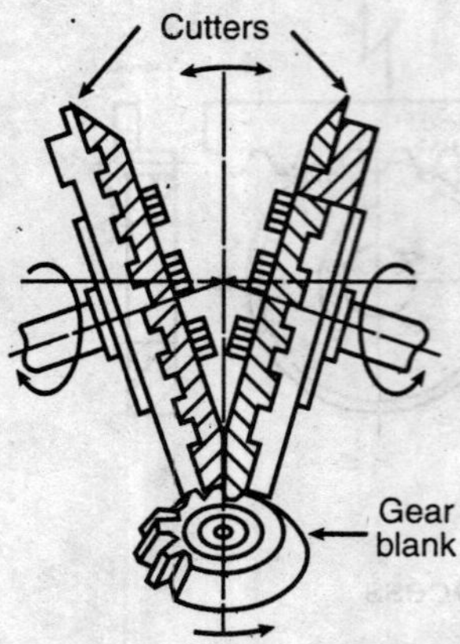

Sketch S-5.27.6

This machine has the following advantages:

1. Saving one handling of the blank.
2. Longer cutter life.
3. Improved gear quality.
4. Less set-up time.

Spiral bevel gears can also be manufactured by the generating method but the cutter is circular and rotates as a face milling cutter.

A5.27.7 : Calculation of Machining Time

Cutting speed, feed and depth of cut in milling operations.

Cutting Speed

It is the continuous length of the chip removing from a workpiece by revolving a rotary milling cutting in one minute. It is expressed in metre per minute (m/min.) and can be evaluated from the formula

Cutting Speed, $V = \frac{\pi DN}{1000}$ m/min, where, D is the diameter of the milling cutter in mm, N is r.p.m. of the cutter.

Cutting speed depends mainly upon the machinability of the material and the actual machining condition.

Workpiece Material	Tool Material	Speed m/min.	Feed, mm per tooth	
			Face Milling	Slab Milling
Aluminium Alloys	H.S.S. / Carbide	70–90 / 60–160	0.45	0.35
Copper Alloys	H.S.S. / Carbide	45–70 / 65–90	0.3	0.2
Mild Steel	H.S.S. / Carbide	30–40 / 90–120	0.2	0.15
Grey Cast Iron	H.S.S. / Carbide	18–28 / 70–90	0.3	0.25
Stainless Steel	H.S.S. / Carbide	16–20 / 35–55	0.15	0.10
Alloy Steel	H.S.S. / Carbide	12–18 / 40–50	0.15	0.10

The above chart shows the different cutting speed and feed of different workpiece and tool materials.

Feed

Feed in milling operations is the rate with which the workpiece advances under the milling cutter. It is expressed in the following methods:

(i) Feed per tooth (f_t) is the linear distance advances by the workpiece in one tooth movement of rotating milling cutter. It is expressed in mm/tooth.

(ii) Feed per revolution (f_r) is the linear distance advances by the workpiece in revolution of the rotating milling cutter.

i.e., $f_r = f_t \times Z$ mm/revolution, where, Z is the number of teeth in milling cutter.

(iii) Feed per minute (f_m) is the linear distance advances by the milling cutter rotation.

i.e., $f_m = f_t \times Z \times N = f_r \times N$ mm/min., where, N is the speed of the milling cutter in r.p.m.

Depth of Cut

Depth of cut is the amount of penetration of the rotating milling cutter into the workpiece in one pass. It is the thickness of the chip or the distance between the original surface and finished surface of the workpiece in one pass and expressed in mm.

Calculation of Machining Time in Milling Operations

Machining time (T_m) is the time required in minute to perform different operations in a milling machine to finish the work.

Machining time,

$$T_m = \frac{\text{Total length of the table travel to complete the workpiece}}{\text{Feed of table per minute}}$$

$$= \frac{L}{f_t \times Z \times N} \times \text{Number of passes}$$

where, L is the length of table travel in one pass in mm
f_t is the feed per tooth of the milling cutter in mm/tooth
Z is the number of teeth of the cutter
N is the speed of the cutter in r.p.m.

Length of table travel in one pass is the summation of approach length (a), length of the workpiece (l) and over-run length (o) in mm.

$$(\text{i.e., } L = a + l + o)$$

Approach length is calculated for face milling and slab milling (or peripheral milling) are as follows:

From right-angled triangle ABC, for face milling operation,

$$a = \sqrt{(D_c/2)^2 - \left(\frac{D_c}{2} - W\right)^2}$$

where D_c is the diameter of the cutter in mm

$AE = W$ is the depth of cut in mm.

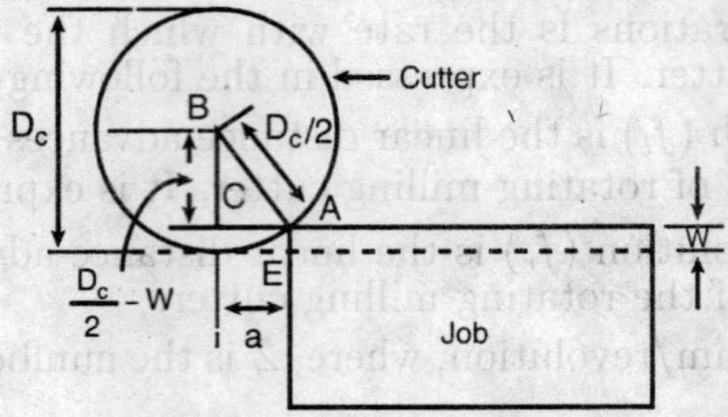

S-5-27.7(1) FACE MILLING OPERATION

For slab milling operation,

$$a = \frac{D_c}{2} - x \text{ in mm}$$

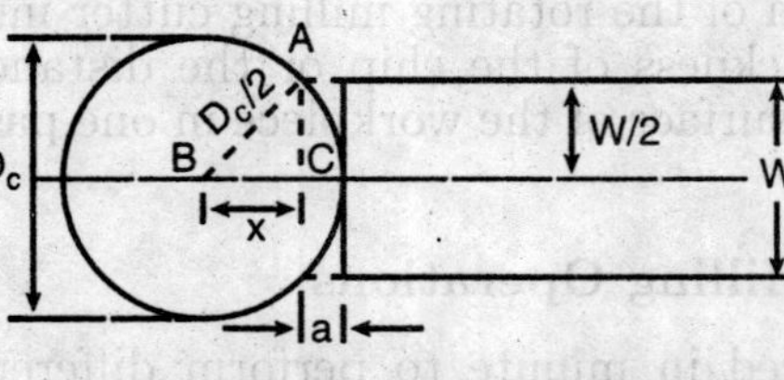

S-5.27.7(2) SLAB MILLING OPERATION

From right-angled triangle ABC,

$$x = \sqrt{\left(\frac{D_c}{2}\right)^2 - \left(\frac{W}{2}\right)^2}$$

where, D_c is the diameter of the cutter in mm

$\frac{W}{2}$ is the depth of cut in mm.

W = width of the workpiece in mm.

Over-run (o) is considered about 2 to 4 mm for removing the last chip from the workpiece and the amount will depend on the size of the workpiece and cutter.

Problem 1. Problems on machining time

Calculate the total machining time required to mill a hexagonal prism on a milling machine from round M.S. bar of 100 mm length and 40 mm in diameter with an H.S.S. 50 mm cutter. Take feed as 20 mm/min. Consider set-up and indexing time as 20 minutes.

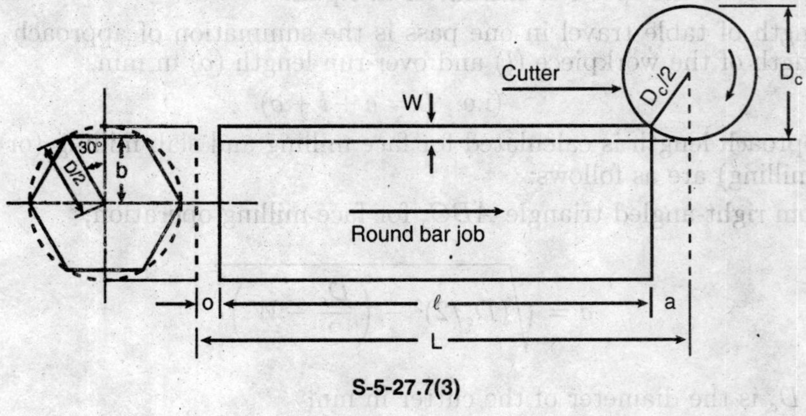

S-5-27.7(3)

Given Data:

Job Length, $l = 100$ mm
Job Diameter, $D = 40$ mm
Cutter Diameter, $D_c = 50$ mm
Feed $= f_m = 20$ mm/min.
Set-up and Indexing Time = 20 min.
$\cos 30° = b/D/2$

$$b = D/2 \cos 30° = \frac{\sqrt{3}}{2}\frac{D}{2}$$

$$W = \left(\frac{D}{2} - \frac{\sqrt{3}}{4}D\right)$$

$$= D\left(\frac{1}{2} - \frac{\sqrt{3}}{4}\right)$$

$$= 2.68 \text{ mm} \quad [\because D = 40 \text{ mm}]$$

$$a = \sqrt{W \times (D_c - W)}$$

$$= \sqrt{2.68(50 - 1.08)} \quad L = a + l + o \text{ in mm}$$

$$= 11.26 \text{ mm} \quad = \sqrt{W(D_c - W)} + l + o$$

$$= 11.26 + 100 + 2 \simeq 115 \text{ mm}$$

$$= 7.27$$

$$\text{Machining time, } T_m = \frac{\text{Length of cut}}{\text{Feed per min.}} \times \text{No. of indexing}$$

$$= \frac{L}{f_m} \times 6 = \frac{115}{20} \times 6 = 34.5 \text{ min.}$$

Total time required $= 34.5 + 20 = 54.5$ min.

Problem 2. A slot 25 mm deep is to be cut on a workpiece of 200 mm length with the help of H.S.S. side and face milling cutter whose diameter is 150 mm and has 10 teeth. The cutting speed is 50 m/min. and feed is 0.25 mm/teeth. Determine (i) Table feed in mm/min., (ii) Total machining time required to machine the slot. Assume suitable approach and over-run.

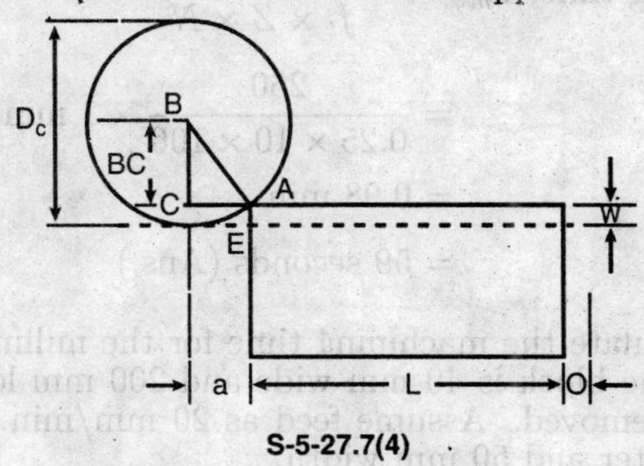

S-5-27.7(4)

D_c = Diameter of cutter = 150 mm

$BA = D_c/2$

Assuming $CA = a$ = approach

$BC = D_c/2 - W$

AE = depth of slot = W = 25 mm

Length of workpiece, l = 200 mm

Cutting speed, $v = (\pi DN)/1000 = 50$ m/min.

or, $50 = (\pi \times 150 \times N)/1000$

or, N = 106 r.p.m.

Feed/Teeth = f_t in mm/tooth = 0.25 mm/tooth

Number of teeth of cutter, $Z = 10$

Answer to first part (i)

Table feed in mm/min., $f_m = f_t \times Z \times N$

$$= 0.25 \times 10 \times 106 \text{ mm/min.}$$

$$\simeq 265 \text{ mm/min.}$$

Answer to second part (ii)

$$\text{Approach, } a = \sqrt{\left(\frac{D_c}{2}\right)^2 - \left(\frac{D_c}{2} - W\right)^2}$$

$$= \sqrt{(150/2)^2 - (150/2 - 25)^2}$$

$$= 55.9 \text{ mm (Ans.)}$$

Assuming over-run, o = 4.1 mm

Therefore, total length of cutter travel, $L = a + l + o$

$$= 55.9 + 200 + 4.1$$

$$= 260 \text{ mm}$$

$$\text{Total machining time, } T_m = \frac{L \times \text{No. of passes}}{f_t \times Z \times N}$$

$$= \frac{260}{0.25 \times 10 \times 106} \times 1 \text{ min.}$$

$$= 0.98 \text{ min.}$$

$$= 59 \text{ seconds (Ans.)}$$

Problem 3. Calculate the machining time for the milling operation of rectangular block. The block is 40 mm wide and 200 mm long. 4 mm layer of material is to be removed. Assume feed as 20 mm/min. and H.S.S. cutter with 30 mm diameter and 50 mm width.

BA = Diameter of the cutter, $D_c = 30$ mm
AE = Layer to be removed, $W = 4$ mm
CA = Approach, $a = ?$

$$BC = BA - AE = \frac{D_c}{2} - W$$

$$CA = a = \sqrt{\left(\frac{D_c}{2}\right)^2 - \left(\frac{D_c}{2} - W\right)^2}$$

$$= \sqrt{\left(\frac{30}{2}\right)^2 - \left(\frac{30}{2} - 4\right)^2}$$

$$= 10.19 \text{ mm} \simeq 10 \text{ mm}$$

Over-run assuming, $o = 3$ mm

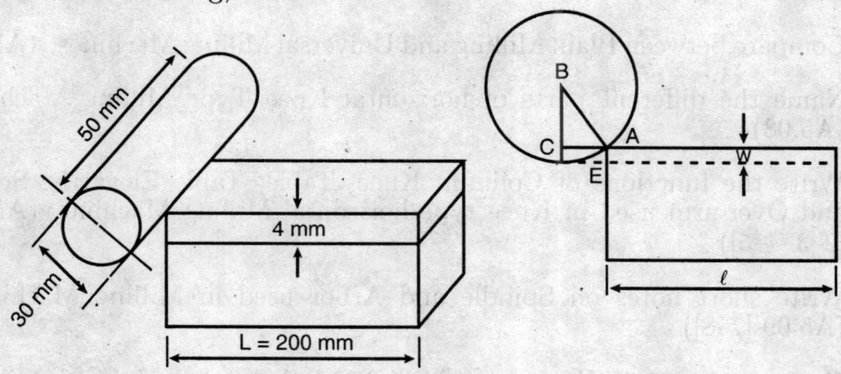

S-5-27.7(5)

$$\text{Machining Time, } T_m = \frac{L \times \text{No. of passes}}{f_t \times Z \times N}$$

$$L = a + l + o$$
$$= 10 + 200 + 3 = 213 \text{ mm}$$

$$\text{Feed/min, } f_m = f_t \times Z \times N \text{ mm/min.}$$
$$= 20 \text{ mm/min.}$$

Due to the greater width of the cutter (i.e., 50 mm) than the workpiece width (i.e., 40 mm) on peripheral milling, the 4 mm layer will be completed in two passes, i.e., one rough cut with 3 mm layer and one finish cut 1 mm layer will make the workpiece finish.

$$\therefore \quad T_m = \frac{L}{f_m} \times \text{No. of passes} = \frac{213}{20} \times 2 = 21.3 \text{ (Ans.)}$$

A5.28 Questions

1. Write a short note on a Milling Machine with respect to its principle, cutters, application and suitability. (A5.01)

2. Classify the numerous types of Milling Machines available. (A5.02)

3. Write the types of Milling Machines. (A5.03)

4. Write brief notes on the following Milling Machines:

(a) Knee Type Milling Machine (A5.04 [2])
(b) Universal Milling Machine (A5.04 [5])
(c) Omniversal Milling Machine (A5.04 [6])
(d) Fixed Bed Type Milling Machine (A5.04 [7])
(e) Planer Type Milling Machine (A5.04 [9])

5. Specify a Horizontal Milling Machine. (A5.05)

6. Compare between Lathe Machine and Universal Milling Machines (A5.06).

7. Compare between Plain Milling and Universal Milling Machines. (A5.07)

8. Name the different parts of horizontal Knee Type Milling Machine. (A5.08)

9. Write the functions of Column, Knee, Table, Table Elevating Screw and Over-arm used in Knee type horizontal Milling Machine. (A5.09 [2–3–4–5])

10. Write short notes on Spindle and Arbor used in Milling Machines. (A5.09 [7–8])

11. Compare between Horizontal Milling Machine and Vertical Milling Machine. (A5.10)

12. Sketch and explain step by step how you will fix arbor on Milling Machine Spindle, fix cutter on arbor and mounting the workpiece on Milling Machine table. (A5.11)

13. Sketch and explain step by step how you will mount cutter(s) on Milling Machine Spindle (Arbor). (A5.12)

14. Sketch and write short notes on any two types of cutter holding devices. (A5.13)

15. Sketch and state about any three types of work holding devices. (A5.14)

16. Sketch, state and compare Up-milling and Down-milling. (A5.15 & A5.16)

17. Sketch and state the angular, Side, Form and End Milling Operations. (A5.17 [3, 4, 6 & 8])

18. Sketch and state Slab, Straddle, Gang and Saw Milling Operations. (A5.17 [1, 5, 7 & 10])

19. Sketch step by step in Milling a Tee-slot. (A5.17 [9])

20. Sketch and state the Face Milling, Milling Keyways and Gear Milling operations. (A5.17 [2, 11 & 12])

21. Sketch two views of a Plain Milling Cutter and show its different angles and nomenclature. (A5.18B)

22. Classify different types of Milling Cutters. (A5.18A)

23. Name the different Milling Machine Attachments. (A5.19)

24. State about Vertical Milling Attachment, Universal Milling Attachment and Rotary Milling Attachment. (A5.19 [1, 2 & 3])

25. State about Slotting Milling Attachment, Rack Milling Attachment and Universal Spiral Milling Attachment. (A5.18 [4, 5 & 6])

26. Sketch, write functions and working of a Universal Dividing Head Attachment used in Gear Cutting Operations. (A5.20)

27. Compare Plain and Universal dividing heads. (A5.21)

28. What do you mean by Indexing? Name the different types of Indexing. (A5.22)

29. Explain step by step the method of Differential Indexing. (A5.22)

30. Sketch and write the steps to be followed in Spur Gear Milling. (A5.24)

31. Write the steps to be followed in Helical Gear Milling with necessary sketches. (A5.25)

32. Sketch and state the different Gear Driving systems. (A5.26.1)

33. Sketch and explain about the Gear Hobbing. (A5.26.2)

34. Sketch and explain about the working of Rack Cutter Generating Process. (A5.26.3)

35. Sketch and state about the Pinion Cutter Generating Process. (A5.26.4)

36. Explain with sketch about the manufacture of Bevel Gear Generators. (A5.26.6)

37. Name the Gear Cutting Machines. Sketch and explain any two of these. (A5.26.2 & A5.26.3)

38. Write brief notes on the elements of Plain Milling Cutter Angles. (A5.18 C)

39. State cutting speed, feed, depth of cut and machining time in Milling Machine Operations. (A5.27.7)

6

SHAPING MACHINE

A6.01 : Introduction

Next to Lathe and Milling Machines, the reciprocating type of machines, like shaping, planning and slotting are equally important in machining of flat surfaces. In these machines either tool or job reciprocates, while the other job or tool gives intermittent linear feed during machining. The Shaping Machine is suitable for small jobs up to 900 mm size cubical block, but in usual practice, maximum length of stroke is possible about 800 mm and in this the Ram holding the tool reciprocates, while the job is giving intermittent linear feed during machining. The Planing Machine is suitable for bigger jobs up to 3 m × 3 m × 18.25 m size block or many small jobs at a time and in which the table holding the job(s) reciprocates while the tool is giving intermittent linear feed during machining. The Slotting Machine is suitable for machining smaller in size, vertical slots, keyways, grooves, etc., and in which the Ram holding the tool reciprocates in vertical axis while the table holding the job gives intermittent linear feed during machining.

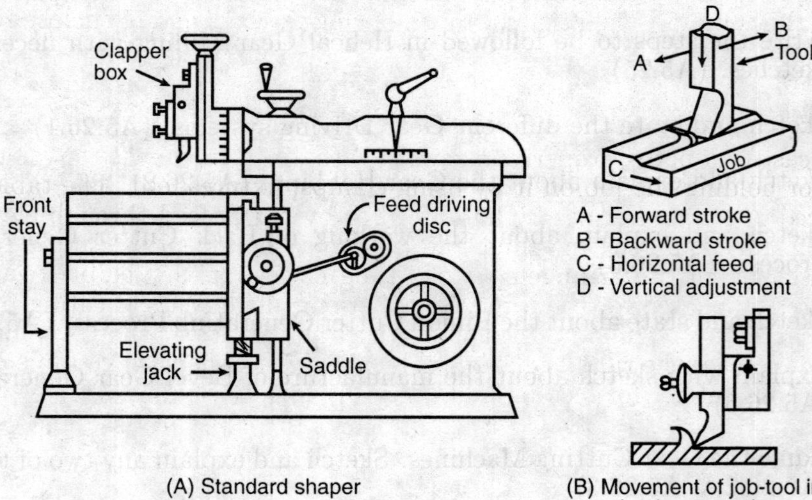

(A) Standard shaper

(B) Movement of job-tool in shaping operation

Sketch S-6.01 (A) & (B)

The Shaping Machine is compact, easy to operate and economical in use of single points cutting tool, less set-up time but mainly for small jobs with less accuracy.

This machine can machine the flat surfaces both horizontal and vertical, angular and contoured surfaces, spur and bevel gears, Tee-slot operations, etc.

A6.02 : Descriptions Of Main Parts

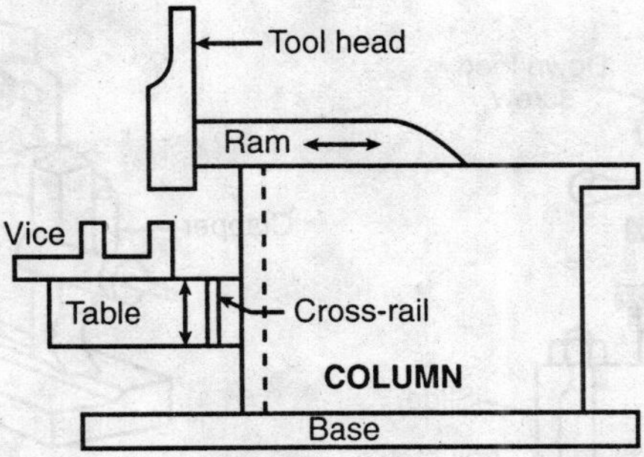

Sketch S-6.02

[1] Base

Base is rigidly bolted to shop floor and machine components rest on it. This base has to take up machine load, both dynamic and impact load, during machining operations. Grey cast iron is generally used as base material to resist vibration and to take up maximum compressive load.

[2] Column

The main part, column of the machine, consists of a hollow casting; a ram reciprocates on the top of the column through two machined horizontal guide-ways and the table moves vertical ways through cross-rail on the front vertical face of the column. The ram and table driving mechanism are enclosed inside the column.

[3] Table

It is a cast iron body, properly machined with Tee-slots on the top, sides and front for holding the job on it by using clamps, fixtures, etc. The table may be raised or lowered either by hand feed or by power feed through elevating jack. In universal type machine, the table may be fitted in either direction and also having the arrangement of tilting either forward or backward. A separate top is used for the table. These two arrangements of tilting provide to machine any angular jobs inclined to either planes. Front stay may be used to increase the rigidity of the table during operations. The table can be moved horizontally for feeding and vertically for positioning of the work.

[4] Ram

It carries a tool head on its front end. It reciprocates (or moves to and fro) on horizontal guideways of machine in the top of the column across the work. The motion of the ram is obtained by quick return mechanism. It machines only in forward stroke, moves at slower speed but runs at higher speed or a minimum time is taken for the idle return stroke. The position of the ram relative to the workpiece can also be adjusted.

[5] Tool Head

It is fixed to the front end of the ram. It consists of tool post, clapper box, down feed screw. Tool slide can be swivelled for the shaping of bevels, angular cuts, etc. The clapper box (or tool box) hinges on a pin, in order to allow the tool to swing up during the return stroke. Down feed screw is operated by hand or by power for required depth of cut and down feed cutting.

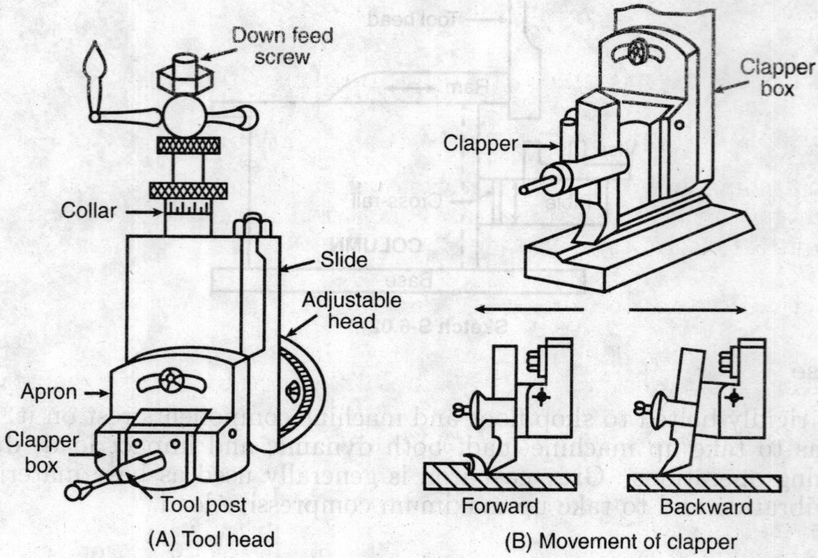

(A) Tool head (B) Movement of clapper

Sketch S-6.02.5 (A) & (B)

[6] Cross-rail and Saddle

Cross-rail slides up and down for positioning the table over the front vertical ways of the column by the cross-rail vertical screw. The work table is bolted on the saddle and the saddle is also mounted on the cross-rail. The work table along with the saddle can be moved horizontally by the table cross-feed screw.

A6.03 : Classification Of Shaping Machines

Shaping Machine may be classified with respect to the direction of ram travel, type of ram drive, design of the table, type of work, type of cutting, etc.

Shaping Machine may be classified as follows:

(i) With respect to the direction of ram travel—

(a) Horizontal, (b) Vertical, (c) Travelling tool head.

(ii) With respect to the type of ram drive—

(a) Mechanical < Crank type / Geared type, (b) Hydraulic type.

(iii) With respect to the type of design of the table—

(a) Plain, (b) Universal.

(iv) With respect to the type of work—

(a) Production work (Plain), (b) Tool room work (Universal),

(c) Special purpose work such as gear cutting, etc.

(v) With respect to the type of cutting—

(a) Push cut, (b) Draw cut.

A6.04 : Statement On Different Shaping Machines

[1] Horizontal Type Shaping Machine

In this machine, ram holding the tool travels in horizontal axis and generally produces flat surfaces. The plain table may travel in horizontal and in vertical directions for intermittent feeding.

[2] Vertical Type Shaping Machine

In this machine, ram holding the tool travels in a vertical direction, while the table is of circular rotary type. The ram of the vertical shaping machine can be adjusted 10° on either side on its vertical axis for machining an inclined surface, in addition to its vertical movement.

[3] Travelling Tool Head Type Machine

In this machine, the ram holding the tool, during reciprocation, can move crosswise for required feeding while the job will remain stationary. Heavy jobs which are not possible to be fixed on machine table, may be put on floor or base and can be machined by this machine.

[4] Mechanical Type Shaping Machine

The ram holding the tool reciprocates and cuts only in one stroke while the other stroke goes on idle. The mechanism adopted for this reciprocation is being done by the following methods: (a) Crank and slotted link, (b) Whitworth quick return and (c) Geared type. Crank and slotted link and Whitworth quick return mechanisms are adopted in such a manner that the idle stroke or return stroke in case of a push-cut type, speed will be higher than during cutting stroke time and hence will reduce the machining time.

[5] In Geared Type Machine

The ram reciprocation motion is made by means of a rack and pinion. The number of gears in the gear train will settle the speed and direction of machine traverse.

[6] Hydraulic Type Shaping Machine

The ram holding the tool reciprocates under the pressure of oil by pumping in a cylinder, from which hydraulic power is transmitted to the ram through piston and piston rod is connected in between cylinder and ram. This type of machine is costly, shock and vibration absorbent, less noise, constant speed and cutting force during cutting stroke, flexibility of speed and feed by controlling the amount of oil, etc.

[7] Plain or Standard Type Shaping Machine

In this type, the table holding job permits only two movements either in horizontal or in vertical for giving desired feed.

[8] Universal Type Shaping Machine

In this type of machine, in addition to the two movements as in the plain type machine, the table can be swivelled and tilted to permit machining at any angle. Table swivelling takes place to an axis parallel to the ram ways and the table top tilting takes place about an angle to the swivelling axis. The versatility of the job holding on the machine table at any angle on both the planes is named as Universal.

[9] Production Work Type Machine

A Horizontal, Plain type and Push-cut type shaper is commonly used for production and general purpose work.

[10] Tool Room Work Type Machine

The horizontal tool travel Universal type table and Push-cut type shaper is mostly suitable for Tool Room and Maintenance works.

[11] Special Type Shaping Machine

This machine is specially prepared to suit the special type of machining, such as gears, pulleys, flywheel, etc. The job is held on the horizontal table and the tool is reciprocated in vertical axis. The feeding is done by job with table adjustments.

[12] Push-cut Type Machine

The metal is removed from the job during the forward motion of the ram, i.e., when ram moves away from the column. This type is mostly used in standard and universal type horizontal shapers.

[13] Draw-cut Type Machine

The metal is removed from the job during the return motion of the ram (i.e., the tool is pulled across the work by the ram instead of being pushed). The machine should be sufficiently rigid. Horizontal draw-cut shaping machines are subjected to heavy cuts, being widely used for large die blocks and machining large railway parts.

A6.05 : Specification Or Size Of A Shaping Machine

1. Type of Machine, i.e., Horizontal/Crank Type/Universal/Tool Room Work/Push-cut Type, etc.

2. Maximum size of a cube (900 mm^3) can be machined, i.e.,

   1st length of a cube is the length of stroke in mm of tool travel,

   2nd length of a cube is the horizontal travel of the table in mm,

   3rd length of a cube is the vertical travel of the table in mm.

3. Vertical travel of tool slides in mm.

4. Ram cutting/Return time ratio.

5. Type of quick return mechanism, i.e., Crank and Slotted Link/Hydraulic/ Whitworth Quick Return Mechanism, etc.

6. Speed range and no. of speeds available in the machine.

7. Feed range and no. of feeds available in the machine.

8. kW of the machine.

9. Weight of the machine in Tonne.

10. Floor space required in Square Metre.

A6.06 : Shaping Tools

Generally, Shaping Tools are manufactured from High Speed Steel, and sometimes cemented carbide is brazed at the tool tips. Single point cutting tool used in shaping machine should be sufficiently rigid and heavier to withstand shock during shearing of metal. To withstand shock and intermittent cutting, negative rakes (ranging 0° to 15°) are generally preferred. Higher negative rakes are used for tougher job materials. Side rake angles, ranging from +3° to −12° are preferred on the machinability of the metals. Side end clearance angles ranging from 2° to 5° are preferred for tougher to harder materials respectively. The shape of the cutting edge mainly depends on the planing or shaping work and type of finish desired.

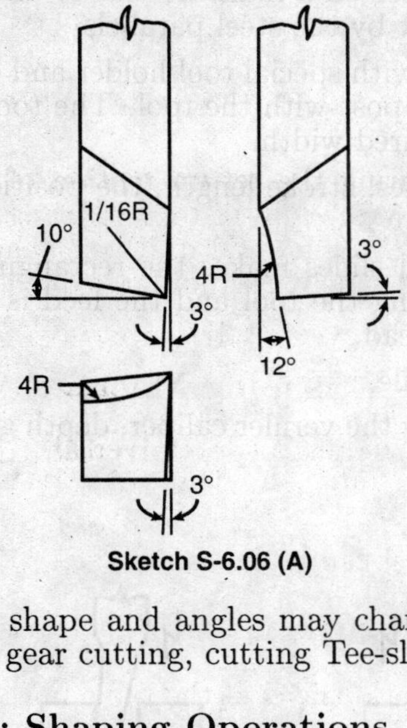

Sketch S-6.06 (A)

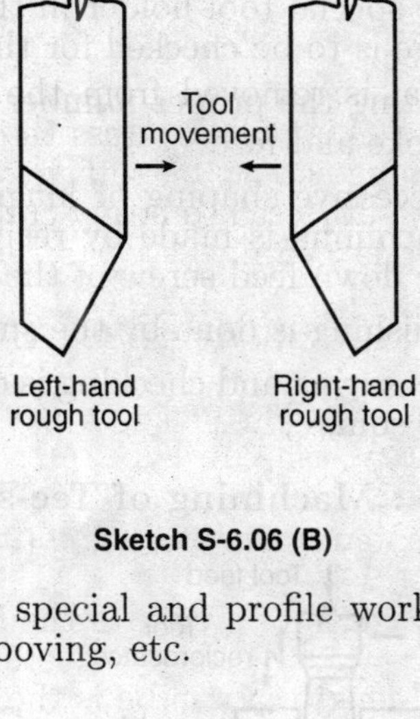

Sketch S-6.06 (B)

Tool shape and angles may change for special and profile work, e.g. for parting, gear cutting, cutting Tee-slots, grooving, etc.

A6.07 : Shaping Operations

A Shaping Machine is to generate the flat surfaces by a single point cutting tool. The generation of flat surfaces is horizontal, vertical, angular, irregular, splines, etc. Grooves, Slots keyways, Tee-slots, Dovetail, etc., may also be machined by the shaper.

A6.07.1 : Machining of Rectangular Slot

For machining a rectangular slot or a keyway on a small workpiece—

Machine Required	: Horizontal Shaping Machine.		
Tools Required	: Cutting Tool	:	Specific Shaping Tool and Smooth File
	: Holding Tool	:	Special Tool Holder, Fixture or Machine Vice for holding job.
	: Measuring Tool	:	Vernier Caliper, Depth Gauge, Steel Parallels, Try Square.
Job Material	: M.S.		
Tool Material	: H.S.S.		

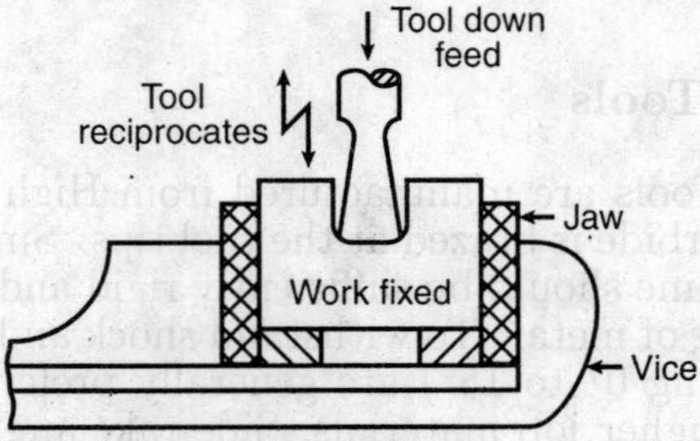

Sketch S-6.07.1

Steps to be followed:

1. Mounting the rectangular workpiece on a machine vice or on a fixture rigidly and checking the alignment by the steel parallels.
2. Selecting the proper shaping tool with special tool holder and clamping the special tool holder in the tool post with the tool. The tool cutting edge is to be checked for the required width.
3. Setting the proper number of cycles, stroke length, the position of the stroke and feed.
4. Successive shaping of longitudinal slides makes the rectangular slots. Machining is made by reciprocating the tool and the feed is given by the down feed screw of the tool head.
5. Finishing is done by the smooth file.
6. Measuring and checking is done by the vernier caliper, depth gauge and try square.

A6.07.2 : Machining of Tee-slot

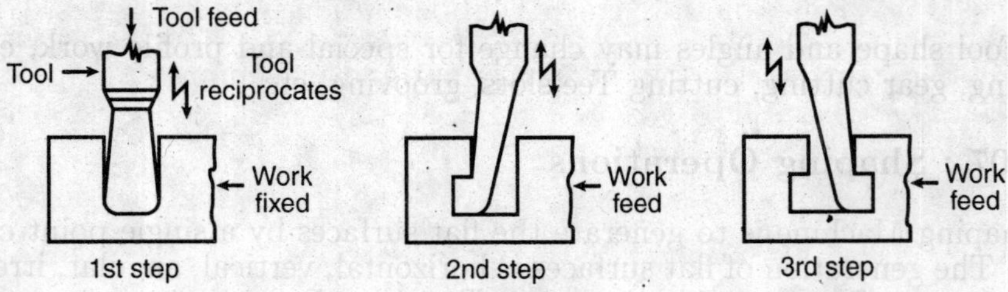

Sketch S-6.07.2

To machine a Tee-slot on a small workpiece—

Machine Required: Horizontal Shaping Machine.

Tools Required: Cutting Tools : (A) Slot Cutting tool.
(B) Suitable Undercutting Tools.
Holding Tools : (A) Suitable Fixture or Vice.
(B) Special Tool Holder with Suitable Spring.
Measuring Tools : (A) Steel Parallels.
(B) Vernier Caliper.
(C) Depth Gauge.
(D) Try Square.

Job Material : Mild Steel.

Tool Material : High Speed Steel.

Steps to be followed:

1. Mounting the workpiece on a suitable fixture or on a vice rigidly and checking the alignment by steel parallels.

2. Selecting the proper shaping tools for machining slots and undercutting with special tool holder and with the arrangement of lifting the tool automatically after each cutting stroke by a suitable spring. Cutting tool edges are to be checked for the required width, clamping the above tools in sequence of operations.

3. Setting the proper number of cycles, length of strokes, the position of strokes and feed.

4. The machining operation of Tee-slots is carried out in three operations. At first, a plain slot is machined to the full depth and width and then left-hand under cut and right-hand undercut operations are machined separately in sequence with suitable undercutting tools. Machining is made by the reciprocating motion of the tools and the required feeding either by the tool or by the table is given as mentioned in the Sketch.

5. Measuring and checking is done by the vernier caliper, depth gauge and try square, etc.

A6.08 : Mechanism Of Ram Drive

Rotary motion of the Bull gear is converted into a reciprocating motion of the ram through the crank arm. The machining is done during forward cutting stroke and no machining (or no metal is removed) is done during return or backward stroke of the ram, (may be called as idle stroke). If by some mechanism, the return (or idle) stroke travel time is reduced to reduce the total machining time, it is known as quick return principle.

This mechanism may be attained by any one of the following methods:

(a) Whitworth quick return mechanism.

(b) Crank and slotted link mechanism.

(c) Hydraulic shaper mechanism.

A6.08.1 : Crank and Slotted Link Mechanism

The driving pinion (DP) which receives power from the individual motor or overhead line shaft, drives the bull gear (BG) at different speeds either directly or through back gears. The crank pin (CP) may be adjusted by a handwheel operating through suitable gears for length of stroke. The slotted link is pivoted at its lower end, while the upper fork end is connected with

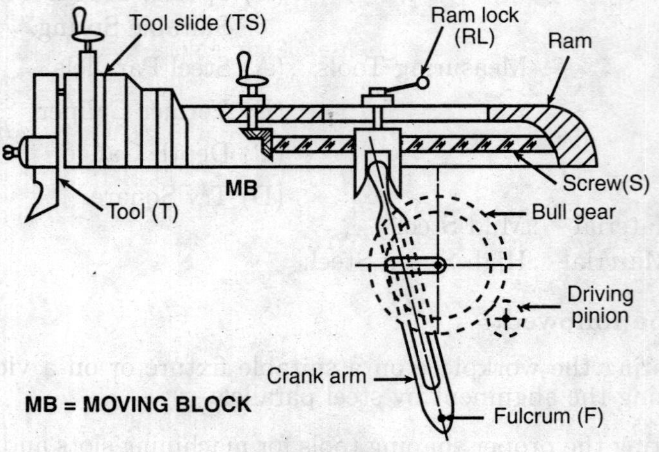

Sketch S-6.08.1

the moving block (MB). Crank pin is the connector between the crank gear and the slotted link. The rotary motion of the crank gear transmits the rocking movements of the slotted link by the crank pin within the slot in the slotted link. This rocking movement of the slotted link is communicated to the reciprocating movement of the ram through the moving block which is locked with the ram screw(S).

A6.08.2 : Adjustment of Length of Stroke

The length of stroke may be adjusted by moving the position of crank pin from the centre of the crank gear. Greater the distance of the crank pin centre from the centre of the crank gear, longer will be the stroke, and vice versa. Crank pin (CP) can be adjusted inward/outward w.r.t. the centre of the crank gear(CG).

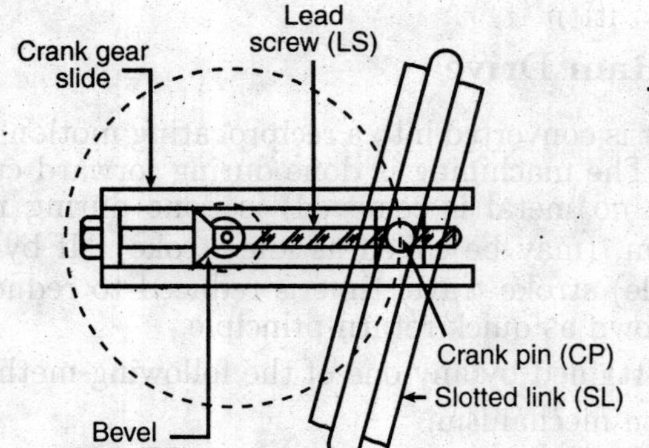

Sketch S-6.08.2

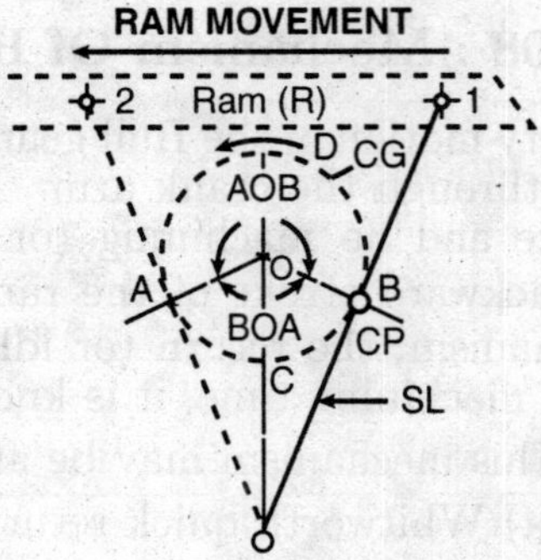

Sketch S-6.08.4

This movement of the crank pin (CP) along with the crank gear sliding block is achieved by rotating the two bevel pinions (BP) and a small lead screw (LS). The adjustment of crank pin (CP) for the length of stroke may be either by hand during machine stopping condition or in some shaper, by a handwheel operating through suitable gears (pinions) during the machine running conditions.

A6.08.3 : Adjustment of Position of Stroke

The position of the ram relative to the job can also be adjusted. For adjusting the position of stroke, the following steps are to be followed:

1. Switching-off the machine.

2. Unlocking the ram lock (RL); this will unlock the moving block (MB) and now it is ready to move axially w.r.t. the ram screw(S).

3. By rotating the handwheel (HW), the bevel pinions will rotate the ram screw(S), which will bring the moving block (MB) to required position by trial and error method.

4. Locking the ram lock (RL); this will lock the moving block (MB) with the ram screw(S).

Now the machine is ready for machining.

A6.08.4 : Principle of the Quick Return Mechanism

The crank pin (CP) is driven at a uniform speed by the crank gear (CG). This rotary movement of the crank pin (CP) rocks the slotted link (SL) as its lower end is fixed at the fulcrum (F). The rocking movement of the slotted link drives the ram to and fro through moving block (MB). The crank pin (CP) moves the ram through the slotted link to its two extreme positions (1) and (2), it only possible when two axes of the slotted link and lead screw of crank gear are at right angle.

During the forward (or cutting stroke) motion of the ram (R), the crank pin (CP) will travel the path BDA and will make the reflex angle AOB and also the ram will travel from position (1) to position (2) in straight line path.

During the backward (return stroke or idle stroke) motion of the ram, the crank pin (CP) will travel the path ACB and will make the obtuse angle BOA and also the ram will travel from position (2) to position (1).

It is seen from the diagram that the path travelled by the crank pin during forward motion is much more than the backward (or return) motion. Supposing the obtuse angle (or return angle) BOA is 144°, then the reflex angle (cutting angle) AOB will be $360° - 144° = 216°$.

$$\frac{\text{Cutting Time}}{\text{Return Time}} = \frac{\text{Cutting Angle}}{\text{Return Angle}} = \frac{AOB}{BOA} = \frac{216°}{144°} = \frac{3}{2} = 3:2$$

From the above equation, it is clear that the ram moves $1\frac{1}{2}$ times faster on the return stroke than on the cutting stroke.

A6.09 : Table Automatic Feed Mechanism

Feed is the movement of the table intermittently in either direction during return stroke only. The driving disc (D) with Tee-slots is driven by the gear shaft and is used to connect the connecting rod (C) by an adjustable bolt (B). The bolt (B), running eccentrically on disc (D), causes the connecting rod to angular motion. Connecting rod (C) connected with the rocker arm at its lower end by a pin (P_N), causes the rocker to rock with respect to its fulcrum (F). This rocker arm is fitted with a reversible spring load pawl (P). This pawl (P) is engaged with teeth of the ratchet wheel, which is also mounted on the spindle of the table.

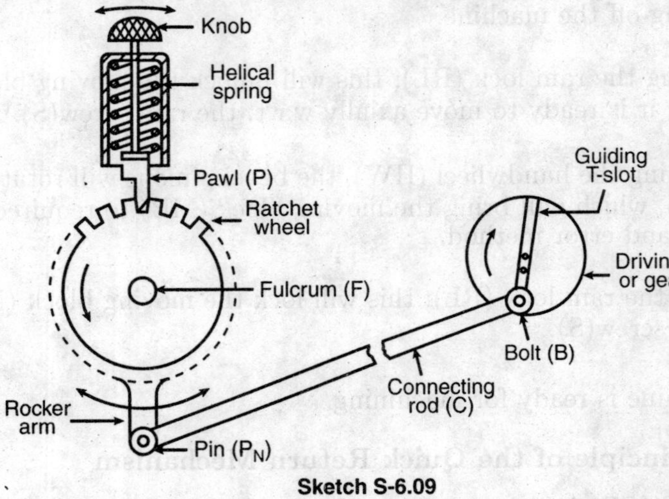

Sketch S-6.09

During forward motion (i.e., during cutting stroke), the ratchet wheel connecting the slant side of the pawl (P) slides over the wheel, imparts no motion to the table and meshes again between the teeth.

During backward motion (i.e., during return stroke), the ratchet wheel connecting the straight side of the pawl (P), imparts a short rotary motion to the ratchet wheel, which transmits the motion to the screw spindle of the table and hence the feed movement of the table.

The feed direction can be reversed by turning the pawl to 180°.

The amount of feed can be adjusted by shifting the bolt (B). Several teeth may be shifted by the pawl during roughing operation and tooth by tooth shifting during finishing operation.

A6.10 : Cutting Speed, Feed And Depth Of Cut

In shaping machine, the cutting speed is the ratio of the length of the forward cutting stroke and the time required during the forward cutting stroke.

$$\text{Cutting speed, } (v) = \frac{\text{Length of cutting stroke } (L) \text{ in m}}{\text{Time for the cutting stroke } (T_c) \text{ in min.}} \text{ m/min.}$$

Number of Cycles

The number of cycles/min. depends on the length of the stroke and the permissible cutting speed.

Cutting speed, v = Length of cycle $(2L) \times$ No. of cycles/min.

$$\text{No. of cycles/min.}, N = \frac{\text{Average speed in m/min. } (v)}{\text{Length of cycle in m } (2L)} \text{ r.p.m. or, } N = \frac{v}{2L}$$

Problem.

Calculate the no. of cycles/min., when the length of stroke is 200 mm and the cutting speed is 15 m/min.

Solution:

$$N = \frac{v}{2L} = \frac{15}{2 \times 200/1000} = 40 \text{ cycles/min.}$$

Number of cycles is directly proportional to permissible cutting speed and inversely proportional to the length of stroke.

Approximate cutting speed in (m/min.) w.r.t. the tool materials and job materials

Job Material / Tool Material	Mild Steel	Gun-metal, Brass	Grey Cast Iron
Tool Steel	10 to 12	20	12
High Speed Steel	14 to 16	30	14
Cemented Carbide	30	50	25

[A] Adjusting the Length of Stroke

l_a = Allowance, length before cutting or approach length approximate = 20 to 30 mm.

l = Length of the job to be machined.

l_o = Allowance, after cutting or over-run length approximate 10 to 15 mm.

L = Length of stroke = $l_a + l + l_o$ in m.

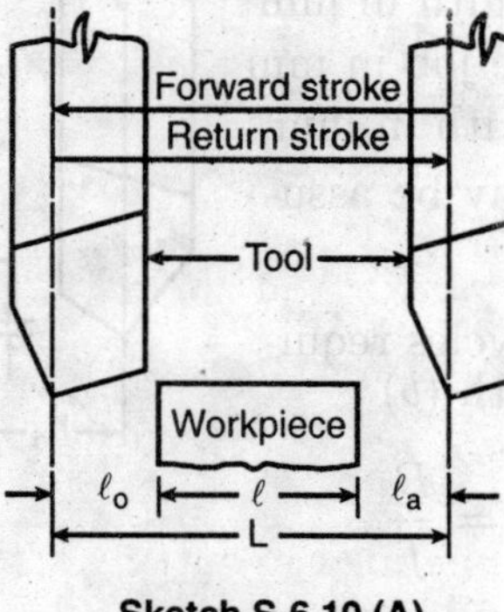

Sketch S-6.10 (A)

The length of stroke is the summation of the length of job, approach length and the over-run length.

[B] Feed and Depth of Cut

Feed (f) in shaper is the movement of the job relative to the tool in a direction perpendicular to the axis of reciprocation of the ram per double strokes (i.e., forward and backward strokes) in mm/double strokes. It depends on the type of machining and the capacity of the machine. Feed may be assumed between 0.25 mm to 1 mm/double strokes.

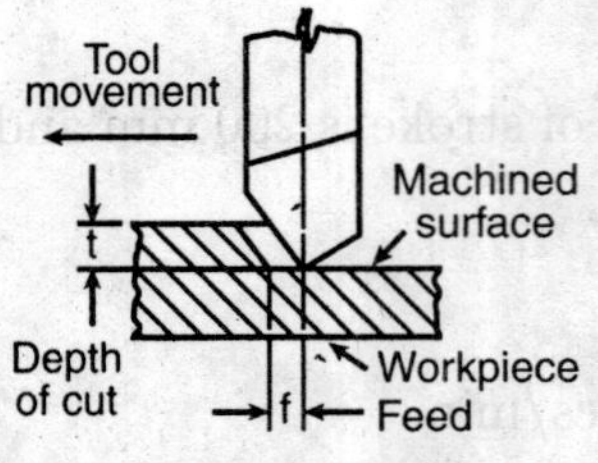

Sketch S-6.10 (B)

Depth of cut (t) is the amount of the tool penetrates into the job or the thickness of metal removed in one cut or the distance between the machined and non-machined surfaces and is expressed in mm.

Cross-section of the chip (A) = Feed × Depth of cut = $f \times t$.

Increase of chip cross-section mainly depends on the capacity of the machine.

Depth of cut may be assumed as 3 to 5 times the feed.

A6.11 : Machining Time Calculation In Shaping Machine

$$\text{Machining time, } T_m = \frac{\text{Distance travelled by the tool in metre}}{\text{Speed in m/min.}}$$

$$\text{Time required for cutting stroke, } T_c = \frac{\text{Length of cutting stroke in m}}{\text{Cutting speed in m/min.}} = \frac{L}{V_c}$$

$$\text{Time required for return stroke, } T_R = \frac{\text{Length of return stroke}}{\text{Return speed in m/min.}} = \frac{L}{V_R}$$

Time for Cutting + Return strokes (or for one cycle), $T = T_c + T_R$

$$\text{where, } L = \text{Length of stroke in metre} = \frac{(l_a + l + l_o) \text{ in metre}}{1000}$$

(Already explained earlier)

Shaping width, $B = b_a + b + b_o$

where, b_a = approach width in mm
b = width of the job in mm
b_o = over-run width in mm.

Approach and Over-run may be assumed as 5 mm.

Therefore, the number of cycles required to machine the job width (b)

$$N_B = \frac{\text{Shaping width}}{\text{Feed}} = \frac{B}{f}$$

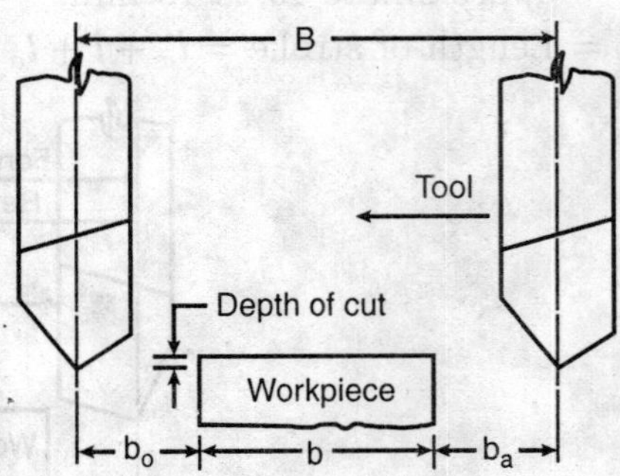

Sketch S-6.11

Machining time, t_m = Number of cycles × Time per cycle = $T \times B/f$

Total time required to complete the cut in one face, $T_m = T \times N_B \times$ No. of passes.

Number of passes depends on the depth of cut and depth of metal to be machined in one face.

$$\frac{\text{Return stroke time}}{\text{Cutting stroke time}} = m$$

$$\text{Total time for cutting + Return strokes or per cycle} = \frac{L}{v} + \frac{mL}{v} = \frac{L(1+m)}{v}$$

$$\text{Total time required to complete the cut} = \frac{L(1+m)}{v} \times \frac{B}{f} \times \text{No. of passes.}$$

Cutting speed may also be calculated from the formula,

where, $v = NL(1+m)$ m/min.

N is r.p.m.

L is the length of stroke in metre

m is the return/cutting time ratio.

A6.11.1 : Work Out Problems

Problem 1.

Calculate the machining time of a face of a rectangular block made of M.S. by a shaper.

Data given: Length of the Rect. Block = 400 mm, width of Rect. Block = 300 mm, $l_a = 20$ mm, $l_o = 10$ mm, $V_c = 10$ m/min., $V_R = 15$ m/min., Feed = 1 mm/cycle, $b_o = b_a = 5$ mm. Machining will complete in two passes.

Solution:

$$\text{Length of Stroke, } L = (l_a + l + l_o)/1000 = \frac{20+400+10}{1000} = 0.43 \text{ metre.}$$

$$\text{Time for the Cutting Stroke, } t_c = \frac{L}{V_c} = \frac{0.43\text{ m}}{10\,\text{m/min.}} = 0.043 \text{ min.}$$

$$\text{Time for the Return Stroke, } t_R = \frac{L}{V_R} = \frac{0.43\,\text{m}}{15\,\text{m/min.}} = 0.0287 \text{ min.}$$

Time required in one cycle, $t = 0.043 + 0.0287 = 0.0717$ min.

Shaping width, $B = (b_a + b + b_o)$ min. $= 5 + 300 + 5 = 310$ mm

$$\text{Number of cycles required, } N_B = \frac{310\text{ mm}}{1\,\text{mm/cycle}} = 310 \text{ cycles}$$

Machining time/cut, $t_m = N_B \times t = 310 \times 0.0717 = 22.28$ min.

Total time required $= t_m \times$ No. of passes

$= 22.28 \times 2 = 44.56$ min. **Ans.**

Problem 2.

A cast iron block of 190 mm length and 100 mm width is to be machined with H.S.S. shaping tool. The permissible cutting speed is 30 m/min.

Maximum quick return ratio is 3 : 2 at the design value of stroke of 400 mm. Calculate the time of machining of the block, assuming total depth of cut to be 8 mm. Assume also a suitable feed rate and any data if required.

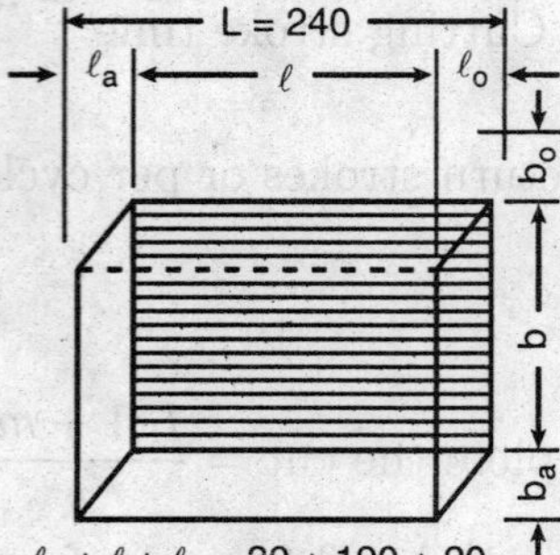

$L = \ell_a + \ell + \ell_o = 30 + 190 + 20$

L = 240 mm

$B = b_a + b + b_o = 5 + 100 + 5$

B = 110 mm

Sketch S-6.11.1(3)

Solution:

1. Machine time, $T_m = \dfrac{\text{Distance travelled by the tool in metre}}{\text{Speed in m/min.}}$

2. Time required for cutting stroke, $t_c = \dfrac{L}{v_c} = \dfrac{0.24}{30\,\text{m/min.}} = 0.008$ min.

3. Time required for return stroke, $t_R = \dfrac{L}{V_R} = \dfrac{0.24\text{ m}}{\frac{30\times 3}{2}\text{m/min.}}$

$$= \frac{0.24}{45} = 0.00533 \text{ min.}$$

4. Time required for one cycle, $t = t_c + t_R = 0.008 + 0.00533 = 0.01333$ min.

5. Shaping width, $B = 5 + 100 + 5 = 110$ mm

6. No. of cycles required, $N_B = \dfrac{110\,\text{mm}}{1\,\text{mm/cycles}} = 110$ cycles

Assuming Feed, $f = 1$ mm/cycle.

7. Machining time/cut, $t_m = N_B \times t = 0.01333 \times 110 = 1.4666$ min.

8. Number passes will be three, considering *two* @ 3 mm depth of cut for rough cut and *one* @ 2 mm depth of cut for finish depth of cut.

8 mm depth of cut $= 2 \times 3 + 1 \times 2 = 8$ mm

9. Total machining time, $T_m = t_m \times$ No. of passes $= 1.4666 \times 3 = 4.4$ min.

Problem 3.

Calculate the machining time required for taking a complete cut by two rough cuts and one finish out on a grey C.I. plate 800 mm ×600 mm, cutting speed = 12 m/min., return/cutting time ratio = 2:3, feed = 1.5 mm/cycle, allowances length-wise = 20 mm, allowances widthwise = 5 mm.

Solution:

No. of Passes = 3

Length of Stroke, $L = 20 + 800 + 20 = 840$ mm = 0.84 m.

Time for Cutting stroke, $t_c = \dfrac{L}{V_c} = \dfrac{0.84}{12} = 0.07$ min.

Time for Return stroke, $t_R = \dfrac{L}{V_R} = \dfrac{0.84}{18} = 0.0466$ min.

$\left[\because \; V_R = \dfrac{V_c}{m} = \dfrac{12 \times 3}{2} = 18 \text{ m/min.}\right]$

Time required for cutting + Return strokes, $t = 0.07 + 0.0466 = 0.1166$ min.

Shaping width, $B = 5 + 600 + 5 = 610$ mm.

No. of cycles required/cut, $N_B = \dfrac{B}{f} = \dfrac{610 \text{ mm}}{1.5 \text{ mm/cycle}} = 406$ cycles.

Time required/cut, $t_m = N_B \times t = 406 \times 0.1166$ min.

Total machining time required for taking a complete cut, $t_m \times$ No. of passes

$= 406 \times 0.1166 \times 3 = 140$ min.

= 2 hrs. 20 min. **(Ans.)**

Problem 4.

Find the time required on a shaping machine for completê one cut only on a plate of 200 mm ×300 mm, if the cutting speed is 15 m/min. The return time to cutting time ratio is 2 : 3, and the feed is 2.5 mm. The clearance; at each end is 40 mm.

Solution:

L = The length of the stroke in mm $= a + l + o = 40 + 300 + 40 = 380$ mm.

B = The breadth of the work in mm $= a + b + o = 10 + 200 + 10 = 220$ mm.

S = Feed in mm/double stroke = 2.5 mm.

m = The ratio of return time to cutting time = 2/3.

v = Cutting speed in m/min. = 15 m/min.

Time taken to complete in double strokes $= \dfrac{L(1+m)}{1000 \times V}$

Total number of double strokes required to complete the job $= \dfrac{B}{S}$

Total time required to complete the cut $= \dfrac{L \times B(1+m)}{1000 \times V \times S}$

$$= \frac{380 \times 220\left(1 + \frac{2}{3}\right)}{1000 \times 10 \times 2.5} \text{ min.}$$

= 5.573 minutes **(Ans.)**

A6.11.2 : Problems for Practice

1. Find the machining time for machining an M.S. Plate of 400 mm × 300 mm by one rough cut and one finish cut with the following data—Tool material = H.S.S., Cutting speed = 15 m/min., Return speed = 25 m/min., Feed = 1.2 mm/cycle. Lengthwise approach and over-run allowances are 30 and 20 mm respectively. Assume, if any data is missing. (**Ans.** 24.8 min.)

2. Calculate the machining time of a face of a square plate 500 mm^2, made of gun-metal with a tool material of H.S.S.; Cutting speed = 25 m/min., Return/cutting ratio 1 : 2, Feed = 0.8 mm/cycle, l_a = 30 mm, l_o = 10 mm, b_a = 5 mm, b_o = 5 mm. Machining time will complete face in three passes. (**Ans.** 62 min.)

A6.12 Questions

1. Write a brief note on a shaping machine. (A6.01)

2. Describe in short the main parts of a shaping machine. (A6.02)

3. Classify the different types of shaping machines. (A6.03)

4. State in short the different shaping machines commonly used. (A6.04)

5. Specify a shaping machine. (A6.05)

6. Sketch left-hand and right-hand shaping tool. (S6.06 A)

7. State, sketch and show the angles of a left-hand rough shaping tool. (A6.06) and (S6.06 B).

8. Sketch and state step by step the operations of a key slot or a rectangular slot in a shaping machine. (A6.07.1)

9. Sketch and state step by step the operations of a Tee-slot in a shaping machine. (A6.07.2)

10. State how many ways a Ram in a shaping machining can drive. (A6.08)

11. Sketch and explain the Crank and Slotted Link Mechanism for ram drive in a shaping machine. (A6.08.1)

12. How length of stroke and position of strokes are adjusted in case of ram drive by crank and slotted link mechanism? (A6.08.2, A6.08.3 and A6.08.1)

13. Sketch and state the principle of quick return mechanism adopted in a shaping machine. (A6.08.4)

14. Sketch, label and explain the automatic table feed mechanism adopted in a shaping machine. (A6.09)

15. State about cutting speed, feed and depth of cut in case of a shaping machine. (A6.10)

16. Write step by step how you will calculate the machining time of a flat surface in a Shaping Machine. (A6.11)

7

SLOTTING MACHINE

A7.1 : Introduction

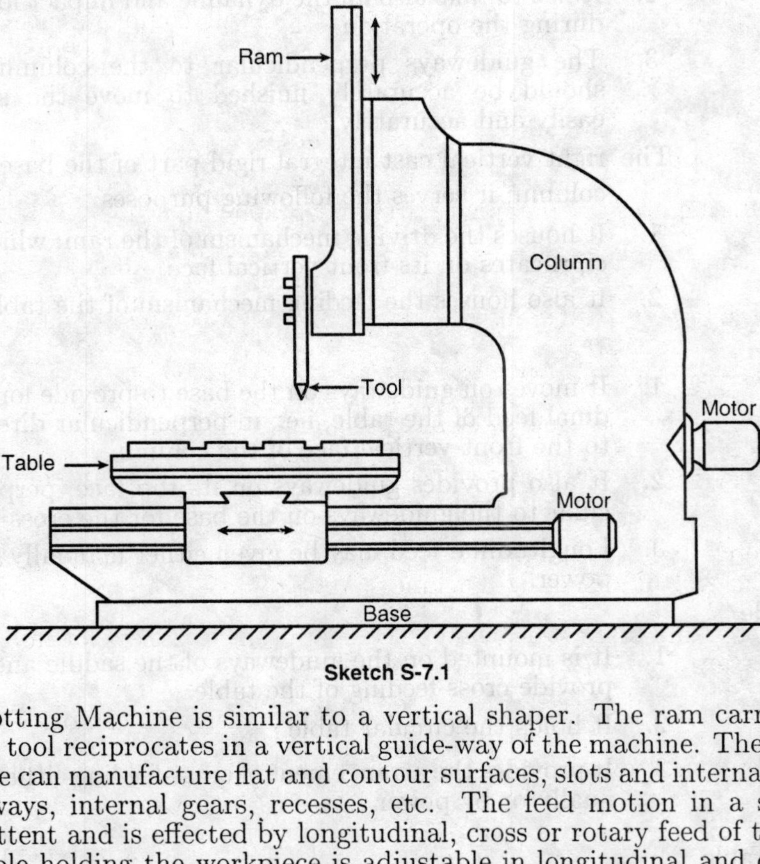

Sketch S-7.1

The Slotting Machine is similar to a vertical shaper. The ram carrying the slotting tool reciprocates in a vertical guide-way of the machine. The Slotting Machine can manufacture flat and contour surfaces, slots and internal grooves or keyways, internal gears, recesses, etc. The feed motion in a slotter is intermittent and is effected by longitudinal, cross or rotary feed of the table. The table holding the workpiece is adjustable in longitudinal and crosswise directions. In small machines, the table is adjustable in vertical direction also.

The main dimensions of a slotter are the maximum length of stroke of the ram and the diameter of the table. The ram of a slotter can reciprocate either by a mechanical crank drive or by hydraulic drive.

In mechanical crank drive, the ram is actuated by the connecting rod and the other end of which is attached to the driving disc in the opening of the column behind the ram. Various stroke lengths of the ram can be set by adjusting the radial position of the connecting rod and on the driving disc. The vertical position of the ram is adjustable by releasing the nut on its front face, which clamps the ram to the upper end of the connecting rod.

The feed drive actuates the longitudinal, cross-wise and circular motion of the table. It is driven by either from the main drive or from an individual motor.

A7.2 : Functions Of Main Parts

Bed

1. It should be sufficiently rigid to take up the entire load of the machine.
2. It has to take also all the dynamic and impact loading during the operation.
3. The guideways perpendicular to the column face should be accurately finished to move the saddle easily and accurately.

Column The right vertical cast integral rigid part of the base is the column; it serves the following purposes:

1. It houses the driving mechanism of the ram, which reciprocates on its front vertical face.
2. It also houses the feeding mechanism of the table.

Saddle

1. It moves on guideways on the base to provide longitudinal feed of the table, i.e., in perpendicular direction to the front vertical face of the column.
2. It also provides guideways on its top face, perpendicular to the guideways on the base for the cross-slide.
3. Longitudinal feed may be given either manually or by power.

Cross-Slide

1. It is mounted on the guideways of the saddle and can provide cross-feeding of the table.
2. It holds the circular table.
3. It provides the arrangement of cross-feeding either manually or by power.

Circular Table

1. It can rotate on cross-slide to provide circular feeding of the work.
2. It can hold the work by clamping, using fixtures, fixing vice, etc.
3. It has the arrangement of rotary feeding either by hand or by power.
4. On some machines, the table may be graduated in degrees for indexing.

Ram and Tool Head

1. Ram slides on the guideways mounted on the front vertical face of the column.

2. Ram provides a slot for changing the position of stroke.
3. Ram carries the tool, which is mounted at the bottom -end of the tool head.

Ram Drive

1. It provides the circular motion of the motor to the reciprocating motion of the ram in vertical plane.
2. It also provides the arrangement of slower speed during the cutting or forward stroke and greater speed during the return stroke to reduce the idle time.

Feed Drive

1. It provides the arrangement of longitudinal and cross-feed of the table either by power or manual.
2. It also provides the arrangement of circular or rotary feeding of the table on a vertical axis.
3. Intermittent feeding of the table is supplied at the beginning of each cutting or forward stroke of the ram.

A7.3 : Specifications Of A Slotting Machine

1. Maximum length of stroke of the ram in mm.

2. Diameter of the table in mm.

3. Maximum vertical adjustment of the ram in mm.

4. Distance between the centre of tool and the axis of the column in mm.

5. Maximum longitudinal and cross travel of the table in mm.

6. Number and range of speeds and feeds available.

7. Floor space required in square metre.

8. kW of motor.

9. Net weight of the machine in tonne.

A7.4 : Slotter Tools

Solid tools of H.S.S. material or inserted bit of cemented carbide with H.S.S. as a tool holder is generally used for slotting. Tools are provided with top rake, front clearance and side clearance angles. The nose of tool slightly projects beyond the shank to avoid chatter. The shape of the cutting edge mainly depends on the operation and shape of the workpiece. For machining flat surfaces, square nose tools are used and for machining contoured or circular surfaces, round nose tools used. Tool holder should be sufficiently large and rigid in section to resist the impact and cutting forces during the operation.

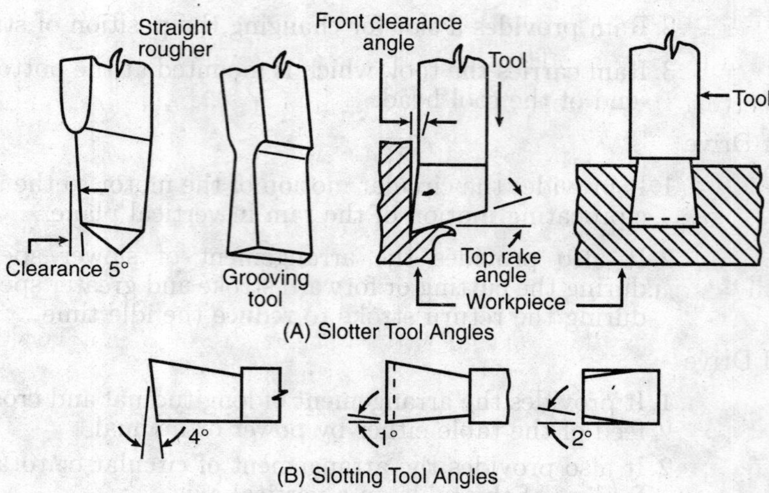

Sketch S-7.4

A7.5 : Slotting Of Keyway

Tools Required

(1) Slotting Machine

(2) Clamps, Parallels, Clamping Screws, etc.

(3) Scriber or Marking Block

(4) Slotting Tool for 8 mm wide keyway.

(5) Plug or Slip Gauge.

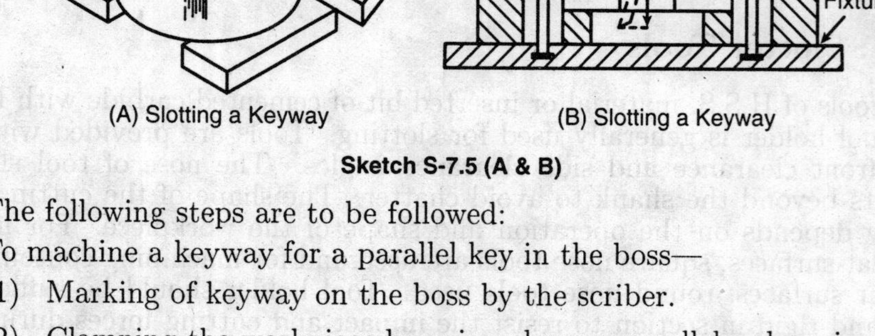

Sketch S-7.5 (A & B)

The following steps are to be followed:

To machine a keyway for a parallel key in the boss—

(1) Marking of keyway on the boss by the scriber.

(2) Clamping the workpiece by using clamps, parallels, clamping screws, etc.

(3) Selecting a keyway tool (square nose) and clamping the tool in tool holder.

(4) Rotating the table until the marking on boss and the tool centre is in line (by hand).

(5) Setting the number of cycles, length of stroke and position of stroke.

(6) Slotting of keyway and feeding gradually by hand.

(7) Checking and measuring the keyway by plug or slip gauge.

A7.6 Questions

1. Sketch a block diagram and write a short note on a Slotting Machine (A 7.1).

2. Write the functions of main parts of a Slotting Machine (A 7.2).

3. Specify a Slotting Machine (A 7.3).

4. Sketch a Slotting Tool; show its different angles and explain in short its action (A 7.4).

5. Sketch and write the systematic operation in machining a keyway (A 7.5).

8

PLANING MACHINE

A8.1 : Planing Machine

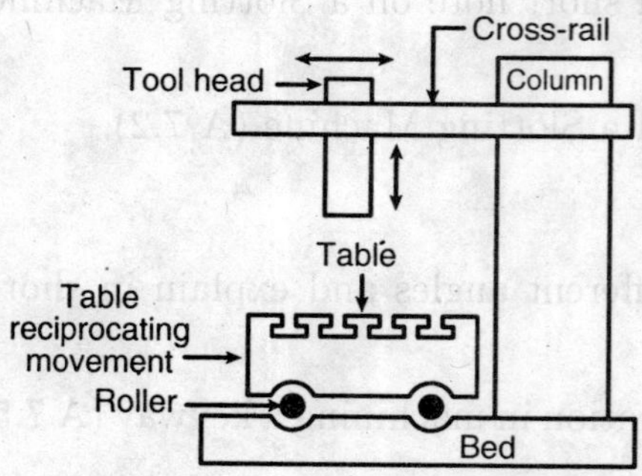

Open Side Planer

Sketch S-8.1-A

Introduction

This is also a reciprocating type of machine like shaper. The table with T-slots for mounting workpiece, slides in the guideways of the machine bed past the stationary tool head. The tool head holding the tool, machines during the forward (or cutting) stroke of the worktable; during the cutting stroke the tool remains stationary. The tool horizontal feed is given during return (or backward) stroke of the worktable.

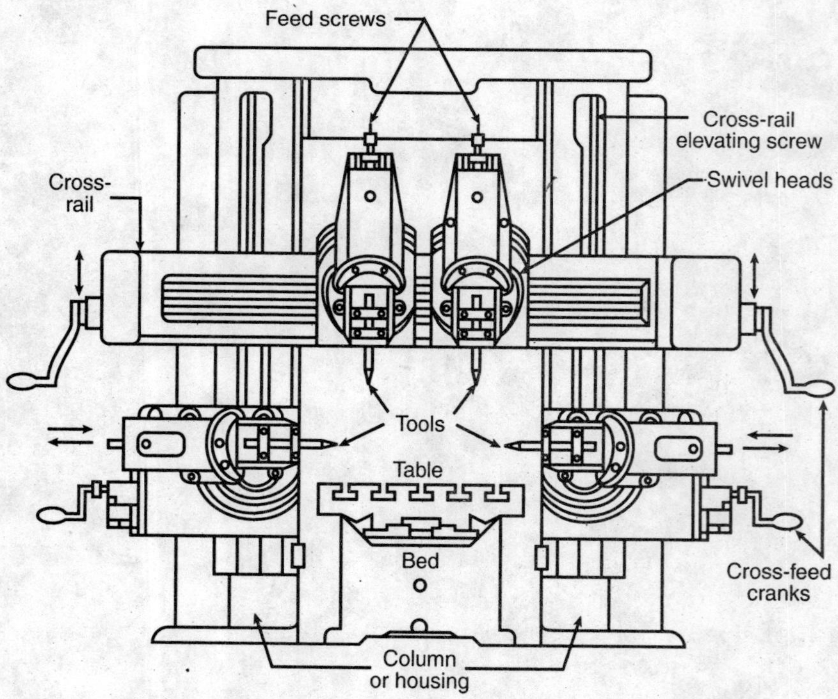

Double Housing Planer

Sketch S-8.1-B

The vertical or angular feed of the tool is given manually by the downward movement of the tool head slide. The relative movement of the work and tool in planer is just reversed compared to shaper. The planer uses one or more single-point tool(s) and a number of work surfaces can be machined at a time. The planer can machine flat (horizontal or vertical), angular and also contour surfaces with a series of straight cuts; it is suitable for the big castings, heavy forgings and also number of small workpieces at a time.

A8.2 : Specifications Of A Planer

The size (or specification) of a planer is as follows:

1. Maximum size of a rectangular block can be machined

(i) First length is the length of stroke (table) in metre.

(ii) Second length is the width of the table in mm.

(iii) Third length is the distance from the table to rail in mm.

2. Type of machine: Double housing/open side/pit/plate or edge/divided table type, etc.

3. Arrangement of table reciprocating mechanism.

4. Number of speeds and range of speeds available in the machine.

5. Number of feeds and range of feeds available in the machine.

6. kW of the machine motor.

7. Floor space required in square metre.

8. Weight of the machine in tonne.

A8.3 : Classification Of A Planing Machine

Planing Machine is classified as follows:

1. According to housing arrangement—

(a) Open side planer, (b) Double housing planer.

2. According to the reciprocating drive of the table—

(a) By open and cross belt drive planer, (b) By variable speed reversing motor drive planer, (c) By hydraulic drive planer.

3. According to method of construction—

(a) Light duty planer, (b) Medium duty planer, (c) Heavy duty planer.

4. According to the general purpose—

(a) Open Side Planer, (b) Double Housing or Plano-miller, (c) Pit Planer, (d) Edge or Plate Planer, (e) Hydraulic Planer, (f) Universal Planer, (g) Divided Table Planer.

A8.4 : Description Of Different Planing Machines

[1] The Open Side Planer

The Open Side Planer has only one housing (or column) on one side of the base. The workpiece of greater width can be accommodated on the extended table. Two tool swivel heads are on the cantilever type cross-rail and one side tool head is on the side rail; it can machine two surfaces at a time.

[2] The Double Housing or Plano-Miller

This type is the most rigid and common type of planer. The table is reciprocated on a long heavy base. There are two housings (columns) at two sides of the base for supporting the cross-rail upon which the tools are fed across the work. There may be two or more tool heads on the cross-rail and one on each side on the housing (or column) can machine three surfaces at a time. Tools are fed manually or by power in either vertical or crosswise directions. The accuracy of the machine is mainly depending on its rigidity and the manner in which the ways in the bed are machined. The controls for operation are all at the housing. The stroke length of the machine is controlled by adjusting the dogs at the side of the bed.

[3] The Pit Planer

The Pit Planer is massive in construction. The table can hold the workpiece of about 4 m width and 10 m length, and will remain stationary during machining, so requires less floor space. The sufficiently rigid column carrying the cross-rail, holding tool head(s) can reciprocate on horizontal rails situated on both sides of the table.

[4] Edge or Plate Planer

This type of planer is intended for the fabrication work of heavy steel plates for pressure vessels and ship-building works, etc. The edge of a long plate is securely mounted to a large bed by air-operated clamps. The carriage which is supported on the horizontal ways of the planer, carrying the cutting tools, moves to and fro past the stationary workpiece (i.e., Plate).

[5] Hydraulic Planer

Hydraulic Planer is highly satisfactory and similar to the hydraulic shaper. This type of planer has certain advantages like uniform cutting pressure, rapid table reversal, quick means of varying the stroke and less noise during operation. The table holding the workpiece can reciprocate by means of hydraulic drive.

[6] Universal Planer

Universal Planer will cut on both forward and return strokes, by using two cutting edged tool, which is pivoted in the fixed holder. The one cutting edge of the tool engages the work during the forward stroke, while the other cutting edge, somewhat longer than the first one, engages the work on the return stroke. Two separately pivoted tools for a two-way planing are used in cutting slots in machine tables.

[7] Divided Table Planer

In this type of planer, the table on the bed is divided into two parts, it may be reciprocated separately or together. This type is mostly suitable for continuous production and thus reduces the machining time by saving the idle time. One of the table is used for setting up large number of identical workpieces while the other one reciprocates against the stationary cutting tool machining the workpieces. Both the tables may clamp together for holding large and heavy workpiece, and can also reciprocate together under the tool.

A8.5 : Table Drive Mechanism

The table reciprocates in the guideways on the machine bed. The mechanism used to drive the table is mainly by gear drive and the reversible mechanism is adopted by any of the following ways:

(a) By having an open and cross belt drive in which the belts are alternately used as drivers, (b) By having a reversible type of D.C. motor drive, (c) By means of a crank as employed on shapers, (d) By reversing the flow of oil from a hydraulic pump.

A8.5.1 : Quick Return Mechanism of a Planer to Drive the Table by Open and Cross Belting

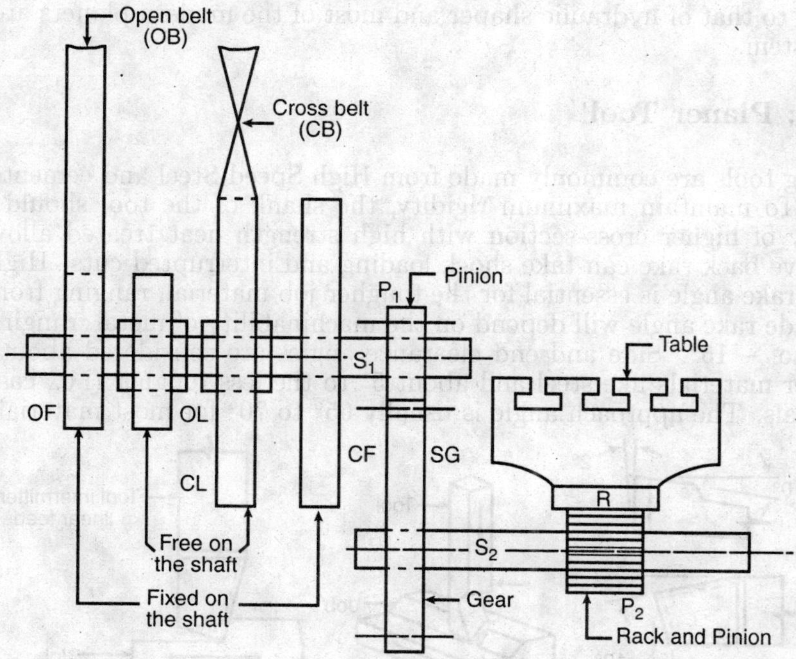

Sketch S-8.5.1

In open and cross belt drive, the belts are alternatively used as drivers to reciprocate the table. The cross belt with a greater arc of contact on the larger pulley is used to drive the table on the cutting stroke or forward stroke. Greater power and less speed is essential during the cutting stroke and is being achieved by connecting the cross belt with larger diameter pulley CF, which is fixed on the shaft S_1. The power from the shaft S_1 transmits to the shaft S_2 through pinion P_1 and spur gear SG. Power again from the shaft S_2

is transmitted to table for cutting stroke through the pinion P_2 and the rack R (which is keyed at the bottom of the table). During cutting stroke, the open belt will be on the smaller pulley OL (free on the shaft S_1). At the end of cutting stroke, the trip dog on the side of the planer table actuates the belt shifting fork, so that the cross belt CB is shifted from the fixed larger pulley CF to the loose larger pulley CL while the open reversing belt OB is shifted from the loose smaller pulley OL to the fixed smaller pulley OF. This causes the fixed smaller pulley OF to change its direction of rotation for non-cutting or return stroke. The table returns at greater speed due to fixed smaller diameter OF or less number of teeth on the spur gear w.r.t. the cutting stroke during the non-cutting or return stroke and hence quick return principle is adopted in planing machine. Return-cutting speed ratio may be 2 : 1 or 4 : 1 or as per design. A gear train may be used to transmit power from the drive shaft to the table.

Quick return motion may also be obtained by D.C. reversible motor in which speed can be changed very quickly from full forward to full backward and the instantaneous reversibility is also possible. The trip dogs change over the supply to the motor at the end of each stroke. Modern industries are mostly using this drive due to its wide range of table speeds and better control.

Quick return motion is also possible by hydraulic system. This system is similar to that of hydraulic shaper and most of the modern planers are using the system.

A8.6 : Planer Tool

Planing tools are commonly made from High Speed Steel and cemented carbide. To maintain maximum rigidity, the shank of the tool should be relatively of higher cross-section with high strength heat treated alloy steel. Negative back rake can take shock loading and interrupted cuts. Higher negative rake angle is essential for the tougher job material, ranging from 0° to 15°. Side rake angle will depend on the machinability of metal, ranging from + 3° to − 15°. Side and end clearance angles are considered about 2° for tougher materials like steel and about 5° to the less tougher (i.e., cast iron) materials. The approach angle is usually 65° to 70° for most materials.

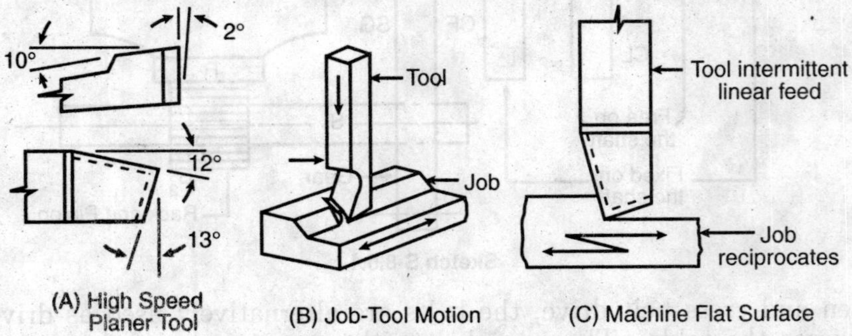

(A) High Speed Planer Tool (B) Job-Tool Motion (C) Machine Flat Surface

Sketch S-8.6 (A, B & C)

The reciprocating motion is performed by the workpiece which is mounted on the planing table. The planing tool gives the linear intermittent cross feed automatically. Down feed or depth of cut of the planing tool is given manually.

A8.7 : Method Of Machining Of A Dovetail In A Planing Machine

Tools Required:

1. Double Housing Planing Machine
2. Proper fixture
3. Dovetail end cutting roughing tool (Left and right handed)
4. Dovetail end cutting finishing tool (Left and right handed)
5. Surface gauge
6. Rule
7. Universal bevel protractor
8. Vernier caliper
9. Depth gauge
10. Try square

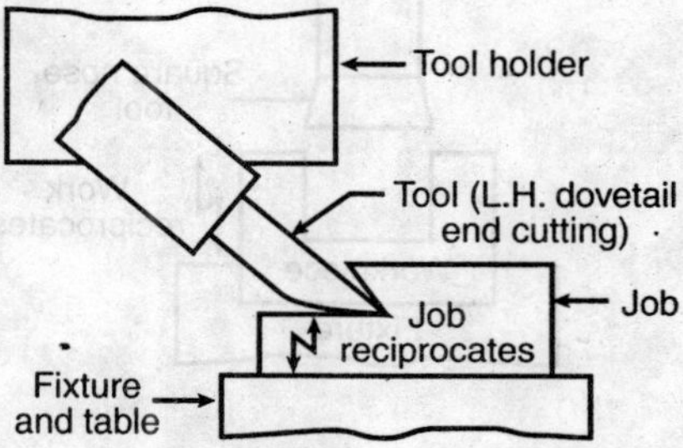

Sketch S-8.7

For machining dovetail operations on Mild Steel workpiece by an H.S.S. Tool in a Planing Machine, the following steps are to be followed:

1. Setting of job on machine table by a proper fixture
2. Choice of proper planing tools for dovetail operations—
   (a) Left-hand dovetail end cutting tool for machining left side dovetail operation
   (b) Right-hand dovetail end cutting tool for machining right side dovetail operation
3. Selection of proper speed, feed, depth of cut and cutting fluid—

Surface Speed	= 20 to 30 m/min.
Roughing Feed mm/rev.	= 0.2 mm/double strokes
Roughing Depth of Cut	= 2 mm

   Cutting fluid requires: Soluble oil, Finishing feed = 0.1 mm/double strokes, Finishing depth of cut = 0.25 mm.
4. Setting of proper tools (as mentioned above) in proper operational sequences. Care should be taken to machine dovetail operation; tool swivel is essential to the desired dovetail angle either to the **right or left side** of the dovetail operations.
5. After arranging the above, the machining is started with the following steps—

   By switching on the machine, the table holding the job reciprocates, the tool holder holding dovetail end cutting tool feeds in crosswise for machining left side and right side in sequences after swivelling the tool to the proper angle to the desired length from right end of the job or left end of the job. But roughing and finishing cuts are done with proper cutting conditions.

6. Measuring and checking the workpiece are necessary with proper inspecting tool(s).

A8.8 : Machining Method In A Planing Machine To Machine A Tee-Slot

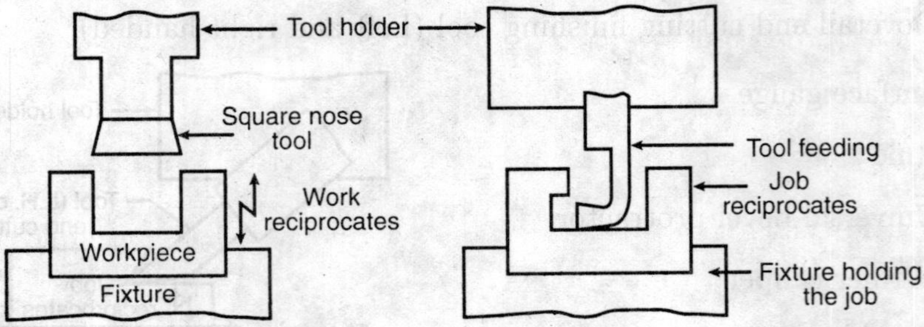

Sketch S-8.8

Tools Required:

1. Double Housing Planing Machine
2. Fixture for holding the job
3. Square nose planing tool
4. Tee-slot Planing tool (Right and left handed)
5. Surface gauge
6. Rule
7. Vernier caliper
8. Depth gauge

For Machining T-slot operations on Mild Steel workpiece by Carbide tools in a planing machine, the following steps are to be followed:

1. Setting up the workpiece on machine table by a proper fixture.
2. Selection of proper planing tools for T-slot operation—
   (a) Square nose planing tool for key-slot operation
   (b) Tee-slot planing tool for T-slot operations
3. Selection of proper speed, feed, depth of cut and cutting fluid—

Surface speed = 120 to 140 m/min.	Roughing depth of cut = 2 mm
Roughing feed = 0.2 mm/double strokes	Finishing depth of cut = 0.2 mm
Finishing feed = 0.12 mm/double strokes	Cutting fluid requires = Soluble oil

4. Setting of proper tools (as mentioned above) in proper operational sequences. T-slot Planing tool may have to be set either to the left side or right side according to the recess operations, to the left or to the right side respectively

5. After arranging the above, the machine is started with the following steps:

(a) By switching on the planing machine, the table holding the job reciprocates, the tool holder holding square nose planing tool feeds vertically for machining key-slot operation. Both roughing and finishing operations are done with proper cutting conditions.

(b) Stopping the machine, changing the tool to T-slot planing tool feeds crosswise directions for T-slot operations (both left side and right side recesses) with roughing and finishing cuts are done with proper cutting conditions. Suitable spring is provided to raise the tool box automatically after each cutting stroke.

6. Measuring and checking the workpiece are necessary with proper inspecting tool(s).

A8.9 : Distinction Among Shaping, Planing And Slotting Machines

The reciprocating type Machine Tools (like shaping, planing and slotting) are primarily used to produce flat surfaces but they differ in job-tool motion, construction, work size, machining time, accuracy, use, etc.

Points	Shaping Machine	Planing Machine	Slotting Machine
1. Job-Tool motion	Tool reciprocates in horizontal axis and Job feeds in intermittent translatory motion	Job reciprocates in horizontal axis and Tool feeds in intermittent translatory motion	Tool reciprocates in vertical axis and Job feeds in intermittent translatory motion
2. Construction and rigidity.	Lighter in construction and less rigid	Heavier in construction and more rigid	Lighter in construction and less rigid but more rigid than vertical shaper
3. Motor Power required	Power is very less; round about 15 kW	Power is very high as compared to the shaping and slotting machines; about 5 to 10 times the shaper	Same as shaper

Contd...

Points	Shaping Machine	Planing Machine	Slotting Machine
4. Job size and fixing time	Smaller jobs can be fixed quickly and simply in a vice Job size: 900 mm cube (max.)	Bigger jobs require fixtures for fixing and hence more fixing time and skillness are required. Max. Job size: 3 m ×3 m ×18.25 m	Same as shaper
5. Number of surfaces that can be machined at a time	Only one surface at a time	Three surfaces can be machined at a time	Same as shaper
6. Machining time and type of job suitability	More time required as one tool works at a time; suitable for jobbing work and few no. of pieces	Less time required as a number of tools can work at a time; suitable for production type of work	Same as shaping machine
7. Number of tools that can be operated at a time	Only one tool can be operated at a time	Three or more tools can be operated at a time	Same as shaper
8. Tool size	Standard tool size is not so robust as used in Planing Machine	Robust tool size is required for machining bigger jobs with higher depth of cut, feed and to take up the impact forces	Same as shaper
9. Range and number of speeds and feeds	Smaller range and smaller number of speeds and feeds	Wide range and more number of speeds and feeds available	Same as shaper
10. Amount of feed and depth of cut	Heavier depth of cut and course feed is not employed	Heavier depth of cut and course feed can be employed	Same as shaper
11. Cutting return ratio	Fixed in a particular machine	Ratio can be changed independently	Same as shaper
12. Arrangement of quick return mechanism	Generally being adopted	Not generally being adopted but quick return is possible by using variable gearing arrangement	Same as shaping machine

Contd.

Points	Shaping Machine	Planing Machine	Slotting Machine
13. Cost of the Machine	Less	Most	Little less than shaper
14. Accuracy	Less accurate as tool reciprocates	More accurate due to more rigidity; the tool remains stationary during machining	Same as shaper
15. Floor space required	Less space is required	More space is required. Space required more than double the length of the machine table	Same as shaper

A8.10 Questions

1. Write a note on Planing Machine (A8.1).
2. Specify a Planing Machine (A8.2).
3. Classify a Planing Machine (A8.3).
4. Write short notes on open side planer and double housing planer (A8.4 [1–2]).
5. State Pit Planer and Edge or Plate Planer (A 8.4 [3–4]).
6. Write a statement of Hydraulic, Universal and Divided Table Type Planers (A 8.4 [5–6-7]).
7. Sketch a Planer Tool and state with its angles (A8.6).
8. Write how many ways the planing machine table can reciprocate (A8.5).
9. Sketch and explain the quick return mechanism of a planer to drive the table by open and cross belting (A8.5.1).
10. Write step by step the method of machining of a Dovetail in a Planing Machine with necessary sketch (A8.7).
11. Write the machining method in a Planing Machine to machine a Tee-slot with necessary sketches (A8.8).
12. Distinguish among Shaping, Planing and Slotting Machines (A8.9).

9

GRINDING MACHINES

A9.01 : Introduction

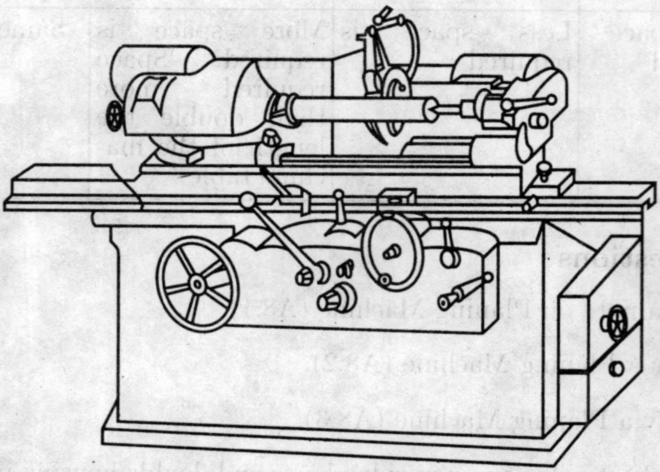

(A) Cylindrical Grinding Machine

Sketch S-9.01 (A)

Grinding is a process of removing materials in the form of ground chips from a workpiece by mechanical action of many small abrasive particles bonded together in a grinding wheel. Each abrasive particle is acting as a small cutting tool. It is a finishing process employed for producing close dimensional accuracies and smooth surface finish.

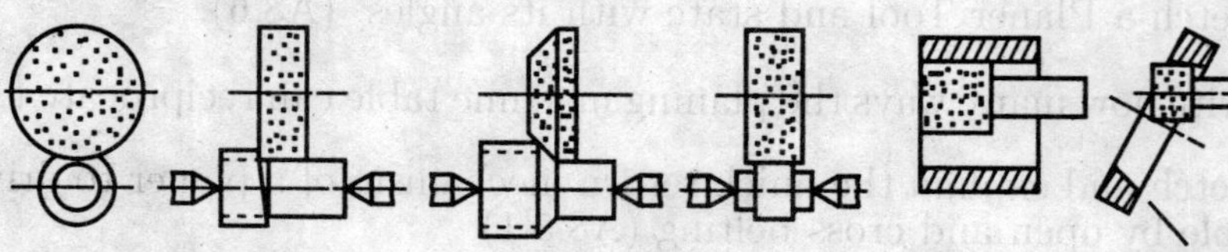

(B) External Cylindrical Surface Grinding

(C) Internal Cylindrical Surface Grinding

Sketch S-9.01 (B, C)

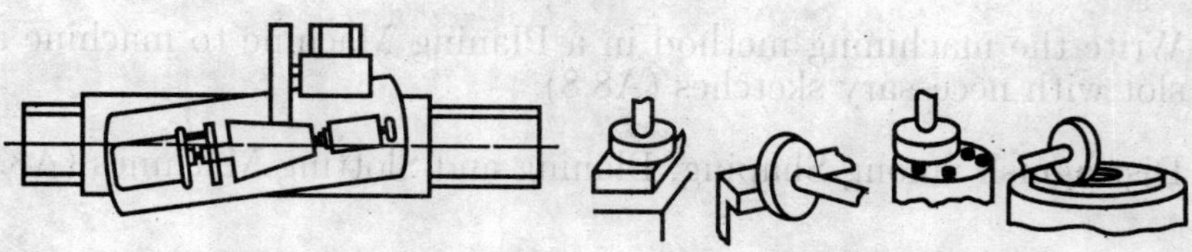

(D) Taper Grinding

(E) Flat Surface Grinding

Sketch S-9.01 (D & E)

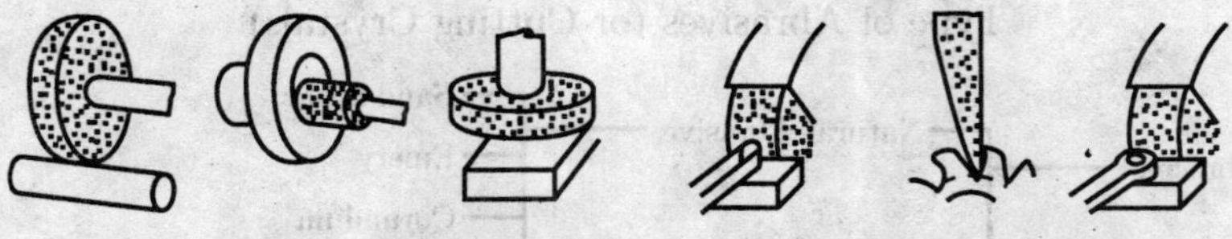

(F) Grinding of Workpieces (G) Sharpening of Tools

Sketch S-9.01 (F & G)

A9.02 : Advantages Of Grinding Over Other Processes

The grinding process has certain advantages that are not found in other cutting processes:

1. Mineral abrasives are much harder than the workpiece materials.

2. Higher cutting speeds than conventional tool materials can be attained as abrasives are less sensitive to heat.

3. By dressing or truing the wearout grinding wheel during abrasive action, new sharp cutting edges can be exposed.

4. Complex profiles can be machined accurately by grinding with relatively inexpensive truing templates.

5. Closer dimensional accuracies and smoother surface finish can be obtained by using a very small amount of depth of penetration of the abrasive grain.

6. In grinding hard skin materials, minimum depth of cut is not a factor as to be considered for conventional tools.

7. Harder materials may easily be machined by grinding which cannot be machined by conventional tools, unless the material is heat treated.

8. Very little pressure is required for the machining by grinding and mostly magnetic chucks are used for holding the work.

A9.03 : Compositions Of Grinding Wheels

Grinding wheels are composed of selectively sized hard and sharp edged abrasive grains which are held together by a suitable bonding agent. The compositions of the grinding wheel are as follows:

(1) Type of Abrasive, (2) Grit or Grain Size, (3) Grade of the Wheel, (4) Structure, (5) Type of Bond.

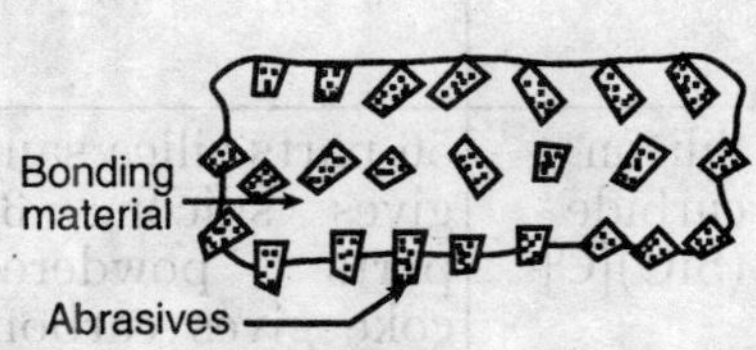

Sketch S-9.03

A9.03.1 : Abrasive

An abrasive is a hard material which can be used to cut or wear away by abrasion to other materials.

Type of Abrasives (or Cutting Crystals)

- Abrasive
  - Natural Abrasive
    - Sand Stone
    - Emery
    - Corundum
    - Diamond
  - Manufactured or Artificial or Synthetic Abrasive
    - Silicon Carbide (C) (SiC)
    - Aluminium Oxide (A) (Al_2O_3)
    - Boron Carbide

A9.03.2 : Comparison among Different Types of Abrasives

Abrasive	Composition	Preparation	Use and Remark
Sand Stone	Natural Abrasive		Rare use, Relatively Soft.
Emery	55% to 65% Aluminium oxide and rest iron oxide and impurities.	Abrasives are crushed and bonded by using vitrified process.	Emery cloth—Greater hardness and better abrasive action than sand stone. Unreliable in modern use because non-uniformity of abrasives.
Corundum	75% to 95% Aluminium oxide and rest iron oxide and impurities.		
Diamond	Ore	Wheels are made with a resinoid bond.	For sharpening cemented carbide tools, ceramics, etc. Slow wear, rapid cutting ability, very low heat generation, proved economical, still having high initial cost.
Silicon carbide (SiC)[C]	56 parts silica sand gives silicon, 34 parts powdered coke gives carbon, 2 parts salt helps to minimise impurities, 8 parts saw dust makes the charge porus to escape gases.	Heat for 36 hours in an electric furnace at a 2000°C temperature; the mass of crystals is then crushed, rolled, screened, graded, cleaned and dried.	Suitable for hard materials of low tensile strength, Grey C.I., Soft bronze, cemented carbide. Black in colour, Harder than aluminium oxide. brittle.

Abrasive	Composition	Preparation	Use and Remark
Aluminium oxide (Al_2O_3) [A]	From mineral bauxite, may contain silica, iron oxide, titanium oxide, coke.	Preparation similar to silicon carbide.	For grinding steel, H.S.S., Tough Bronze, Ferrous alloys with high tensile strength. Aluminium oxide is slightly softer than silicon carbide but much tougher.

A9.03.3 : Grit or Grain Size

It indicates the size of the cutting teeth or the size of the abrasive particles used in making a grinding wheel. Grain size or number is indicating the number of meshes per linear inch (25.4 mm). This number varies from 8 to 600. Grain size 60 means that the abrasive grains will pass through a screen having 60 openings per linear inch. The choice of grain size depends on the nature of grinding operation, the work material, material removal rate and surface finish. The table shows the different types of grains with their applications.

Type of Grains	Grain Number	Applications
Coarse	8–24	Heavy material removal for soft and ductile materials.
Medium	30–60	Average material removal and finish.
Fine Very fine	80–180 220–600	For finish work and also used to grind hard and brittle materials.
Special	1000	Finer, very high surface finish.

A9.03.4 : Grade of the Wheel

Grade is the degree of hardness of the bond and is designated by the letter of alphabet. 'A' is the softest and 'Z' is the hardest grade.

Soft – A to H – Soft wheels for hard materials.

Medium – I to P –

Hard – Q to Z – Hard wheels for soft materials.

A9.03.5 : Structure

Abrasive grains are distributed through the bond. The relative abrasive grain spacing is known as structure and denoted by the number of cutting edges per unit area of wheel face as well as by number and size of void spaces between grains. The purpose of structure is to provide chip clearance. The structure may be open or dense.

Dense or close – 1-2-3-4-5-6-7-8 – Suitable for heavy cut, soft, tough and ductile materials.

Open – 9-10-11-12-13-14-15 – Suitable for finishing cut, hard and brittle materials.

An open structure is more porous than a dense one.

A9.03.6 : Bond

The hardness of the abrasive wheel depends on the amount of bonding material used, which cements the grains of abrasives. Degree of hardness caused by the bond is called the grade of the wheel.

A9.03.7 : Comparative Study of Different Types of Bonding Materials

The table below shows the comparative study of different types of bonding materials with their composition, preparation, properties and applications.

Bond Type	Represented by Alphabet	Composition	Preparation	Properties	Applications
Vitrified Bond	V	Ceramic clays, Porcelain and Water	Bond with abrasive mixed, moulded, dried for several days and fused for 12 to 14 days at 1320°C	Strong, rigid and high strength, well porosity, better for high stock removal. Maximum possible diam. 915 mm	Suitable for maximum metal removal and 75% of the total use
Silicate Bond.	S	Silicate of soda water or glass	Bond with abrasive mixed, moulded and dried and then baked in a furnace for 20 to 80 hours at 260°C	Maximum possible diameters 1525 mm of any sizes	Suitable for small metal removal as knife, tool, etc.
Shellac Bond.	E	Powder shellac.	Bond with abrasive mixed, rolled in heated moulds and baked for few hours at 150°C.	3 mm or less thickness sheets are available, greater elasticity.	Suitable for light work, used in chilled C.I., steel rolls, steel cams, Al-piston, etc.
Rubber Bond.	R	Sheets of pure rubber and sulphur as vulcanized agent, spreading abrasive on it	The mixtures are rolled into sheets and wheels are punched out from sheets	More resilient, less heat resistant, extremely thin wheels can be made, elastic	Used for good surface finish, roll finish
Synthetic Resinoid Bond.	B	Synthetic resins and other compounds.	Mixture is moulded and heated at 200°C.	Strong and high speed wheels are available.	Rapid stock removal used for precision grinding cams, rolls for high finish, and high speed grinding.

A.9.03.8 : Grinding Wheel Shapes and Faces

The following are some of the standard grinding wheel shapes and faces that are in common use:

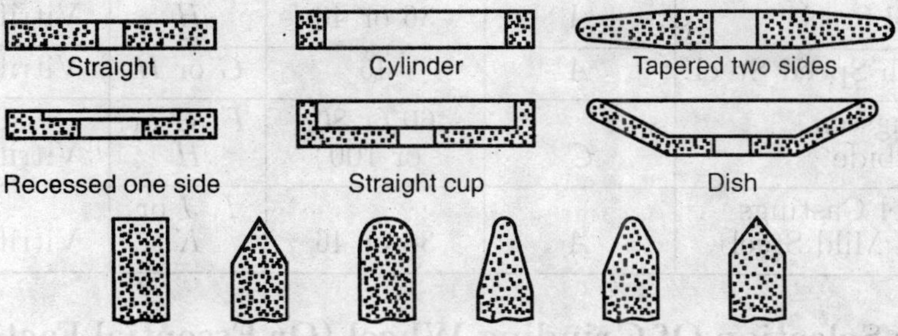

Sketch S-9.03.8

A9.04 : Specification (Or Designation) Of A Grinding Wheel

A grinding wheel is designated or specified by a line with letters and figures which show the characteristics of a wheel.

Sequence:	1	2	3	4	5	6	7
	Prefix	Abrasive	Grain Size	Grade	Structure	Bonding	Suffx
	44	A	60	L	8	V	16

1. Prefix: Manufacturer's abrasive type symbol. (44 Alundum).

2. Abrasive: C stands for Silicon carbide and A stands for Aluminium oxide.

3. Grain Size:

Course grain no.	:	8-10-12-14-16-20-24.
Medium grain no.	:	30-36-46-54-60.
Fine grain no.	:	80-100-120-150-180.
Very Fine grain no.	:	220-240-280-320-400-500-600.
Special	:	1000
(Medium grain 60)		

4. Grade: Soft: A to H, Medium: I to P and Hard: Q to Z. (Medium-L).

5. Structure: Dense: 1-2-3-4-5-6-7-8

and Open: 9-10-11-12-13-14-15, etc. (Dense-8).

6. Bonding: Vitrified (V), Resinoid (B), Silicate (S), Rubber (R), Shellack (E), etc.

7. Manufacturer's wheel shape and size type symbol (16).

A9.04.1 : Specification of Grinding Wheel Composition of Different Tool Materials

Job Material	Abrasive	Grain Size	Grade	Bonding
Tool Steel	*A*	36 or 46	*H*	Vitrified
High Speed Steel	*A*	46	*G* or *H*	Vitrified
Tungsten Carbide	*C*	60 or 80 or 100	*F*, *G* or *H*	Vitrified
Steel Castings and Mild Steel	*A*	36 or 46	*I*, *J* or *K*	Vitrified

A9.05 : Selection Of Grinding Wheel (Or Essential Factors For Consideration In Choosing A Grinding Wheel)

Proper care should be taken in selecting a grinding wheel towards its abrasive, grain size, grade, structure, bonding, etc.

1. Selection of abrasive (Aluminium oxide and Silicon carbide)

(a) Aluminium oxide for materials of high tensile strength, i.e., steel and steel alloys.

(b) Silicon carbide for materials of low tensile strength, ductile and non-ferrous metals, i.e., Brass, Copper, Soft Bronze, etc.

(c) Diamond for extremely hard material, i.e., sintered carbide.

2. Selection of grain size (coarse, medium, fine and very fine)

(a) Coarse and medium grain for fast removal of materials and ductile materials.

(b) Fine and very fine grain for fine finish and hard and brittle materials.

(c) Medium and fine grain for tool grinding and precision work.

3. Selection of grade (soft, medium and hard)

(a) Harder wheel for softer materials.

(b) Softer wheel for harder materials.

4. Selection of structure (dense and open)

(a) Dense structure for heavy cuts of soft and ductile materials.

(b) Open structure for finishing cuts of brittle and hard materials.

5. Selection of bonding (vitrified, resinoid, rubber, shellac, silicate)

(a) **Vitrified** (*V*) bond for general work (about 75% of total use), and diameter up to 915 mm and maximum metal removal at a surface speed of about 2,000 m/min.

(b) **Silicate** (*S*) bond for diameter up to 1,525 mm and small amount of metal removal as knife, tool, etc.

(c) **Shellac** (E) bond has greater elasticity, used for light work to remove material 3 mm or less thickness sheets, at surface speed over 2,000 m/min. For machining chilled C.I., steel rolls, steel cams, Al-pistons, etc.

(d) **Rubber** (R) bond for extremely thin sheets, best finish, and for serious bending at high surface speed.

(e) **Synthetic Resinoid** (B) bond for high speed snagging of castings for precision grinding cams, rolls for high finish.

6. Selection on amount of stock removal

(a) Fast cutting: More area of contact, coarse grain, wide spacing, vitrified bond, slower wheel speed.

(b) Finish cutting: Less area of contact, fine grain, close spacing, resinoid-rubber-shellac bond, higher wheel speed.

A9.06 : Manufacture Of Grinding Wheels

Silicon carbide and Aluminium oxide grinding wheels are manufactured in the following steps:

1. Crushing the required percentages of materials to small sizes by passing through rolls and jaw crushers.

2. Screening the crushed materials.

3. Using magnetic separators to remove iron compounds.

4. Washing for removing all dust and foreign materials.

5. Grading the grains by passing over vibratory standard screens.

6. Mixing the grains with proper and requisite amount of bonding material.

7. Moulding in an electric furnace at 2,000°C for 36 hours.

8. Preparing to proper size and shape of the wheel required.

9. Lastly, wheels are bushed, trued and tested.

A9.07 : Specifications Of A Grinding Machine

1.	Type of Grinding Machine	: Centre Type
2.	Size of grinding wheel	: External diameter in mm Internal diam. in mm Width of the wheel in mm
3.	Wheel specification	: Abrasive, Grain Size, Grade, Structure, Bonding
4.	Distance between two grinding wheels in mm	
5.	Distance of the wheel centre from the base plate in mm	
6.	Spindle diameter in mm	
7.	Spindle speed in r.p.m.	
8.	kW of the driving motor	
9.	Weight of the machine in tonne	
10.	Floor space required in square metre	

A9.08 : Classification Of Grinding Machines

The Grinding Machines may be classified according to the—

1. Type of operations:

   (a) Tool grinders, (b) Cut-off grinders, etc.

2. Quality of surface finish:

   (a) Precision grinders, (b) Rough grinders
   - Bench
   - Floor Stand
   - Portable
   - Flexible Shaft
   - Swing Frame
   - Abrasive Belt
   - Disc

3. Type of Surface Generated:

   (a) Cylindrical Grinders
   - Centre type
     - Plain
     - Universal
     - Plunge
   - Centreless
   - Chucking type
   - Tool post

   (b) Internal Grinders
   - Chucking type
   - Vertical type
   - Planetary
   - Centreless

   (c) Surface Grinders
   - Planer type (Reciprocating table)
     - Horizontal spindle
     - Vertical spindle
   - Rotating table
     - Horizontal Table
     - Vertical Table

   (d) Tool Grinders
   - Universal
   - Special

   (e) Special Grinders
   - Swinging frame - Snagging
   - Cutting off - Sawing
   - Portable - Off hand grinding
   - Flexible shaft - General purpose
   - Honing & Laping - Accurate finishing
   - Etc.

   (f) Surface finishing Grinders
   - Buffing Machine
   - Flexible Band

A9.09 : Description Of Different Types Of Grinders

1. Bench and Floor Stand Grinders

Small size grinding machine mounted on a bench is known as bench grinder. The relatively larger grinding machine mounted on base or pedestal is known as floor stand grinder. Motor is used to drive the grinding wheel mounted on horizontal spindle. These machines may be used for snagging and off-hand grinding of tools, etc.

2. Portable Grinder

This type of grinding machine is very handy for cleaning castings, welded works, sheet metal works, rough works, etc., in which horizontal spindle is being operated either by electric motor or by pneumatically.

3. Flexible Shaft Grinding Machine

The grinding wheel is mounted at the end of a long rotating flexible shaft driven by a motor on a relatively stationary stand. This machine may be used in removing comparatively small amount of stock from widely separated areas. Small wheels is used for burring and die work and heavy wheels for roughing and snagging operations.

4. Swing Frame Grinder

It has about 3 m long horizontal frame suspended at its centre so as to move freely within the area of operation. This is used for cleaning of castings or smoothing large castings.

5. Two Wheels or Pedestal Grinder

Two wheels, one for rough grinding and the other for finish grinding, are mounted at the two ends of the horizontal spindle, it is driven by motor and placed on base or pedestal. This machine is suitable for freehand tool bit grinding.

6. Abrasive Belt Grinder

The endless abrasive belt is slipped over two drums, one of which is driven at high speed. The work may be fed against the open belt to grind various curve and flat surfaces. Dry belt, wet belt or combination type may be used for heavy stock removal or for light polishing work.

7. Disc Grinder

It is driven either manually or mechanically and used to grind small flat surfaces. The workpiece can be fed towards the abrasive disc.

8. Precision Grinding Machines

The workpiece and the grinding wheel are mounted rigidly but they are adjustable according to the requirement to the method of machining, for grinding to shape and size quickly and accurately.

A9.09.1 : Cylindrical Grinder

These machines are used for grinding the external surfaces of cylindrical, contoured cylinders, tapers, shoulders, fillets, cams, crank-shafts, etc. Grinding operations are accurate, easier, quicker and cheaper as compared to finishing operation in Lathe Machines.

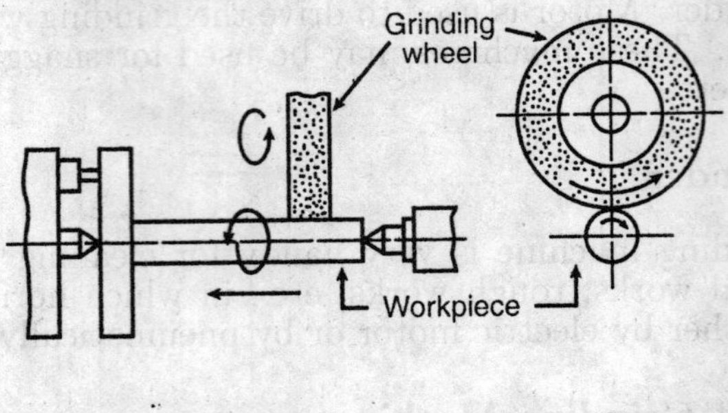

(A) External Cylindrical Surface Grinding (B)

Sketch S-9.09 (A & B)

Finishing an external surface by cylindrical grinding:

In finishing the external surface of a shaft by cylindrical grinding, the following steps are to be followed:

1. The shaft (workpiece) is clamped between the centres.
2. Selection of proper grinding wheel, i.e., straight grinding wheel (soft wheel preferable).
3. Mounting of grinding wheel in its spindle with proper alignment with the workpiece.
4. Selecting and setting of proper cutting speed of the grinding wheel, peripheral speed of the workpiece, longitudinal (lateral) table feed and down feed of the grinding wheel.
5. Adjusting and making the proper alignment between the workpiece and the grinding wheel.
6. Selection, use and proper functioning of required coolant for the purpose.
7. Checking the machine to be ready to grind.
8. Workpiece and grinding wheel are rotated in the same direction of rotation with their requisite speed. Longitudinal (lateral) feed is given by the workpiece table, using the proper down feed by the grinding wheel.
9. Using number of cuts to attain the required diameter of shaft with proper depth of cut.
10. Measuring and checking the shaft by using micrometer, snap gauge and dial gauge.

A9.09.2 : Internal Grinders

These are used for grinding the internal surfaces of holes of any size, straight or taper in parts like gears, cutters, bushings, etc.

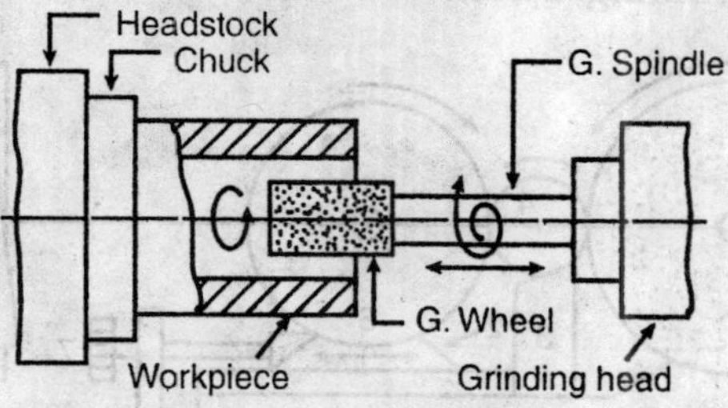

Sketch S-9.09.2

Finishing an Internal Surface by Cylindrical Grinder

In grinding the internal surface of a bush by cylindrical grinding, the following steps are to be followed:

1. Clamping the bush on a cross-slide by chuck. The headstock receives power from the motor, rotates the chuck of the table and the table also can give lateral feed.
2. Selecting the proper internal soft grinding wheel and mounting the same in grinding head.
3. Dressing of grinding wheel.
4. Aligning the workpiece with respect to the grinding wheel.
5. Selecting and setting of proper cutting speed of the grinding wheel, peripheral speed of the workpiece, longitudinal (lateral) table feed and down feed of the grinding wheel.
6. Selection, use and proper functioning of required coolant for the purpose.
7. Checking, the machine is ready to grind.
8. Finishing the bore of a bush is being done with proper speed of the grinding wheel and workpiece (wheel and workpiece rotate in opposite directions) lateral and down feed by the wheel.
9. Using proper number of cuts to finish the bush to the required bore dimension with proper depth of cut.
10. Measuring and checking the bush bore by micrometer, plug gauge and dial indicator.

A9.09.3 : Finishing an External Surface by Centreless Grinder

In centreless grinding, work is not secured between centres, or in a chuck or on any fixture, but the workpiece is supported by using two wheels and a work rest. The larger wheel is the grinding wheel and the smaller one the regulating wheel or pressure wheel. The fine grained resinoid bonded abrasive regulating wheel having frictional characteristics is used to rotate the workpiece. The regulating wheel is having same direction of rotation as the grinding wheel. The axial movement of the workpiece is obtained a longitudinal feed, by tilting the regulating wheel at a slight angle of 1 to 8 degrees relative to grinding wheel.

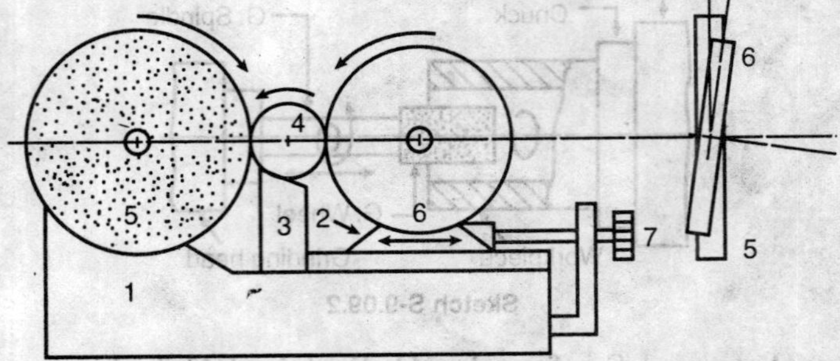

1. Grinding head 2. Regulating wheel head 3. Work rest
4. Workpiece 5. Grinding wheel 6. Regulating wheel 7. End stop

Sketch S-9.09.3

Rate of longitudinal feed of the work, $f = \dfrac{\pi D_R N_R \sin\alpha}{1000} = V_R \sin\alpha$ m/min.

where, feed, f in mm/min
peripheral speed of regulating wheel, V_R in m/min.
diameter of regulating wheel, D_R in mm
revolution of regulating wheel, N_R in r.p.m.
angle of inclination of Regulating wheel, α in degrees
speed of rotation of the workpiece, $V_w = V_R \cos\alpha$ m/min.

Peripheral speed of the grinding wheel (30 to 50 m/s) is 70 to 80 times that of the peripheral speed of the regulating wheel (10 to 50 m/min.). Thus the speed of rotation of the work is a function of the regulating wheel.

After completion of grinding, the regulating wheel is backed out, and the work is replaced by a new one.

A9.09.4 : Finishing an Internal Surface by Centreless Grinder

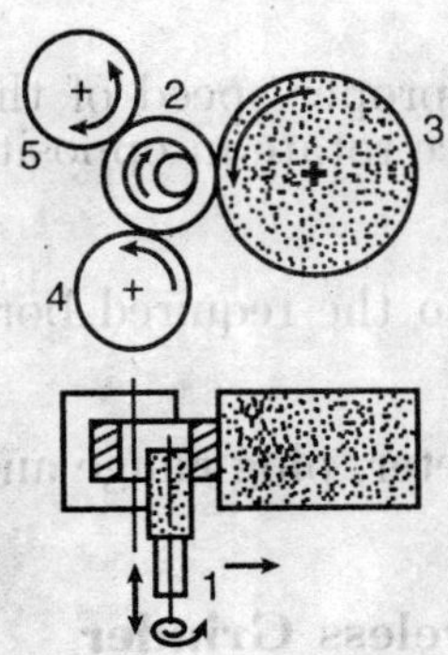

1 = Grinding wheel
2 = Workpiece
3 = Driving wheel
4 = Roller rest
5 = Pressure roller

Sketch S-9.09.4

In centreless internal surface grinding, three rollers are used to support and drive the work. They are (i) driving rollers 3, (ii) work-rest roller 4 and (iii) pressure roller 5. Automatic loading and unloading can be arranged by swinging the pressure roller out of the way at the end of the cycle. This system can eliminate the use of work-holding fixtures and can grind both straight and tapered holes.

Workpiece 2 is rotated by the driving wheel 3. The grinding wheel 1 rotates at a specified speed and effects the longitudinal feed 3 and cross-feed motion 4. The workpiece rests on work rest roller 4 in a predetermined position but the pressure roller 5 is oscillated in a manner to provide a fluctuating grinding pressure. In grinding of a work having diameter less than 30 mm, the pressure roller is replaced by a work-rest blade.

A9.09.5 : Finishing Flat Surfaces by Surface Grinder

Surface grinding is the production of finishing of flat and plane surfaces. Cup wheel or a disc wheel can be used in surface grinding. Modern flat surface grinders are equipped with electromagnetic chucks which greatly reduce the time for loading, mounting and unloading the workpieces; screws and straps are used for clamping the bigger workpieces. Flat surface grinders are classified by the principle of operation (i.e., by the periphery or by the sides of the wheel or both), by the shape of the table (rectangular or circular), by the motion of table (reciprocating or rotary), by the axis of rotation of the grinding wheel (in horizontal axis or in vertical axis). Hydraulic power is mostly used in driving the table.

[a] *Grinding by Horizontal Spindle on Reciprocating Table*

In horizontal spindle surface grinding machine with reciprocating table arrangement, the workpiece mounted on magnetic chuck or on fixture, reciprocates under the rotating grinding wheel, in its horizontal axis. The table or the wheel can feed axially (cross-feed) between passes to produce the flat surface. Depth of cut is given by the down feed of the grinding wheel. This type is well adopted for reconditioning dies, grooves, ways on machine tools and other long surfaces.

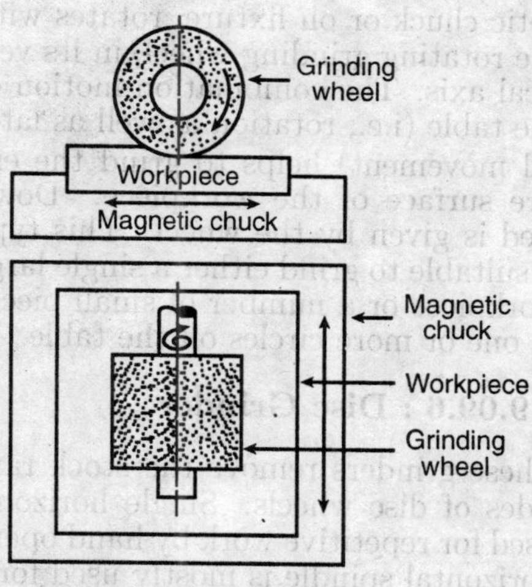

Sketch S-9.09.5 (a)

[b] *Grinding by Horizontal Spindle on Rotary Table*

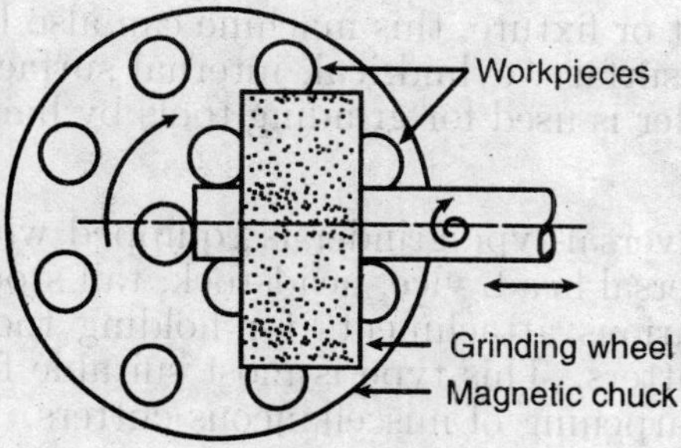

Sketch S-9.09.5 (b)

In horizontal spindle surface grinding machine with rotary table arrangement, the workpieces are mounted on magnetic chucks or on fixture rotates slowly under the rotating grinding wheel in its horizontal axis. The circular table rotates at specified speed and the wheel can feed axially (cross-feed). Depth of cut or down feed is incorporated by the grinding wheel. This type is suitable for grinding circular works such as milling cutters, piston rings, valves, etc.

[c] *Grinding by Vertical Spindle on Reciprocating Table*

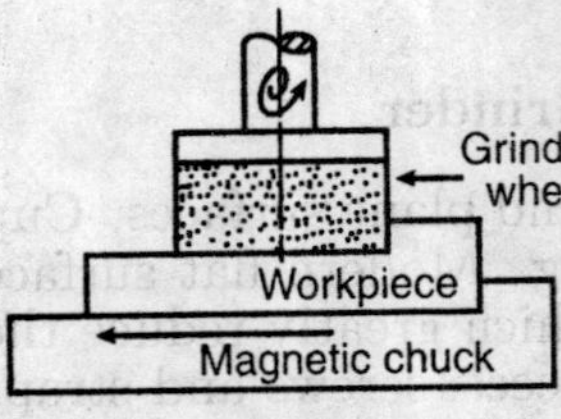

Sketch S-9.09.5 (c)

In vertical spindle surface grinding machine with reciprocating table arrangement, the workpiece, mounted on magnetic chuck or on fixture, reciprocates under the rotating grinding wheel, in its vertical axis. Vertical and cross-wise adjustment for the grinding wheel is incorporated in wheel head assembly. This type is specially adopted for grinding large surfaces.

[d] *Grinding by Vertical Spindle on Rotary Table*

In vertical spindle surface grinding machine with rotary table arrangement, the workpiece, mounted on magnetic chuck or on fixture, rotates with the rotating grinding wheel, in its vertical axis. The combination motion of the table (i.e., rotation as well as lateral movement) helps to grind the entire surface of the workpiece. Down feed is given by the wheel. This type is suitable to grind either a single large workpiece or a number of small pieces in one or more circles on the table.

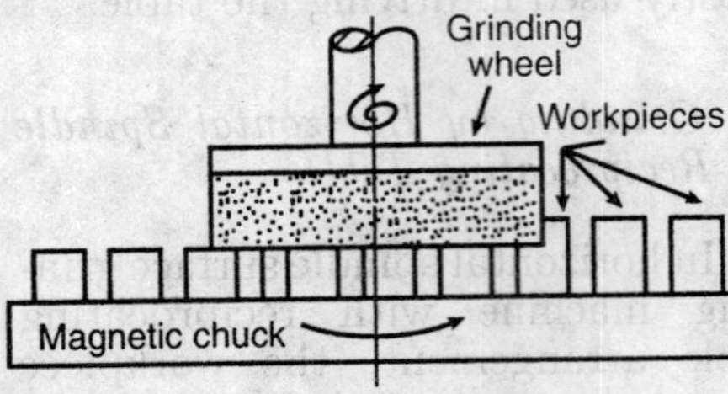

Sketch S-9.09.5 (d)

A9.09.6 : Disc Grinder

These grinders remove the stock rapidly and finish the flat surfaces by the sides of disc wheels. Single horizontal or vertical spindle disc grinders are used for repetitive work by hand operation or with simple fixture. The double horizontal spindle is mostly used for production operations.

A9.09.7 : Tool and Cutter Grinder

As the name implies, this type is mostly used to sharpen and reconditioning various types of cutters like single-point tools, milling cutters, drills, taps, hobs, etc. By using requisite attachment or fixture, this machine can also be used for grinding die, jig, fixture, light surface, cylindrical, internal surface, etc. A bench or pedestal type tool grinder is used for grinding tools by hand (known as off-hand grinding).

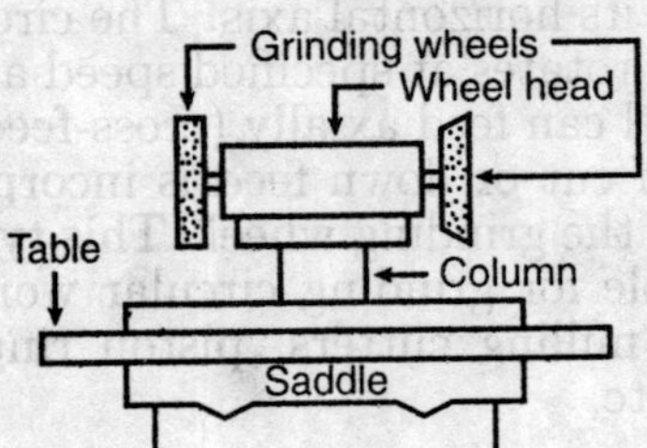

Sketch S-9.09.7

Universal-type grinder is equipped with a universal head, vice, headstock, tail stock and various attachments for holding tools and cutters. This type is most suitable for the sharpening of miscellaneous cutters.

The heavy box-type base maintains the machine's stability and rigidity. Saddle mounted on the base provides the arrangement for moving the work forward and backward. The column is mounted on saddle. The column is supporting the wheelhead. The wheelhead can move up and down, and swivelled to either side of the column. The work table is mounted on sub-table, which can be swivelled to grind tapers.

The headstock and tailstock are essential in cylindrical grinding, mounted on either side of the table for holding the workpiece in between the centres.

A9.09.8 : Comparison between Centreless Grinding and Cylindrical Grinding

Points	Centreless Grinding	Centre Type or Cylindrical Grinding	Advantages of Centreless over Centre Type
1. Work rests on centres	No (True floating condition)	Yes (Mounted)	–
2. Mounting of work by holding devices, i.e., on Chuck, Mandrels, etc.	Not essential	Essential	Advantage
3. Rigidity of the work-holding and time required for holding	More rigid and less time required	Less rigid and more time required	Advantage
4. Tendency of work deflection and chatter during operation	Very minimum	Reasonable	Advantage
5. Skillness of the operator required	Semi-skilled	Skilled	Advantage
6. Process suitable for	Production work	General purpose	–
7. Process is continuous or intermittent	Continuous	Intermittent	Advantage
8. Machining time required	Less	More	Advantage
9. Control of work size	Easy	Difficult	Advantage
10. Maintenance required	Less	More	Advantage
11. Grinding other than cylindrical surface, i.e., flats, keyways, etc.	Not possible	Possible	Disadvantage

Contd.

Points	Centreless Grinding	Centre Type or Cylindrical Grinding	Advantages of Centreless over Centre Type
12. Handling of different diameters of work	Not so easy	Not so difficult	Disadvantage
13. Constant wall thickness of the hollow work	Not possible	Possible	Disadvantage

A9.10 : Mounting Or Changing New Grinding Wheel

While changing or mounting a new grinding wheel on machine spindle, the following steps are to be considered:

1. The wheel to be mounted should be carefully checked against damage, any cracks at or below the surface, by ringing test.
2. Make sure that the wheel is compressed against soft paper or card packings on both sides.
3. The two flanges must be machined true and of equal outside diameter of just half that of the outside wheel diameter. Put the left flange on the spindle and keyed to it.
4. The lead bush forming the wheel arbor hole is to be examined for any burrs and then be placed on the spindle for an easy sliding fit.
5. Place the wheel on the spindle.
6. Then place the outer flange on the spindle.
7. Put the spindle nut (with washer) and tighten it firmly but not excessively for protection of the wheel.
8. Place and fix the wheel guard in position.

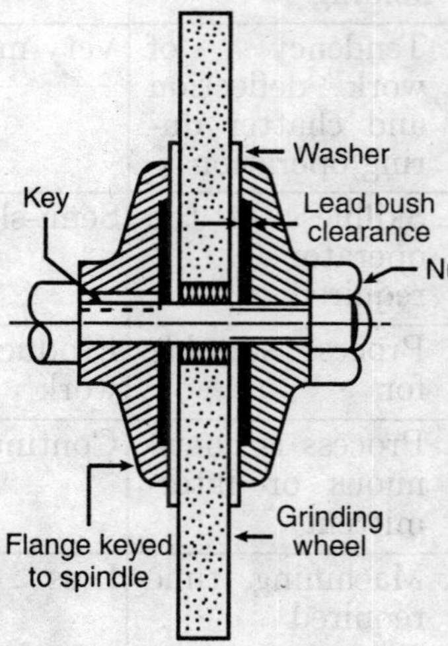

Sketch S-9.10

9. Rotate the wheel by hand and make sure the wheel is clear from the housing.
10. Run the wheel for five minutes at working speed without any one in its proximity.
11. True the wheel by using diamond tool before operation.

A9.11 : Comparison Between Truing And Dressing Of A Grinding Wheel

Points	Truing of a Grinding Wheel	Dressing of a Grinding Wheel
Definition	Truing a wheel is the operation to create a true and concentric surface with the wheel spindle axis.	Dressing a wheel is the operation to remove glazed or other adhesion particles and to expose fresh cutting particles.
When necessary	Truing is necessary to newly manufactured wheels taken from kilns for truing to shape and dimensions	Dressing is necessary to newly manufactured or the dulled crystals, clogged with some foreign materials after it is being used for sometime on the wheel face, for the purpose of exposing new cutting particles
Tool used	Commercial diamond tool or conical point tungsten carbide inserted on a shank of cold rolled steel	Same as Truing Tool
Why necessary	To rotate the wheel in a true and concentric path	To remove the blunt surface on wheels necessary for efficient cutting
Cost of machining	More	Less
Sketch	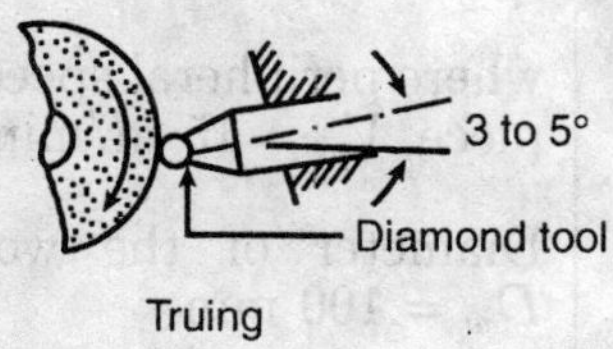	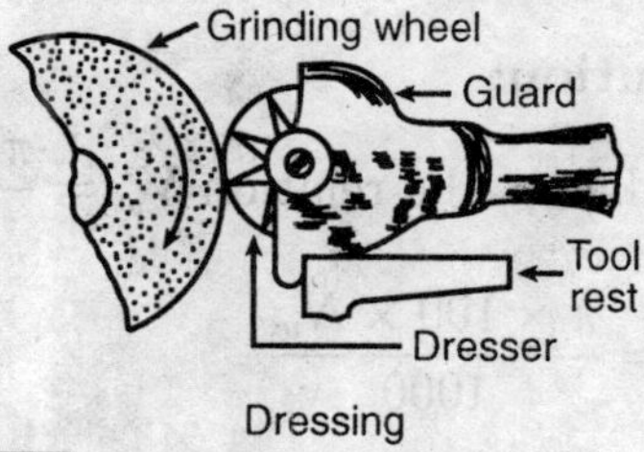

Sketch S-9.11

A9.12 : Cutting Speed, Feed And Depth Of Cut

Grinding Wheel Speed

The cutting speed of a grinding wheel is the peripheral speed (or surface speed), expressed in metres/min. and can be calculated by the following expression:

$$\text{Grinding Wheel Speed in metres/min.,}\ V_w = \frac{\pi D_w N_w}{1000}$$

where, D_w is the diameter of the grinding wheel in mm

N_w is the revolution per minute of the grinding wheel.

Work Speed. Similarly, surface speed or peripheral speed of the work can be determined by the following expression:

$$\text{Work Speed} = \frac{\pi \times \text{Diameter of the work in mm} \times \text{Rev. per min. of the work}}{1000}$$

Grinding wheel speed will vary with respect to the type of grinding operation. Surface speed of grinding wheel may be considered between 1,500 to 2,000 metres/minute. Surface speed of work may be considered 12 to 25 metres/ minute, depending on work materials and type of abrasives.

Feed: Feed is the linear movement of the work per revolution in mm/revolution of the work.

For rough cut, feed may be considered between 0.02 to 0.1 mm.

For finish cut, feed may be considered between 0.01 to 0.05 mm.

For longitudinal feed may be considered between 0.5 to 0.8 times the width of the cutter.

Depth of Cut

Depth of cut is the extent of tool penetrates into the work or the thickness of metal removed in one pass of the work w.r.t. the tool and is expressed in mm, may be considered between 0.005 to 0.05 mm.

A9.13 : Work-Out Problems

Problem 1.

Calculate the wheel speed of a shaft with a diameter of 100 mm which is to be grinded. Assume peripheral speed of the workpiece as 15 m/min.

Solution:

We know, the relation $V_w = \frac{\pi D_w N_w}{1000}$

$$15 = \frac{\pi \times 100 \times N_w}{1000}$$

or, $N_w = \frac{15 \times 1000}{\pi \times 100} = 47.7 \simeq 48$ r.p.m. **Ans.**

where peripheral speed of the workpiece, $V_w = 15$ m/min.

Diameter of the workpieces, $D_w = 100$ mm

Let, wheel speed, N_w in r.p.m.

Problem 2.

Calculate the machining time for cylindrical grinding the entire length of an M.S. shaft from 50.5 mm dia. to finish to 50 mm in diameter for a length of 500 mm. Width of the cutter is 40 mm. Assume peripheral speed of the workpiece as 12 m/min., the lateral feed 0.5 times the grinding wheel width per revolution of workpiece and 0.025 mm as down feed of the grinding wheel.

Solution:

1. RPM of the workpiece, N_w

$$N_w = \frac{V_w \times 1000}{\pi \times D_w} = \frac{12 \times 100}{3.142 \times 50} = 76 \text{ r.p.m.}$$

where, Peripheral speed of the workpiece,

$V_w = 12$ m/min.

2. Longitudinal feed, f_l in mm per revolution of the workpiece

f_l = Cutter width × Lateral feed = 40 × 0.5 = 20 mm.

Diam. of the workpiece, $D_w = 50$ mm

Grinding length, L in mm = approach + workpiece length + over-run = 5 + 500 + 5 = 510 mm

3. Number of cuts, N_c(50.5 − 50) = 0.5 mm in diameter to be reduced)

Grinding allowance applied to the radius 0.5 : 2 = 0.25 mm.

$$N_c = \frac{\text{Grinding allowance}}{\text{Down feed}(f_d)} = \frac{0.25 \text{ mm}}{0.025 \text{ mm}} = 10$$

4. Machining time, $M_T = \dfrac{2 \times \text{ Grinding length} \times \text{No. of cuts}}{\text{Longitudinal feed} \times \text{Speed of the workpiece}}$ min.

$$= \frac{2 \times L \times N_c}{f_l \times N_w} = \frac{2 \times 510 \times 10}{20 \times 76} \simeq 6.71 \text{ min. } \textbf{Ans.}$$

A9.14 : Problems For Practice

1. Calculate the r.p.m. of the workpiece of diameter 50 mm and the peripheral speed of 14 m/min. [**Ans.** 89 r.p.m.]

2. Determine the machining time for cylindrical grinding of the entire length of the workpiece from diameter 40.2 mm to 40 mm for a length of 200 mm. Assume the width of the cutter as 30 mm, peripheral speed of the workpiece = 15 m/min., lateral feed = 0.3 times the grinding wheel width per revolution of the workpiece, and 0.02 mm as down feed of the grinding wheel. [**Ans.** 2 min.]

3. Calculate the machining time for surface grinding for plate of 600 mm length. The grinding allowance of 0.5 mm has to be ground down with 3 cuts. Speed of the table is 2 m/min.

[**Hint.** L = Length of workpiece + Over-travel (Diameter of the wheel) in mm and assume diameter of the wheel = 100 mm and use formula for machining time, $M_T = \dfrac{2 \times L \times N_c}{V_t \times 1000}$ min. where table speed, V_t is in m/min.] [**Ans.** 0.915 min.]

4. A die plate, 250 mm long, 150 mm wide has to be grinded by circumferential grinding. Grinding allowance = 0.5 mm; number of cuts = 5; width of the grinding wheel = 30 mm; lateral feed = 10 mm/stroke. Speed of the table = 2 m/min. Calculate the machining time.

[**Hint.** $M_T = \dfrac{2 \times L \times B \times N_c}{V_t \times 1000 \times f_l}$ where $L = 250 + 5 + 5 = 260$ mm and $B = 5 + 150 + 5 = 160$ mm.] [**Ans.** = 20.8 min.]

A9.15 Questions

1. Write a short note on Grinding Machine (A 9.01).

2. Write the advantages of grinding machine over other conventional machining processes (A 9.02).

3. Name the composition of a grinding wheel (A 9.03).

4. What do you mean by abrasive (A 9.03.1).

5. Compare among different types of abrasives (A 9.03.2).

6. Write short notes on grit or grain, grade and structure of grinding wheel (A 9.03.3-5).

7. What is bonding material (A 9.03.6)?

8. Compare different types of Bonding Materials with respect to their composition, preparation, properties and applications (A 9.03.7).

9. Sketch some standard types of grinding wheel shapes and faces (A 9.03.8).

10. Specify a grinding wheel (A 9.04).

11. Suggest the grinding wheel compositions of grinding different tool materials (A 9.04.1).

12. Write the points to be considered in selecting a grinding wheel (A 9.05).

13. Explain the steps to be followed in manufacturing of a grinding wheel (A 9.06).

14. Specify a grinding machine (A 9.07).

15. Classify the different types of grinding machines (A 9.08).

16. Write short description of different types of grinders (A 9.09).

17. Sketch, label and explain its working (in steps if possible) of the following:

    (a) Finishing an external surface by cylindrical grinder (A 9.09.1).
    (b) Finishing an internal surface by cylindrical grinder (A 9.09.2).
    (c) Finishing an external surface by centreless grinder (A 9.09.3).
    (d) Finishing an internal surface by centreless grinder (A 9.09.4).
    (e) Finishing the flat surface by a surface grinder (A 9.09.5).

18. Compare between centreless grinding and cylindrical grinding (A 9.09.8).

19. Sketch, label and explain the steps in mounting a new grinding wheel or to change a grinding wheel (A 9.10).

20. Compare with sketch the truing and dressing of a grinding wheel (A 9.11).

21. State about the term speed, feed, and depth of cut used in grinding operations (A 9.12).

22. Write a short note on Disc Grinder (A 9.09.6).

23. Sketch and explain the Tool and Cutter Grinder (A 9.09.7).

10

BROACHING MACHINES

A10.1 : Introduction

Broaching Machine is used for machining external, internal and contour surfaces in mass and large lot production on small and medium-sized workpieces. The machine is designed to manufacture workpieces at a high rate and with satisfactory accuracy and finish.

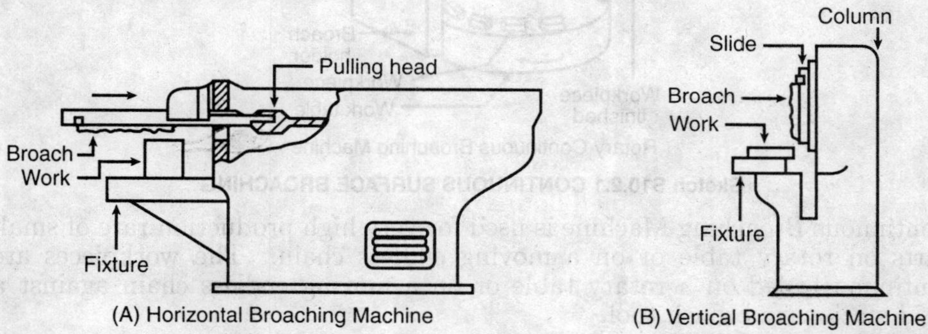

Sketch S-10.1 (A & B)

Broaching is the operation of machining metal by means of a broach tool having a number of teeth gradually increasing in size.The broach moves in a fixed path along the surfaces of the workpiece to be machined. Machining is completed in one stroke of the broach tool or workpiece. Generally, the broach is moved in a straight line motion past the workpiece. In some machines, the workpiece may be moved while the broach tool will remain stationary. Different broach tools are required for different shapes of the workpieces. This machine is limited for mass production due to the high cost of the broach tool. The straight line motion of the tool or workpiece is actuated by rack, hydraulic or electro-mechanical means.

Horizontal Broaching Machine is more versatile, cheaper, smaller output, and requires more floor area than the vertical one.

A10.2 : Classification

The Broaching Machine may be classified as follows:

(a) According to the nature and direction of primary cutting motion: Horizontal, Vertical and Continuous.

(b) According to the purpose: Internal and External Surface.

(c) According to method of operation: Pull or Push.

(d) According to the construction of the broach tool: Solid, Inserted tooth, Progressive Cut, Built-up, etc.

(e) According to the function: Keyway, Burnishing, Spline, Round-hole, Surface, etc.

(f) According to the number of main slides or stations: Single, Double or Multiple Slides.

(g) According to the motion of the broach tool relative to the work: Straight line motion or Stationary broach tool.

A10.2.1 : Continuous Surface Broaching Machines

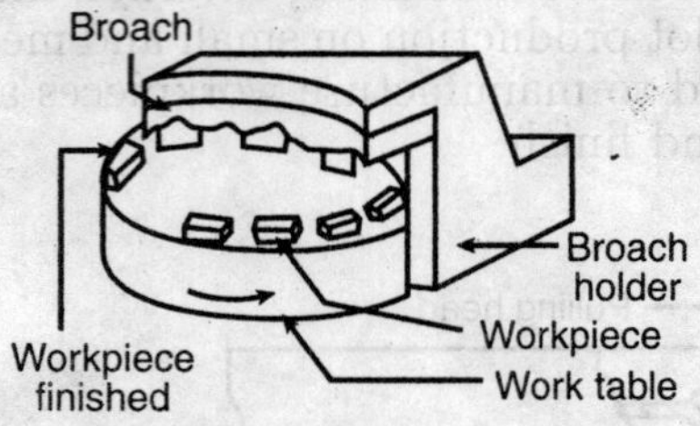

Rotary Continuous Broaching Machine

Sketch S10.2.1 CONTINUOUS SURFACE BROACHING

Continuous Broaching Machine is used for very high production rate of small parts on rotary table or on a moving endless chain. The workpieces are continuously fed on a rotary table or on a moving endless chain against a rigid stationary broach tool.

Grinding : Grinding is a process of removing small amount of material to finish a hardened hole in a drilling machine. The grinding wheel bounded with abrasive particles revolves with the drilling spindle and is fed up and down. The accuracy of grinding operation is from 0.9 to 5 microns (1 micron = 0.001).

Lapping : It is also a precision grinding with loose, dust type abrasives (known as lapping compound) to produce geometrically true surfaces on small diameter hardened hole. The lap tool fits in the spindle of drilling machines, revolves with the spindle and is fed up and down. The accuracy of lapping operation is from 0.08 to 0.25 microns.

A10.2.2 : Compare Pull Broaching and Push Broaching

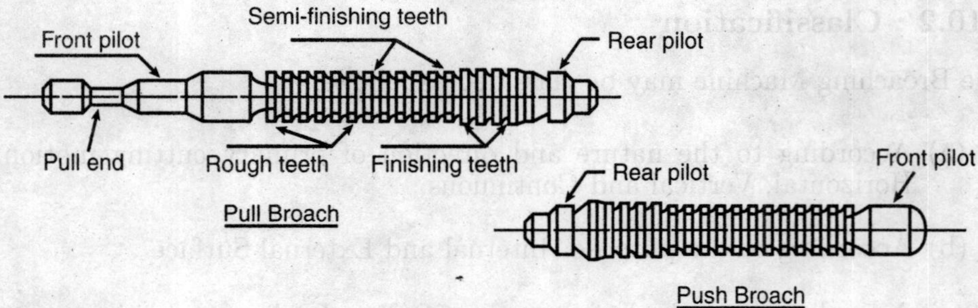

Sketch S10.2.2 PULL AND PUSH BROACHING

Points	Pull Broaching	Push Broaching
1. Sketch	Sketch	Sketch
2. Job-Tool Movement	The broaching tool is pulled through the work while the work is stationary	The broaching tool is pushed through the work while the work is stationary
3. Application	This type is mostly used in internal broaching and can also be done on surface broaching	It can be done on simple utility processes such as sizing holes cutting keyways, etc.
4. Fixing of Tool	Broaches are usually long and so fixed with a special tool head	Hydraulic and hand arbor presses are well known for push broaching
5. No. of teeth of the cutter	More number of cutting teeth	Less No. of cutting teeth
6. Rate of Metal Removal	More metal removal	Less metal removal
7. Preference	Pull type broaches are preferred over push type due to high production rates and capable of completing the operation in one pass	

A10.3 : Specification Or Size Of A Broaching Machine

Main specifications of a Broaching Machine are:

1. Maximum length of stroke in mm
2. Maximum force developed by the slide in tonnes

The subsidiary specifications are:

1. Type of drive for straight line motion, i.e., Electro-mechanical or Hydraulic, etc.
2. Power rating of electrical motor in kW
3. Speed and feed range
4. Weight of the machine in tonnes
5. Floor space required in square metre

A10.4 : Advantages And Limitations Of Broaching

A10.4.1 : Advantages

1. Roughing, semi-finishing and finishing cuts are completed in one pass of the broach.

2. Broaching can be used for either external or internal surface finish.

3. Rapid loading and unloading arrangement of the workpieces by using fixture and machining the surface in one pass, give high rate of production.

4. High accuracy and surface finish of about 0.001 mm; suitable to interchangeable manufacture.

5. Semi-skilled worker can perform a broaching operation.

6. Any form that can be reproduced on a broach can be machined.

7. Broaching is mostly adopted for mass production.

8. Cutting fluid can be applied more easily and effectively.

A10.4.2 : Limitations

1. High initial cost of the broach tool compared to other tools.

2. Job work or batch work is not advisable due to the high tool cost.

3. Obstruction surfaces or blind holes cannot be broached.

4. The workpieces should be rigidly supported to withstand the broaching forces.

5. Broaching is not suitable for removal of large amount of stock.

6. Almost all the workpieces will need fixture for rigid support.

7. Tool grinding is difficult and expensive, needs special type grinding machine.

8. Broach is a single-purpose tool; it requires different broaches for different shapes and sizes.

A10.5 : Broach

Broach is a tool provided with multi-cutting edges, generally made of high speed steel. The Broach is moved along the surface of the workpiece to be machined. It has three cutting zones: roughing teeth, semi-finishing teeth and finishing teeth. The teeth are gradually increasing in height from the first roughing tooth to the first of the finishing tooth, according to the required profile of the workpiece. Normally, the finishing teeth are of the same diameter. Rise per tooth amounts to approximately from 0.02 mm to 0.12 mm. The cutting speed depends on work material and ranges from 2 to 10 m/min. Sufficient coolant must be used to perform cooling of workpiece and tool, to reduce friction and flushing of the chips.

Pull Type Internal Broach

Elements of pull type internal broaching from a previously drilled or bored hole consists of the following:

Pull End : Pull end of the broach is to engage with the broaching machine through a puller head.

Front Pilot : This ensures the alignment of the broach in the hole before machining.

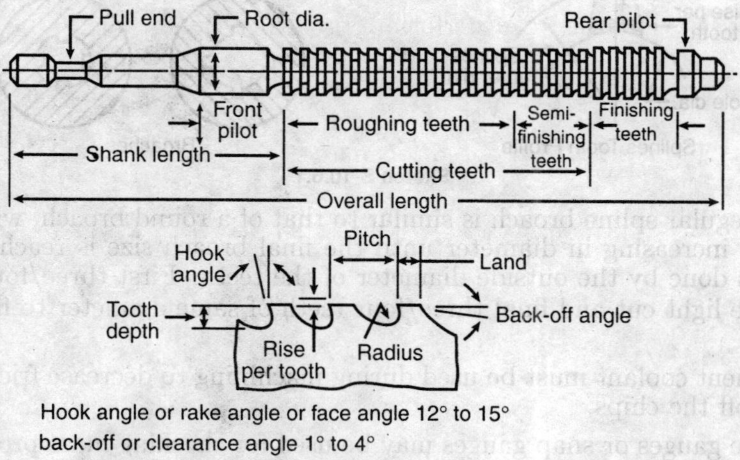

Sketch S-10.5

Roughing and Semi-finishing Teeth

These teeth remove the stock to the required profile of the workpiece.

Finishing Teeth

These teeth are for sizing and maintaining accuracy of the workpiece.

Rear Pilot and Follower Grip

These are supporting the broach even after the last tooth.

Pitch : It is the linear distance from the cutting edge of one tooth to the same point on the next tooth. Pitch may differ for roughing semi-finishing and finishing teeth.

Land : Land is the width of the tooth and may be allowed slight clearance.

Back-off or Clearance Angle : It is the relief angle on the land and may vary from 1 to 4 degrees.

Rake or Hook or Face Angle : It is similar to the rake angle used in Lathe Machine and the angle may vary from 12° to 15°.

A10.6 : Broaching Operations

Broaching is applied for machining a keyway in a hole, an internal/external, or round/contoured surface or a spline hole, etc.

A10.6.1 : Broaching Splines

Straight-sided splines can be broached by the normal keyway broaching procedure, except that the top side will be radiused in relation to the outside diameter by the mating shaft. For broaching the next slot, the horn will be required to permit indexing of the workpiece. The horn should be as many keyways as splines for greater accuracy the key being inserted into each, as the work is indexed.

Sketch S-10.6.1

The regular spline broach is similar to that of a round broach, with teeth gradually increasing in diameter until the final broach size is reached. Machining is done by the outside diameter of the teeth. First three/four round teeth take light cut and final three/four teeth of same diameter to finish the hole.

Sufficient coolant must be used during machining to decrease friction and flushing off the chips.

Profile gauges or snap gauges may be used for checking mass production.

A10.6.2 : Broaching a Keyway

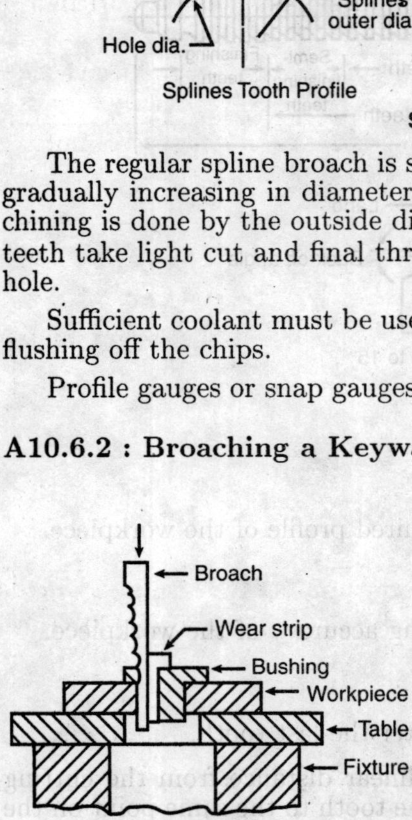

Sketch S-10.6.2

Broaching of internal keyway can be done easily on arbor or hydraulic presses. There is a standard guide bushing with a rectangular slot to support and guide the keyway broach. Bushing is placed in the bore of the work and the front pilot of the broach is pushed in the slot until the first tooth gets in contact with the edges of the work. The required force is applied to push the broach through the work. Several passes are required to machine to the required depth. Plenty of coolant must be used during machining to decrease friction and flushing off the chips.

A10.7 : Distinguish Between Broaching A Key-Way And Milling A Key-Way

Sl.Nos.	Points	Broaching a keyway	Milling a keyway
1.	Cost of the Tool	More	Less
2.	Job Holding	Difficult and more rigidity is essential	Easy and may even be fixed on vice, or on fixture

Contd.

Sl.Nos.	Points	Broaching a keyway	Milling a keyway
3.	Job-Tool Motion	Tool reciprocates and Job gives intermittent linear feed	Tool rotates and the Job gives linear feed
4.	Metal Removal Rate	Less	More
5.	Machining Time	Less, as machined in one pass only	More, as feed rate is slow
6.	Surface Finish	Better surface finish; about ±0.0075 mm	Less; about 0.01 mm
7.	Type of Work	Mainly Productive	Either Productive or Non-productive
8.	Rough, Semi-finishing and Finishing Cut	Done in one stroke only	Done in several number of passes

A10.8 : Machining Time Calculation For The Broaching Process

Machining time in broaching process mainly depends on the length of the broach travel and the cutting speed.

$$\therefore \text{ Machining time, } \quad T_m = \frac{\text{Length of broach travel in metre}}{\text{Cutting speed in metre/min.}}$$

$$T_m = \frac{L}{V} \text{ min.}$$

Cutting speed depends on Job-Tool materials. Broaches are mostly made from High Speed Steel. With this tool material for machining hard materials, cutting speed ranges from 1 to 2 m/min. And for machining soft steel, cast iron, brass, etc., cutting speed ranges from 2 to 10 m/min.

A10.8.1 : Work-out Problems on Machining Time Calculations

Problem: Calculate the machining time to broach a hole on an M.S. job with length of broach = 0.8 m and cutting speed of 2 m/min.

Solution:

$$\text{Machining time, } \quad T_m = \frac{\text{Length of broach travel}}{\text{Cutting speed}} = \frac{L \text{ in metre}}{V \text{ in m/min.}}$$

$$= \frac{0.8}{2} = 0.4 \text{ min. } \textbf{Ans.}$$

[Data Supplied
Length of broach travel, $L = 0.8$ metre

Cutting speed, $V = 2$ m/min.

To be determined

Machining time, T_m in minute]

A10.9 Questions

1. Write a short note on Broaching Machine (A10.1).
2. Classify the Broaching Machines (A10.2).
3. Specify a Broaching Machine (A10.3).
4. Write the advantages of using a Broaching Machine (A10.4.1).
5. Write the limitations of using a Broaching Machine (A10.4.2).
6. State a broach, sketch and show the angles and state its different components (A10.5).
7. Name the operations that may be performed by a Broaching Machine (A10.6).
8. Sketch and explain how you will broaching a Keyway (A10.7).
9. Write the procedure in calculating the machining time of broaching a hole (A10.8).
10. State Rotary type continuous surface broaching (A10.2.1).
11. Compare Pull and Push Broaching (A10.2.2).

11

BORING MACHINES

A11.1 : Introduction

The Boring Machine is one of the most versatile machine tools. This machine is designed for machining of large and heavy workpieces in mass production work of engine frame, cylinder, machine housing, etc. These machines can perform many operations like drilling, boring, counter-boring, spot facing, internal and external thread cutting, peripheral and face milling, facing, turning cylindrical surfaces, gear cutting, etc. Although the above operations can also be done on Lathe, drilling and milling machines but the heavy construction of boring machine is justified for these and is economic in holding and machining the work more accurately.

A11.2 : Types Of Boring Machines

BORING MACHINE
- Horizontal Boring Machine
  - Table Type
    - Smaller Size
    - Medium Size
    - Planer type
  - Floor Type
    - Heavy
    - Universal
- Fine Boring Machine or Precision Boring Machine
- Vertical Boring Machine
  - Standard
  - Vertical Turret Lathe
- Deep Hole Boring Machine
- Vertical Jig-Boring Machine
  - Open Side Construction or Vertical Milling Machine Type
  - Rail or Planer type
- NC (Numerical Control) Boring Machine

A11.3 : Different Boring Machines

A11.3(A) : Horizontal Boring Machines

A11.3(A-1) : Horizontal Boring Machine

Horizontal Boring Machine is the most productive type boring machine. The spindle holding the boring tool rotates in high precision bearings while workpiece mounted on table remains stationary. The work table can be adjusted so that the centre of the bore is in line with the axis of the spindle. The boring tool may also be adjusted and fed through the work by means of the feeding mechanism. Feed motion is imparted either to the work or the tool, depending on the type of machining being done.

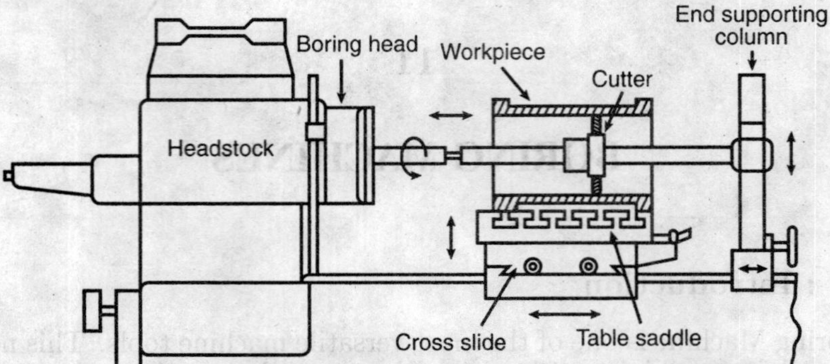

Sketch S-11.3 (A-1)

Horizontal Boring Machines are of three principal types, viz., Table type of smaller size (50 to 125 mm spindle diameter) and medium size (100 to 200 mm spindle diameter), Floor type of heavy (125 to 320 mm spindle diameter), Universal, and a Planer type.

A11.3(A-2) : Smaller and Medium sized Table Type Horizontal Boring Machine

Smaller and Medium sized table type horizontal boring machines differ in their size and spindle diameter, as mentioned above. The machine table travels in one direction only, either longitudinally or cross-wise. The headstock and end-support, both travel along runways on the bed, either in the longitudinal or cross-wise direction.

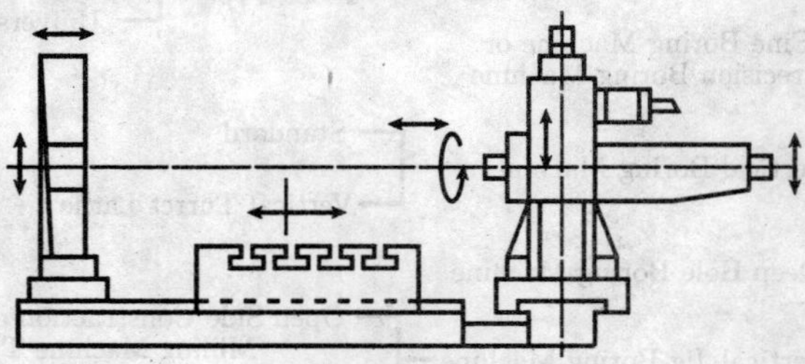

Sketch S-11.3 (A-2)

The planer type horizontal boring machine, in which the table travels cross wise on the bed, has no saddle and the columns travel longitudinally on runways. This type is suitable for supporting long workpieces.

A11.3(A-3) : Floor Type Heavy Horizontal Boring Machine

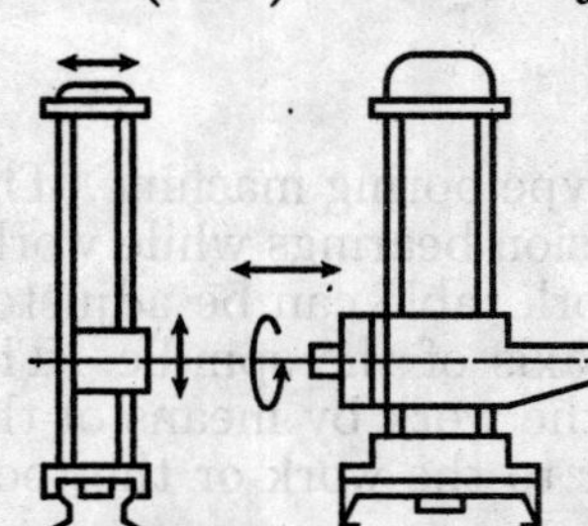

Sketch S-11.3 (A-3)

Floor type heavy horizontal boring machine has no table; the workpiece is mounted on the floor plate.

Floor type universal design horizontal boring machine is one in which the headstock column travels in the longitudinal direction along the ways of the saddle and also in the cross-wise direction with its saddle along a runway. Head-stock column may also travel at a speed equal to working feed. But the end support column travels only in cross-wise direction. This type is suitable for holding large and heavy workpieces.

A11.3(B) : Precision Boring Machine (Fine Boring Machine)

Precision boring machine is constructed for precision boring of cylinders of aircraft and automobile engines, in repair works, certain components of pumps and machine tools, by using diamond tool to within 3 to 5 microns (0.003 to 0.005 mm) accuracy. Precision boring machine may be classified as vertical or horizontal, single or multi-spindle, single-end (holes bored from one end) or double-end machines. Feed motion may be imparted either by the tool or by the workpiece.

A11.3(C) : Vertical Boring Machine

In vertical boring machine, the work rotates on a horizontal table about a vertical axis and the cutting tool is mounted on the adjustable height cross-rail, remains stationary, except for feed motion. This machine is similar to a rotary planer or a vertical Lathe. This machine can hold large heavy parts easily and rigidly for accurate machining, and requires very little floor space. This machine is suitable for holding and machining large, heavy and cumbersome workpieces like flywheels, large gear blanks, locomotive and rolling stocks tyres, large pulleys, vertical housing for pumps and motors, etc.

Standard and vertical Turret Lathes are the types of vertical boring machine.

A11.3(C-1) : Standard Vertical Boring Machine

This machine consists of a heavy cast iron bed which carries a circular table. The machine is provided with two vertical columns; carries a cross-rail holding two tool heads, may be adjusted along cross-rail and in vertical direction. Tool heads may also be swivel to a certain angle for taking an angular cut. Another two tool heads are mounted on the two vertical columns. The size of the workpiece is limited by the size of the table. The workpiece, mounted on the machine table, rotates about its vertical axis and fed against stationary tools. The machine is suitable for large cylindrical and symmetrical workpieces like, turbine castings, machine tools tables, locomotive tyres, pressure vessels, etc.

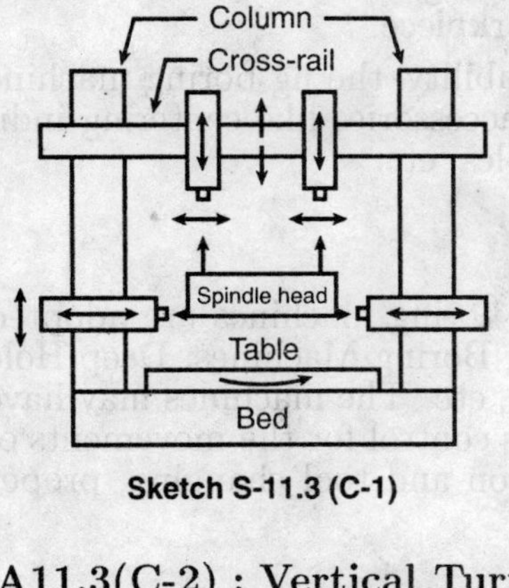

Sketch S-11.3 (C-1)

A11.3(C-2) : Vertical Turret Lathe Boring Machine

The workpiece clamps on rotary table by adjustable jaws and is identical to Lathe chuck; rotates in its vertical axis. It has an arrangement of a multiple tooling turret (indexible) mounted on the cross-rail. The cross-rail can be adjustable vertically and the saddle holding the turret may be moved cross-wise by hand or power. The machine is not only

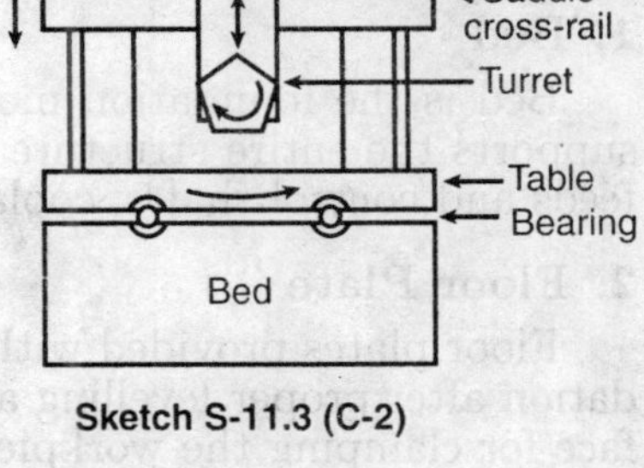

Sketch S-11.3 (C-2)

suitable for boring, but also for turning of gear blanks, piston rings, rail-road wheel, etc.

A11.3(D) : Deep Hole Boring Machine

Deep hole boring machine is specially prepared for production work, for obtaining accurate deep holes in gun boring and where the ratio of depth to diameter increases. This process is similar to deep hole drilling machine but with higher rigidity and accuracy.

A11.3(E) : Vertical Jig Boring Machine

Jig boring machine is suitable in precision tool-room or where highest accuracy is desired (i.e., 2 to 3 microns) for fine finishing of jigs, fixtures, tools, gauges, precision co-ordinate measuring machines, etc. The vertical jig boring machine employs a precision vertical tool spindle, extremely rigid to resist deflection and to minimize vibration. A co-ordinate work table and the tool spindle employs with a high degree of positioning accuracy. Modern jig boring machines are available with numerical control systems.

Open type construction (Vertical Milling Machine type) and Rail or Planer type Boring Machines are the types of Jig Boring Machines.

Open side construction is similar in construction to a vertical milling machine, in which a single column with guideways supports the spindle head in vertical direction. The compound movements of the table help to position the work mounted on the table.

Rail or Planer Boring Machines are constructed with a cross-rail supported on two vertical columns, holding the vertical spindle. The longitudinal movement of the table, cross-movement of the spindle along the cross-rail and the vertical movement of the cross-rail along the guideways of the columns can provide the correct hole location of the workpiece.

To maintain its greater accuracy and workability, the jig boring machine may have the arrangement of fixing different accessories like centering indicator, plain and inclinable precision rotary tables, etc.

A11.3(F) : N.C. Boring Machines

Now-a-days, precision and productive type of boring machines are adopted with numerical controls. The machines are Jig Boring Machines, Deep Hole Boring Machines, Production Boring Machines, etc. The machines may have the arrangement of two, three, even fourth axis control for the movements of spindle and work table, automatic tool selection and tool changing, proper indexing, etc.

A11.4 : Components Of A Boring Machine

1. Bed

Bed is the foundation member of the machine. It is a box like casting, supports the entire structure and consists of columns, table, saddle, various feeds and control shafts, coolant tank, etc.

2. Floor Plate

Floor plates provided with T-slots, are permanently set on a sound foundation after proper levelling and dowelled together to form a continuous surface for clamping the workpiece.

3. Base

Base is an integral part of the bed or may be separated from the bed. It supports and secures the column and also houses the driving mechanisms.

4. Saddle

Saddle should be a very rigid structure to support the table. It can provide a compound movement of the table either longitudinally over the bedways or cross-wise direction to the bedways or both. Movement, are performed by hand power or machine power.

5. Table

The table is provided with T-slots for clamping and supporting the workpiece(s) or for holding various devices. It can travel at right angles to the axis of the spindle.

In vertical boring machine, the table is a circular casting, mounted on the top of the bed. It is provided with T-slots or chuck jaws for clamping and holding the work and has the arrangements of rotational movement.

6. Column

It supports the head stock and also guides it up and down accurately by the guideways. It is a heavily ribbed hollow rigid housing for the counterweights which balance the headstock. It may have the arrangement of sliding along the bed or remain stationary.

7. Head-stock

Head-stock is mounted on the column and may be moved up and down on the column for setting the tool at proper height with respect to the work. The head-stock spindle holding the boring cutter revolving within a quill may provide longitudinal movement for feeding.

8. End Supporting Column

It is used for supporting a long boring bar and is mounted on the other end of the bed. It houses a bearing block adjustable at different heights and side ways of the bed towards or away from the spindle for supporting different lengths of boring bars.

9. Cross-rail

It is essential in planer type horizontal boring machines and vertical boring machines. It is a horizontal member of the rectangular casting mounted on the two vertical members (known as housings). It is used to accommodate one or more head stocks and may be moved up and down for adjusting different heights of the workpiece.

A11.5(A) : Specifications Of Horizontal Boring Machine

Main specification (or the size) of a Horizontal Boring Machine is designated by the following:

1. Type of Machine—Floor type heavy Horizontal Boring Machine.
2. Maximum size of boring spindle diameter in mm (Diameter range may vary from 50 mm to 320 mm).
3. Maximum spindle travel in horizontal and vertical direction in mm.

4. Maximum travel of the table in longitudinal and cross-wise directions in mm.

5. Range and number of spindle speeds in r.p.m.

6. Range and number of spindle feeds in mm/rev.

7. Floor space required for the machine in square metres.

8. Weight of the machine in tonnes.

9. kW of the Machine Motors.

A11.5(B) : Specification Of Vertical Boring Machines

Main specification of a vertical boring machine is the maximum diameter of the work that can be accommodated on the machine table or the diameter of the table or chuck in mm. Diameter of the table may vary from 600 mm to 6000 mm. Other specifications like type of machine (open side or planer type), spindle travel, table movement, speed, feed, weight, floor space and kW required for the motors (similar to Article A11.5 (A)] are to be considered.

A11.6 : Comparison Of Horizontal Boring Machine And Vertical Boring Machine

Points	Horizontal Boring Machine (H.B.M.)	Vertical Boring Machine (V.B.M.)	Adv. of V.B.M. over H.B.M.
1. Principle of operation	Tool is rotated on horizontal axis and work is fed either longitudinally or cross-wise in a horizontal plane	Work is rotated on vertical axis and the tool is fed longitudinally in a vertical plane	—
2. Mainly specified by	Main specification of H.B. Machine is the spindle diameter in mm, generally varies from 50 to 320 mm	Main specification of V.B. Machine is the diameter of the table or chuck in mm, generally varies from 600 mm to 6000 mm	—
3. Design of the Machine	Complicated	Simple and the table being horizontal, table size may be designed as large as possible	Adv.
4. Fixing of work on Machine Table	Difficult to fix the work on smaller table and to feed either longitudinally or cross-wise	Very easy to fix the work on big horizontal table to be rotated only in a vertical plane	Adv.

Contd.

Points	Horizontal Boring Machine (H.B.M.)	Vertical Boring Machine (V.B.M.)	Adv. of V.B.M. over H.B.M.
5. Size of work to be machined	Machining of small and medium sized work may be suitable	Small, medium as well as very large size of work can be machined	Adv.
6. Rigidity of the Machine during operations	Less Rigid	More Rigid	Adv.
7. Tool Rigidity	Less rigid as the tool rotates during operations	Tool is stationary during operations and hence more rigid except for feed movement	Adv.
8. Accuracy	Less accurate work, due to less rigidity of the machine and tool	Very accurate work is possible due to extreme rigidity	Adv.
9. Swivelling arrangement	Work swivelling is possible	Tool Head swivelling is possible	—
10. Application	Heavy and irregular smaller and medium sized work	Large castings, large cylinders, large pulleys, flywheel, etc.	—
11. Operations	Boring, turning, facing, reaming, milling, grooving, etc.	Machining of flat surfaces, external turning, internal boring, shoulder turning, forming, taper boring and taper turning, etc.	—
12. Tool indexing arrangement	No.	Yes, there is a turret type vertical boring machine	Adv.

A11.7 : Boring Tools

Boring Tool is used to finish holes to correct size or on odd size holes in which reaming is unsuitable. Tool in Boring Machine is mounted in either a heavy bar or a boring head. The boring bar or the boring head is connected to the main spindle of the machine. Single-point cutter is used in most of the boring operations. Boring bar is the media to transfer power from the machine to the cutter end and it also maintains the rigidity during the cutting operations. In horizontal boring operations, workpiece generally remains stationary while the rotating cutter is fed through the hole.

Micrometer adjustment tool is used in precision boring or jig boring machine to ensure the true boring of a hole.

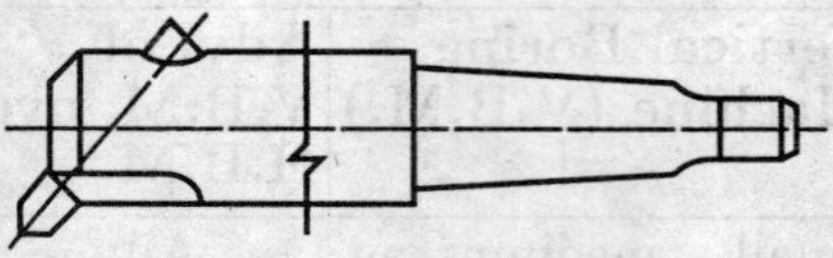

(A) Boring bar with single-point cutter

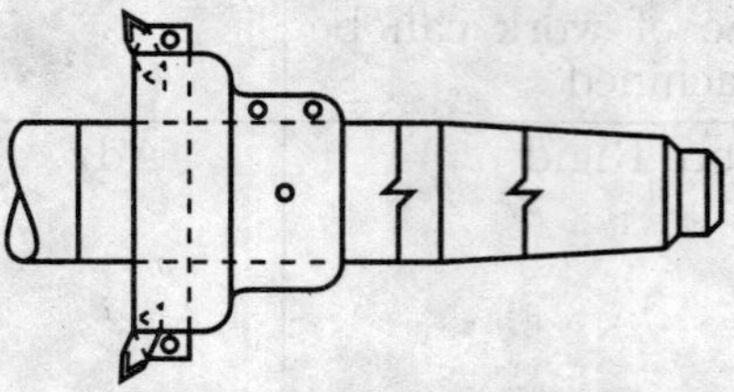

(B) Boring head with double-point cutter

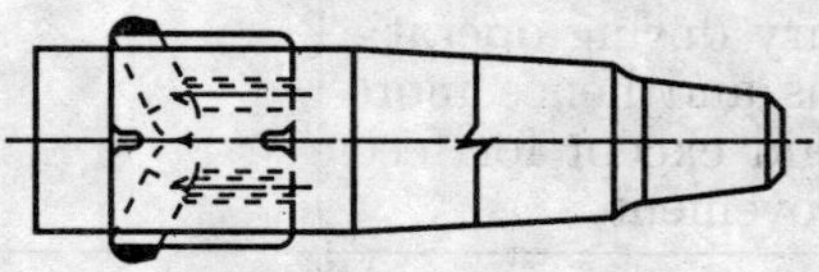

(C) Block type boring cutter

Sketch S-11.7 (A, B & C)

The most useful block type double cutters are placed in grooves on the block and screwed to lock the cutters in position. The entire assembly fits in the boring bar and is keyed in proper place. Multiple cutting edges are widely used in production work. The cutters with inserted (cemented carbide tips) tooth cutters may be adjusted radially to compensate for wear and variation in diameter, which having, longer life and are more economical for production works.

Tool geometry for boring tools is similar to single-point turning tool used in Lathe Machine; although the clearance, rake and cutting angles cannot be used for smaller holes. Greater end clearance and negligible back rake are necessary, owing to the curvature of the whole surface. Suitable side rake and side clearance angles can be used.

A11.8 : Operations In Horizontal Boring Machine

In horizontal boring operations, workpiece generally remains stationary while the rotating cutter is fed through the hole. Several operations like drilling, boring, counter-boring, spot facing, etc., can be done simultaneously on a job with high accuracy. Many operations like drilling, reaming, counter-boring, tapping, spot facing, boring may also be performed by mounting the proper cutter with proper adjustment on a boring bar and the same is fitted to the spindle. The required speed and feed are adjusted according to the job-tool materials, finish desired and available cutting conditions and then operations are performed. Milling operations, like face milling or end milling operation require face milling cutter or end milling cutter are mounted on boring spindle for finishing vertical flat surfaces, grooves, key-slots, etc.

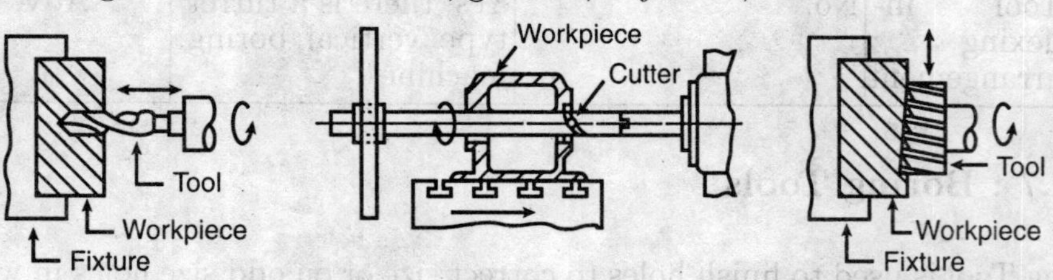

(A) Drilling operation by boring tool | (B) Horizontal boring operation | (C) Machining vertical flat surface

Sketch S-11.8 (A, B & C)

A11.9 : Operations In Vertical Boring Machine

In Vertical Boring Machine, workpiece, mounted on the table, revolves and the ram and cross-rail is locked at the required position while the saddle is fed cross-wise. Depth of cut is given by the ram. Vertical boring machine may produce cylindrical turning, taper turning, boring, taper boring, horizontal flat surface, grooving or shoulder turning, etc.

For producing cylindrical turning, the saddle is clamped to resist any horizontal movement of the ram and down feeding is done by the ram.

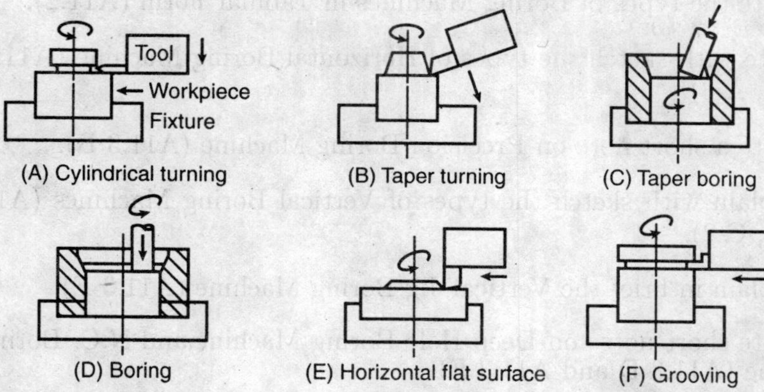

Sketch S-11.09

For producing taper turning and taper boring, tool head is swivelled to the required angle; combined down and cross-feeds are given simultaneously on the tool.

For producing boring, the boring bar or tool head may be fed directly within the rotary workpiece.

In producing horizontal flat surfaces, the cross-rail and the ram are locked at the required position and the saddle is fed cross-wise while the workpiece rotates on the table. Depth of cut may be applied by the ram.

In producing grooving or shoulder turning, forming, cutting-off, necking, etc., operations are performed by the cross-feed movement of the table.

A11.10 : Problems On Machining Time

A casting body having 40 mm hole to a depth of 20 mm is to be brought in alignment by boring in two passes. Assume cutting speed 25 m/min. and feed 0.1 mm/rev. Determine the machining time in boring operation.

[**Data given:**

Length of cut, $L = 20$ mm

Feed, $f = 0.1$ mm/rev.

No. of passes = 2

$$\text{RPM, } N = \frac{1000 \times \text{Cutting speed } (v)}{\pi \times \text{Diam. of hole } (D)} = \frac{1000 \times 25}{\pi \times 40}$$

where $D = 40$ mm $\quad v = 25$ m/min.]

$$\text{Machining time in boring, } M_T = \frac{\text{Length of cut } (L) \times \text{No. of passes}}{\text{RPM}(N) \times \text{feed/rev.}(f)} \text{ in min.}$$

$$= \frac{20 \times 2}{\frac{1000 \times 25}{\pi \times 40} \times 0.1} \text{ min.} = 2 \text{ minutes}$$

Ans. (Approx.).

A11.11 Questions

1. Write a brief note on Boring Machine (A11.1).
2. Write the types of Boring Machines in Tabular Form (A11.2).
3. State with sketch the types of Horizontal Boring Machine (A11.3 A-1, A-2, A-3).
4. Write a short note on Precision Boring Machine (A11.3 B).
5. Explain with sketch the types of Vertical Boring Machines (A11.3 C, C-1, C-2).
6. Explain in brief the Vertical Jig Boring Machine (A11.3 E).
7. Write short notes on Deep Hole Boring Machine,and N.C. Boring Machine (A11.3 D and A11.3 F).
8. State in brief the main components of a Horizontal Boring Machine (A11.4).
9. Compare Horizontal Boring Machine and Vertical Boring Machine (A11.6).
10. Specify Horizontal Boring Machine (A11.5A).
11. Specify Vertical Boring Machine (A11.5 B).
12. Sketch any suitable boring tool and give a statement on Boring Operation (A11.7).
13. Name Horizontal Boring Operations and sketch at least two operations with Job-Tool Motion (A11.8).
14. Name the Vertical Boring Operations and sketch at least four operations with statement (A11.9).

12

PRESS MACHINES

A12.1 : Introduction

Press is a metal forming machine tool used for most cold working operations to shape the sheet metal by applying mechanical or hydraulic pressure. The machine consists of Bed, Frame, Ram, a source of power and a mechanism to move the Ram. In this machine rotary motion power source is transformed into the vertical motion of the Ram. The tools Die and Punch are placed by suitable means on top face of the bed and bottom end of the Ram respectively. Press can perform the operation very quickly, i.e., the time necessary for one stroke of the Ram and the time required to feed the work.

Presses can perform many operations, in which piercing, blanking, shearing, stamping, bending, forming, embossing, coining, etc., are common.

Presses are widely used to mass-production in the field of Automotive, Kitchen appliances, Telephones, Electrical Goods, Sewing Machines, Aircraft, Toys, Radio, etc.

A12.2 : Classification

The Press Machines may be classified under the following:

1. **According to the source of power:**

   (a) Manually operated — Hand Press or Fly Press or Ball Press (For thin sheets and non productive).

   (b) Power operated — Power Presses (Mostly productive).

2. **According to the General design of the frame:**

   (a) Gap Press, (b) Inclinable Press, (c) Adjustable Bed Press, (d) Horn Press, (e) Arch Press, (f) Straight Side Press, (g) Piller Press.

3. **According to the number of actions:**

   (a) Single Action Press, (b) Double Action Press, (c) Triple Action Press.

4. **According to the power transmitting arrangements from source to Ram:**

   (a) Crank Press, (b) Eccentric Press, (c) Rack and Pinion Press, (d) Hydraulic Press, (e) Pneumatic Press, (f) Cam Press, (g) Toggle Press, (h) Knuckle Joint Press, (i) Power Screw Press.

5. **According to the main purpose:**

   (a) Shear Press, (b) Punching Press, (c) Blanking Press, (d) Bending Press, (e) Drawing Press, (f) Embossing Press, (g) Extruding Press, (h) Straightening Press, (i) Coining Press, (j) Brake Press, (k) Seaming Press, (l) Trimming Press, etc.

6. **According to the type of frame:**

   (a) Cast Frame, (b) Welded Frame.

A12.3 : Different Types Of Presses

A12.3.1 : Hand Press Or Fly Press Or Ball Press

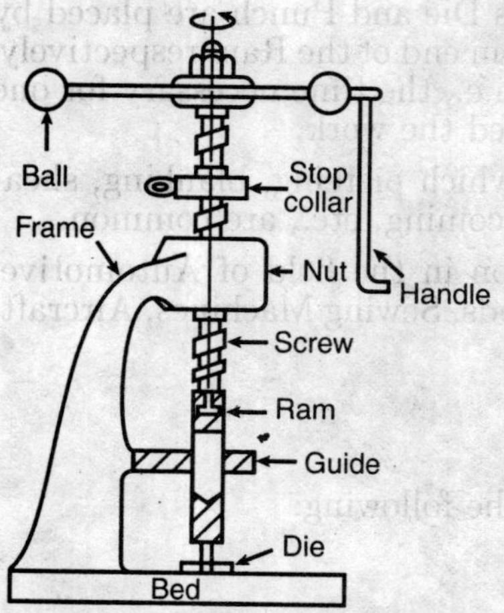

Sketch S-12.3.1

This simplest Fly Press, having balls being operated by hand, is also known as hand press or ball press. The heavy iron balls are rotated by the handle to transform into vertical motion of the Ram by a screw. The kinetic energy of the moving balls is utilized to do the useful work. The stroke length of the Ram can be adjusted and clamped by a collar to the required depth. The Punch is attached to the lower end of the Ram and the die is bolted on the bed of the press. The sheet metal is placed in between the punch and die. The sheet metal may be sheared by moving the Ram, which gets motion from the rotating fly balls.

Kinetic Energy of the two fly balls is equal to the useful work of the Ram (neglecting frictional losses).

The equation becomes,

$$2 \times \frac{1}{2} \frac{W}{g} V^2 = F \times H$$

where, Two fly balls weigh, W in kg,
Acceleration due to gravity, $g = 9.81$ m/sec^2,
Velocity of the rotating balls, V in m/sec,
Force exerted by the Ram, F in kg,
Distance travelled by the Ram, H in metre.

Problem. Calculate the force necessary to be exerted on the Ram to travel a distance of 10 mm. Two fly balls weigh 20 kg each and are moving with a velocity of 10 m/sec. **Ans.** 20.4 tonnes.

A12.3.2 : Power Press

In power press, the Ram is driven by the crank-shaft and connecting rod. This crank-shaft gets power either by mechanical or hydraulic means.

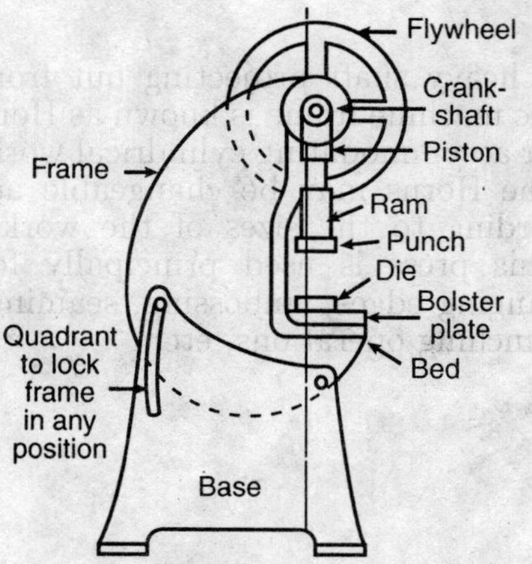

Sketch S-12.3.2

In mechanical press, the rotary motion of the crank-shaft is obtained from an electric motor and is transformed into reciprocating motion of the Ram by using different mechanical arrangements like Rack and Pinion, Eccentric, Crank, Cam, etc. The hydraulic press has a large cylinder and piston, coupled to a hydraulic pump. The piston and Ram are one unit. The cylinder contains fluid under high pressure, drives the piston, and transforms this power for reciprocating movement of the Ram. The working of the power press is at par to that of the hand press.

A12.3.3 : Gap Press

The Gap Press is provided with an opening in the C frame for feeding long and wide sheet metals from front to back, specially suitable in stamping operations. It is more rigid than the inclinable press.

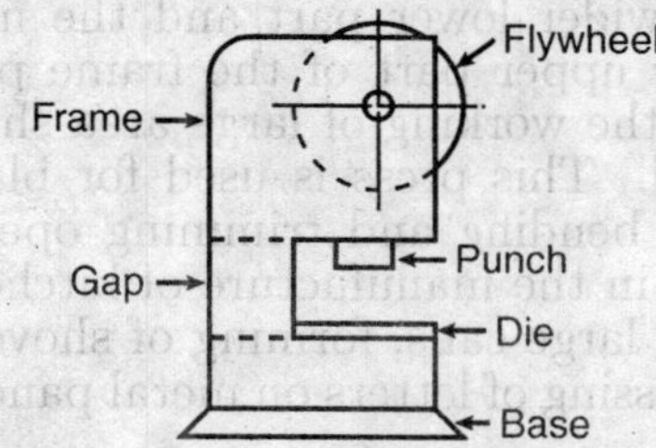

Sketch S-12.3.3

A12.3.4 : Inclinable Press

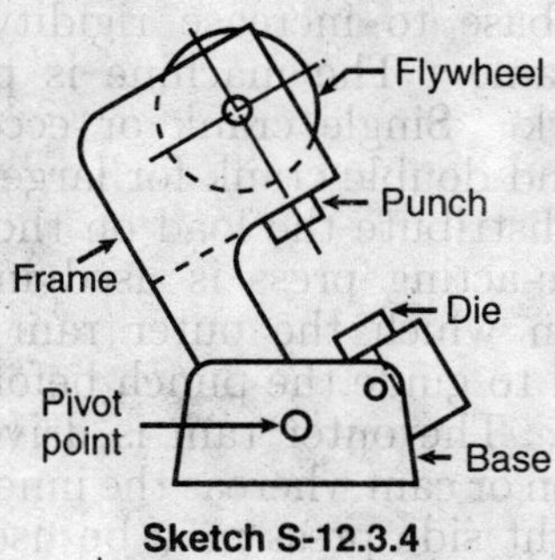

Sketch S-12.3.4

The press frame has the arrangement of tilling back with respect to its base. This helps to discharge the work and scrap from the press by gravity. This press is suitable for the production of small parts, involving blanking, punching, bending and similar operations.

A12.3.5 : Adjustable Bed Press

There is an arrangement, either mechanical or hydraulic, to raise or lower the table with respect to the frame on which the die is fitted. The arrangement helps to accommodate different sizes of work and dies on the table.

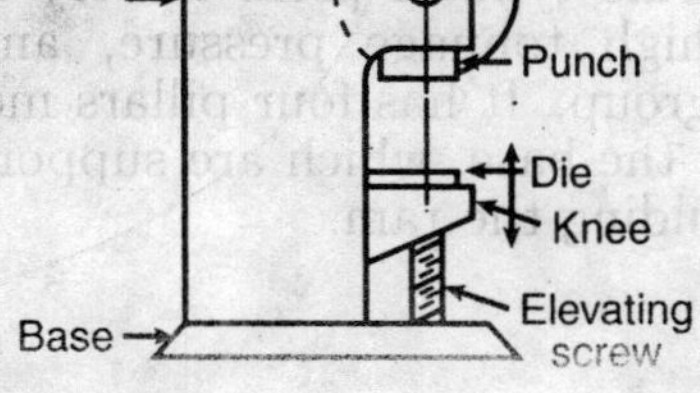

Sketch S-12.3.5

A12.3.6 : Horn Press

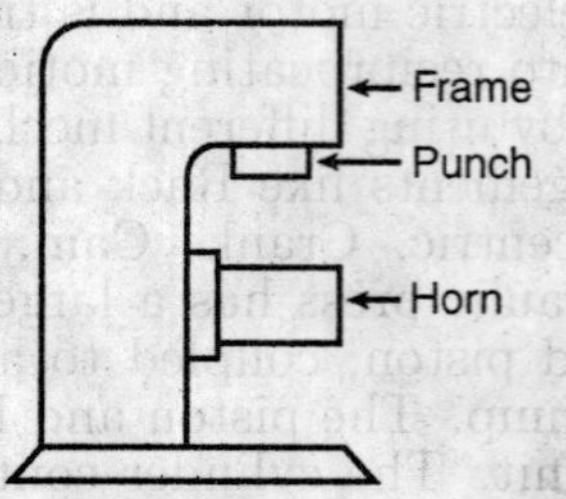

Sketch S-12.3.6

A heavy shaft projecting out from the machine frame is known as Horn for accommodating cylindrical work. The Horns may be changeable according to the sizes of the works. This press is used principally for flanging edges, embossing, seaming, punching operations, etc.

A12.3.7 : Arch Press

The space available for accommodating the work looks like an arch. The wider lower part and the narrower upper part of the frame permits the working of large area sheet metal. This press is used for blanking, bending and trimming operations in the manufacture of kitchenware, large cans, forming of shovels, embossing of letters on metal panels, etc.

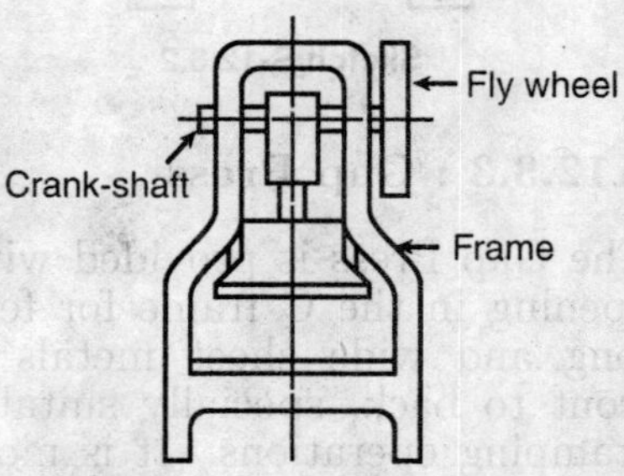

Sketch S-12.3.7

A12.3.8 : Straight Side Press

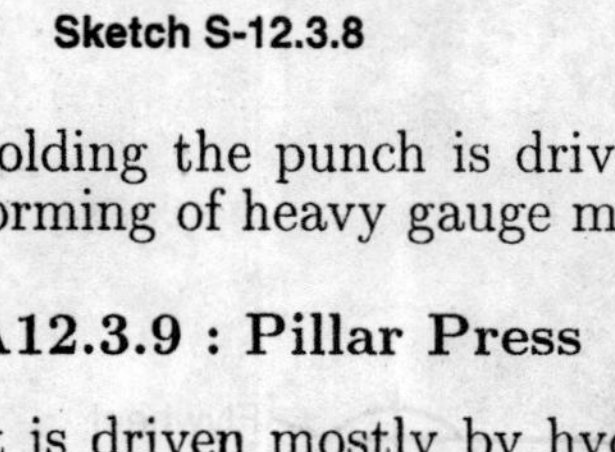

Sketch S-12.3.8

Two vertical rigid side frames are mounted on two sides of the base to increase rigidity and strength of the frame. The machine is prepared for heavy work. Single crank or eccentric for small presses and double crank for large presses are needed to distribute the load on the slide uniformly. Double-acting press is used in drawing operation, on which the outer ram is to hold the blank and to guide the punch before the drawing operation. The outer ram is driven by a special link motion or cam whereas the inner ram holding the punch is driven by crank. Straight side press may be used for forming of heavy gauge material, press forging, deep drawing, coining, etc.

A12.3.9 : Pillar Press

It is driven mostly by hydraulic pressure. This type of press is very rigid with high tonnage pressure, among press group. It has four pillars mounted on the base, which are supporting and guiding the ram.

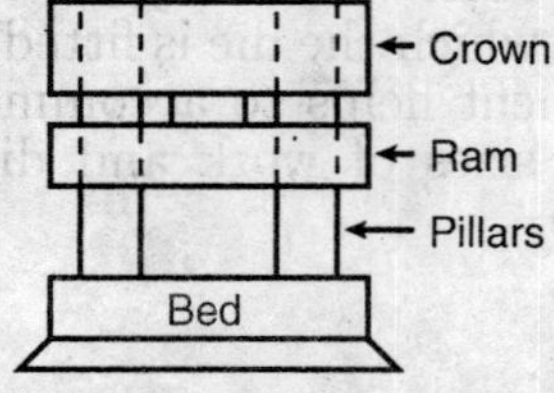

Sketch S-12.3.9

A12.4 : Number Of Actions

Number of actions means the number of rams on the press in function. Single-action press is having only one ram or slide whereas two rams or slides in double-action and three rams or slides in triple-action presses. A double-action press has two slides or rams, an inner slide carries the punch and outer slide carries the blank holder for drawing operation. Triple action press is similar to double-action press with the addition of a third ram, placed in the press pad. The descending of inner and outer slides and ascending of third slide are properly synchronized for triple action for drawing, redrawing and forming operations.

A12.5 : Main Parts Of Power Press

1. Base: Base is the main supporting member and has the arrangement of tilling and clamping the frame in the case of the inclinable press.

Bed: It is a stationary table and usually horizontal part of the press in which a bolster plate is mounted.

2. Bloster Plate: It is a flat plate from 50 to 120 mm thick secured to the base for locating and supporting the die assembly.

3. Ram: It is also named as slide, which reciprocates within the press frame, the lower end of which is fastened the punch or upper die assembly.

4. Flywheel: It absorbs energy from the motor continuously and delivers the stored energy intermittently during pressing the ram into the workpiece.

5. Clutch: It is a device used to start or stop the movement of the ram by connecting or disconnecting the driving shaft with the flywheel respectively.

6. Pitman: It is the connector (or connecting rod) to convey the motion and pressure from crank-shaft or eccentric to the press slide (i.e., Ram). The length of the pitman (or connecting rod) can be varied by changing the position of stroke of the ram.

7. Brake: It helps to stop the movement of the driving shaft just after disconnecting it from the flywheel.

A12.6 : Driving Mechanisms

The mechanism is used to convert the rotary motion of the crank-shaft or driving shaft into the reciprocating movement of the ram. The mechanism may be of any type, i.e., the crank and connecting rod, the cam, the eccentric, the rack and pinion, the hydraulic, the power screw, etc.

A12.6.1 : Crank Press

The most common and simplest drive is the single crank and connecting rod mechanism. In this mechanism, the crank-shaft is being rotated by engaging the clutch with driving flywheel. The flywheel gets power from the motor by belting. The rotary motion of the crank-shaft and crank transforms the reciprocating motion of the ram approaching simple harmonic motion. Most press operations occur near the middle of the stroke where the slide velocity is maximum.

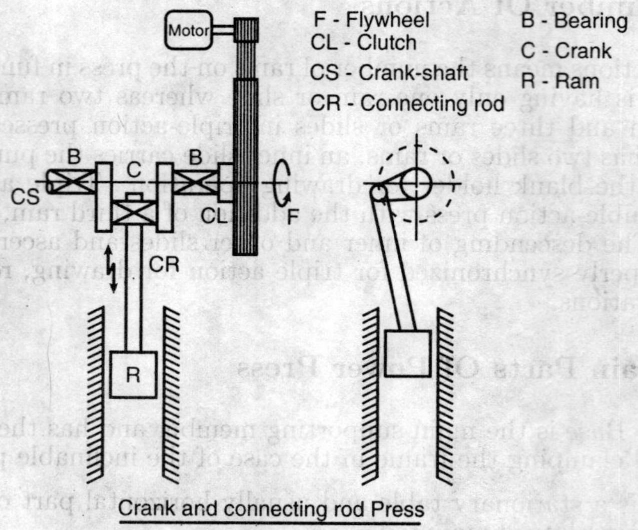

Sketch S-12.6.1

A12.6.2 : Eccentric Press

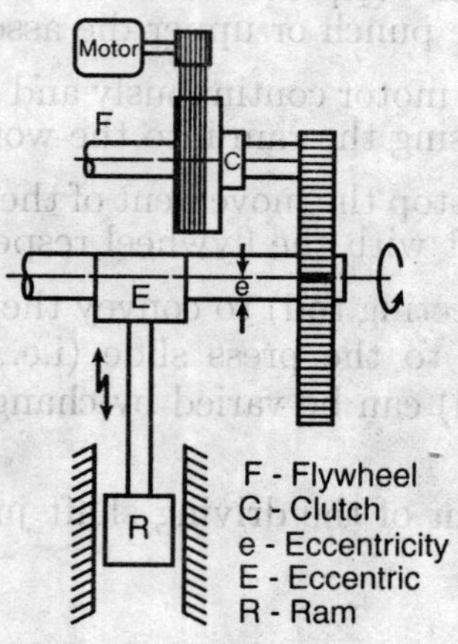

Sketch S-12.6.2

This mechanism is used where shorter stroke length of the Ram is required. The motion is similar to that of a crank drive but having greater rigidity and less tendency for deflection.

A12.6.3 : Rack and Pinion Press

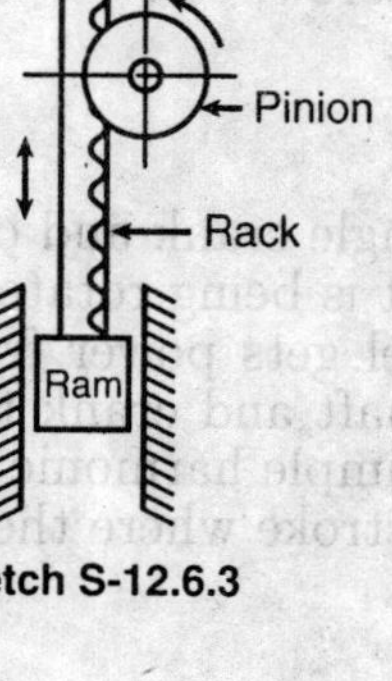

Sketch S-12.6.3

This mechanism is used where a very long stroke length of the Ram is essential. The movement of the slide is much slower and uniform. Arbor Press is an example of this type.

A12.6.4 : Hydraulic Press

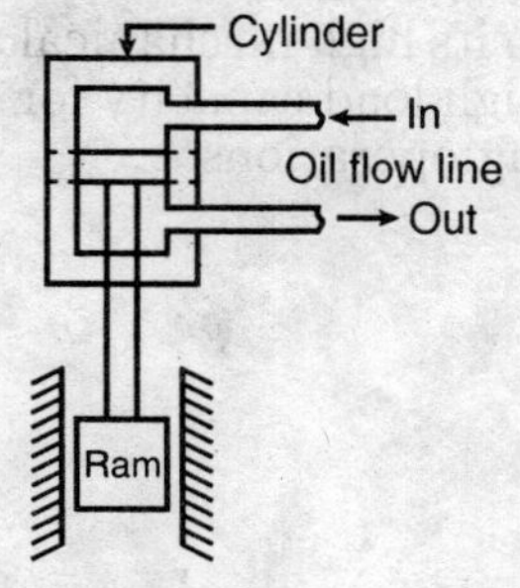

Sketch S-12.6.4

This drive is widely used specially where high pressure and slow speed are essential for forming, drawing and pressing operations.

A12.6.5 : Cam Press

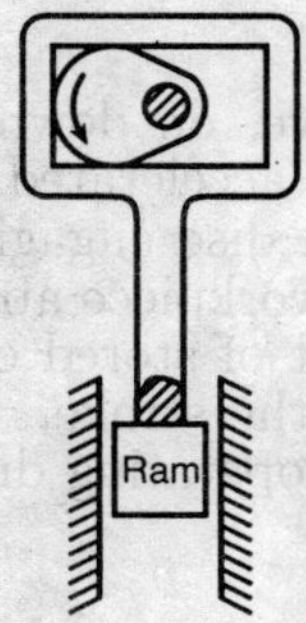

Sketch S-12.6.5

This mechanism is used where some special movement of the Ram is desired, in which the Ram will remain idle at the bottom of the stroke for some time.

A12.6.6 : Toggle Drive

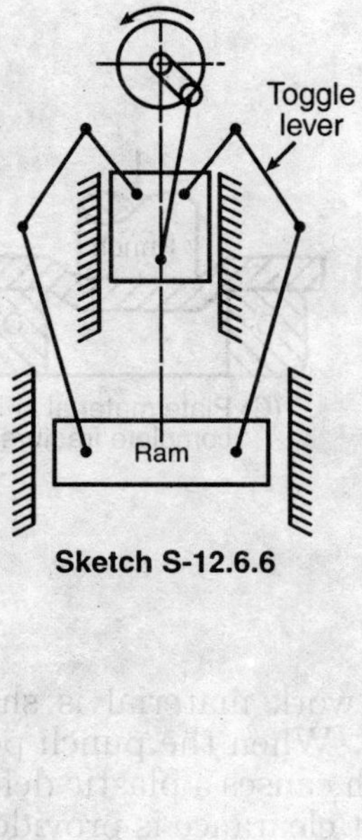

Sketch S-12.6.6

This drive is most suitable in double action press, in which outer slide will hold the blank and inner slide will slide the punch on a drawing operation.

A12.6.7 : Knuckle Press

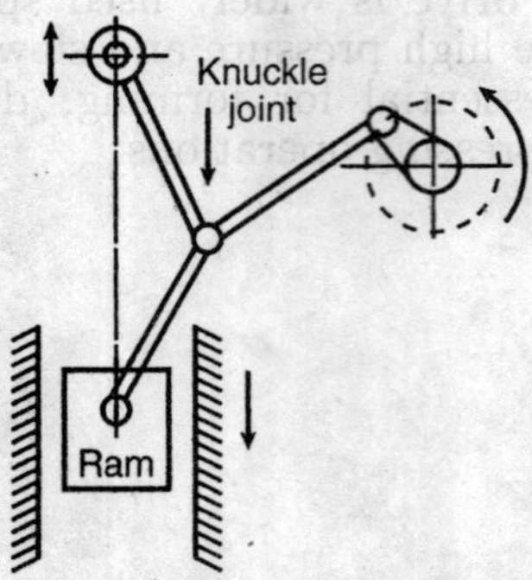

Sketch S-12.6.7

This mechanism is very much used due to its high mechanical advantage and high load capacity for sizing and coining operations.

A12.6.8 : Power Screw Press

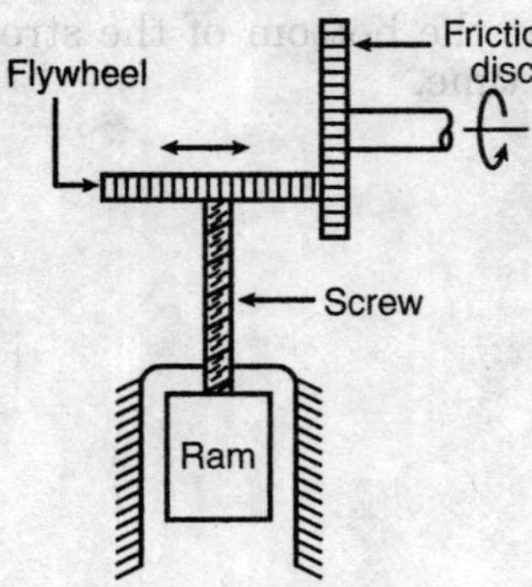

Sketch S-12.6.8

In this drive, on downward stroke, the Ram is accelerated by means of the friction disc engaging the flywheel. The workpiece absorbs the entire amount of stored energy at the bottom of the stroke. This drive is suitable in operating drop hammer.

A12.7 : Press Operations

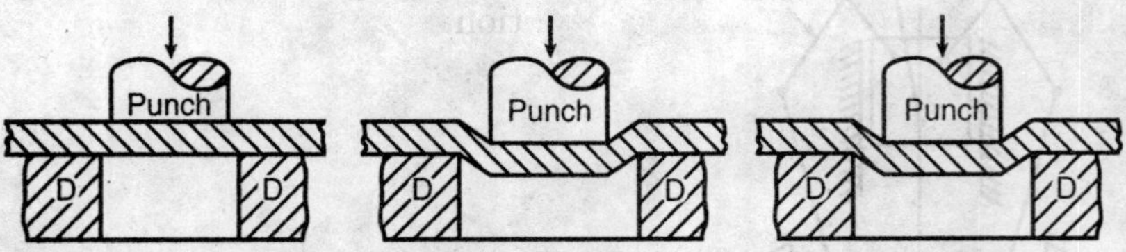

(A) Plate in contact with die and punch (B) Plate material in plastic state (C) Plate material in complete fracture

D = Die

Sketch S-12.7(1)

1. Shearing Operation

Shearing operation is similar to scissors, the work material is sheared off by the action of two blades, i.e., punch and die. When the punch penetrates into the work material, the pressure of the punch causes a plastic deformation and then causes complete fracture, if the correct clearance is provided. Shear force is reduced by providing proper shear angle.

2. Blanking Operation

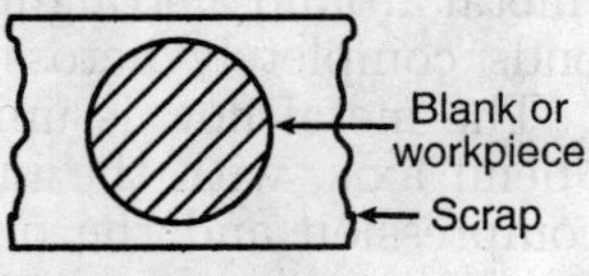

Sketch S-12.7(2)

Blanking is the operation of cutting out to some desired shape from a metal strip or plate. The cutting-out portion is known as blank and the portion left in the strip is known as scrap.

3. Piercing Operation

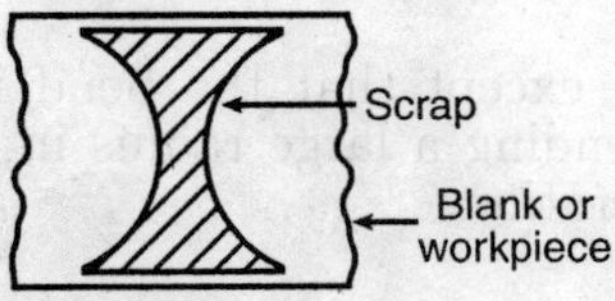

Sketch S-12.7(3)

Piercing (or Punching) is the operation of making a hole of any desired shape in the blank. The cut piece is the scrap and the blank is the workpiece.

A12.7C : Compare Blanking and Piercing

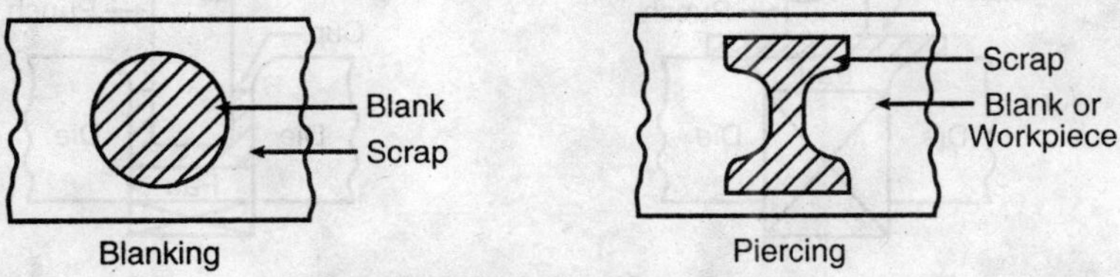

Sketch S-12.7C: Blanking and Piercing

Sl. No.	Points	Blanking	Piercing
1.	Sketch	Sketch S12.7C Cutting-out portion is the blank and the remaining portion is the scrap.	Sketch S12.7C Piercing is the operation of making any cut-out shape on the workpiece and the cut-out portion is the scrap.
2.	Definition	The metal punched out to the required shape from the workpiece is the blanking operation or the product and the remain portion with the hole is the wastage.	The metal punched out to the required shape on the workpiece is the wastage and the remain portion with the hole of required shape is the product.
3.	Clearance provided on	Clearance provided on punch, Diameter of the punch $= D - t_c$, $D =$ Diameter of blank, $t =$ thickness of the plate, 10% clearance on harder material, 6 to 12% on work material composition. where, $t_c = \frac{10}{100} \times t = 0.1t$ for 10% clearance	Clearance provided on Die, Diameter of Die $= D + t_c$, $D =$ Diameter of Punch, $t =$ thickness of the plate, 6% clearance on softer material, 2 to 20% on work material composition. where, $t_c = 0.06t$ for 6% clearance

4. Bending Operation

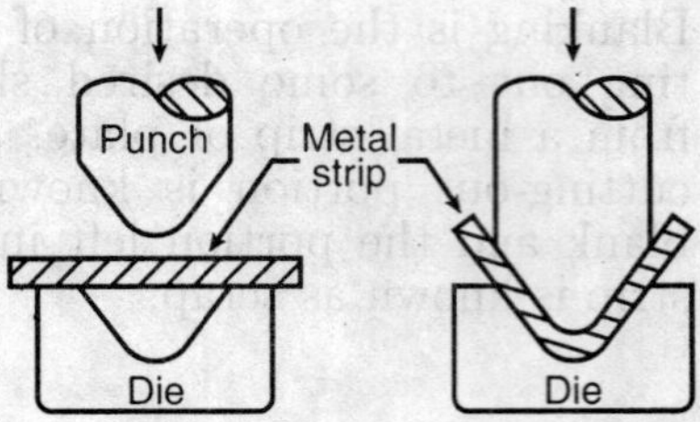

Sketch S-12.7(4)

Bending is the operation of shaping the metal around a straight axis which extends completely across the material. The metal flow is uniform along the bend axis, with the inner surface in compression and the outer surface in tension within the elastic strength and ultimate strength of the material. Other bending operations are curling, seaming, folding, etc.

5. Forming Operation

Forming operation is similar to bending, except that the bend is along a curved axis instead of a straight axis. Bending a large radius in a straight line may also be referred as a forming operation.

6. Drawing Operation

Drawing operation is the process of forming cup or shell from a circular piece of flat blank. The blank material is subjected to severe plastic deformation.

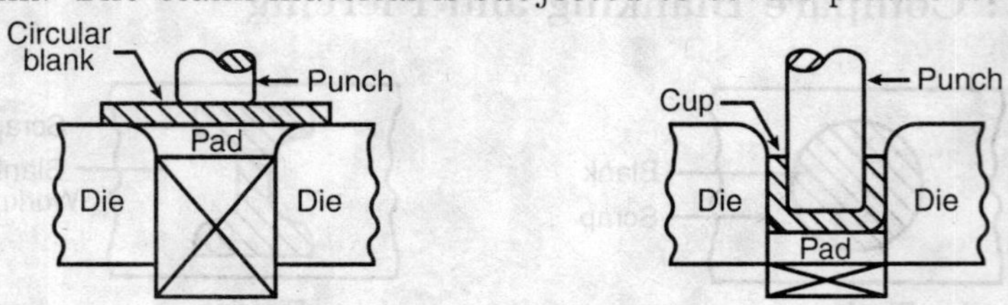

Sketch S-12.7(6)

7. Embossing Operation

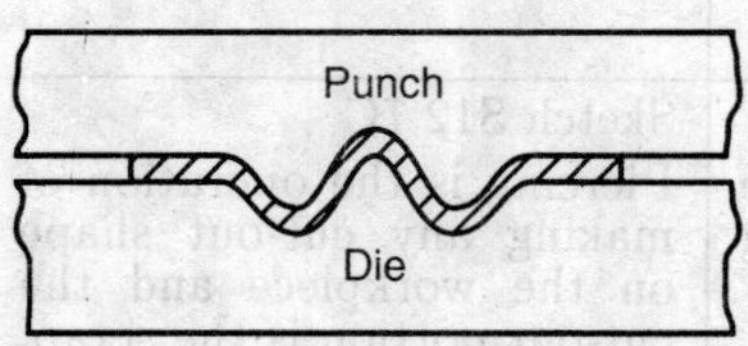

Sketch S-12.7(7)

It is the operation to impart a raised or depressed fancy design on the surface of the blank material. Circular grooves on sheet metal container, metal military buttons, with insignia (identifying mark) raised on the surface, are the examples of embossing.

8. Coining Operation

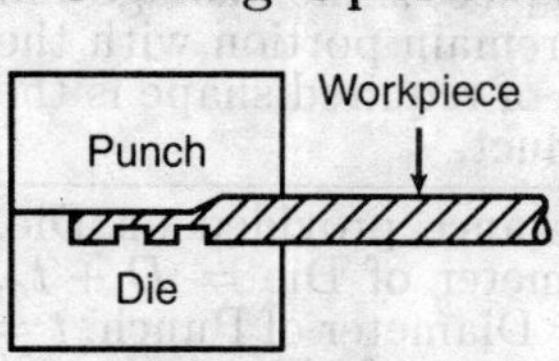

Sketch S-12.7(8)

Coining is the squeezing operation of forming of shapes on the thick metal blank for producing coins, medals, etc. This operation requires severe compressive loads to cause metal flow in cold state.

9. Trimming Operation

Trimming is the operation of removing distorted excess metal from drawn-shaped blank and provides a smooth edge.

10. Notching Operation

Sketch S-12.7(10)

It is the operation of removal of metals of any required shape from the edges of a plate. The size of notching is much smaller than the blanking but the operation is similar to piercing.

11. Latching Operation

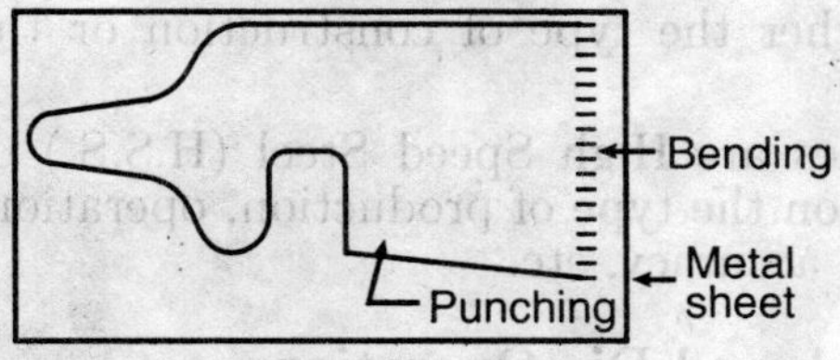

Sketch S-12.7(11)

Latching is the combined operation of bending and cutting along a line in the work material. If a desired shape is partially punched and then bending down the non-cut side, the combined action is called the latching.

Example: Mosquito Coil Stand.

A12.8 : Preparation Of Press Frames

Cast Frames : This type is quite stable, rigid, easy to construct but expensive.

Welded Frames : Heavy steel plates are welded to construct a frame. This type is less expensive, more resistant to shock loading and greater toughness of steel plates.

A12.9 : Specifications Of A Press Machine

1. Maximum pressure or force applied on blank in tonnes :

(a) This force for Mechanical presses may be range from 5 to 4,000 tonnes.

(b) And for Hydraulic presses may be extended to 50,000 tonnes.

2. Available bed area or die space in mm^2.

3. Maximum Ram travel (i.e., stroke length of the Ram) in mm. It is constant on a mechanical press but adjustable on hydraulic press.

4. Maximum opening between the Ram and Bed in mm.

5. Range and number of Ram speeds in m/min.

6. Maximum Ram adjustment into the Pitman in mm.

7. kW of the driving motor.

8. Floor space required by the press in sq. m.

9. Weight of the machine in tonnes.

A12.10 : Press Tools

Die and Punch are known as Press Tools.

Die is the lower part of the press tools which is clamped on the bolster plate and remains stationary during the operation. The Die has a cavity (either with clearance or without clearance) to receive the Punch.

Punch is upper part of the press tools, which is attached on the lower end of the Ram, slides with the ram during the operation and is forced into the die cavity.

Die and Punch must be in perfect alignment for proper operation.

Dies are classified according to either the type of construction or the operation to be performed.

Materials used for punches and dies are High Speed Steel (H.S.S.) or stellite or cemented carbide depending on the type of production, operation, sheet metal composition and thickness, accuracy, etc.

A12.10.1 : Cutting Action in Punch and Die Operation

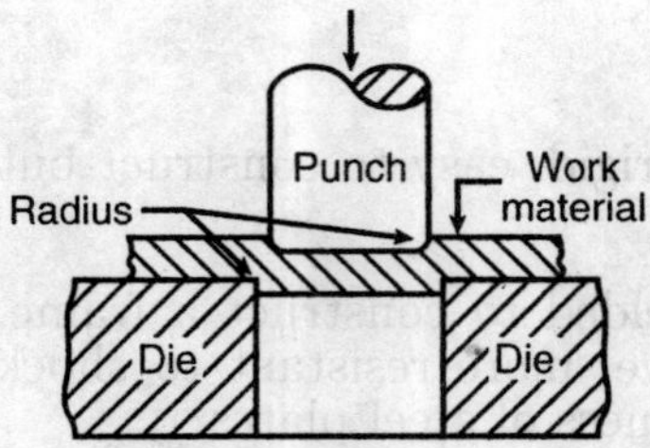

Sketch S-12.10.1

When the punch moves vertically downward, it presses the work material supported by the die and the work material begins to flow plastically. The punch penetrates into the work material and the blank is displaced into the die opening. A radius is formed on the top edge of the hole and the bottom edge of the blank.

Penetration is generally expressed as a percent of work material thickness and depends on properties of work material. Harder the work material, lesser will be the penetration.

A12.10.2 : Die Clearance

Clearance is the space between the punch-cutting edge and die-cutting edge. It is essential to make fracture. It depends on the work material thickness, properties and type of operations.

Clearance is expressed as the amount of clearance per side.

Insufficient clearance increases pressure on Die and Punch, and also reduces the Die life.

In blanking operation, the diameter of the Die will be same as the diameter of the Blank and the clearance per side is provided on the punch (about 6% to 12%) on work material thickness and the composition of the work material. Harder the work material, greater will be the clearance.

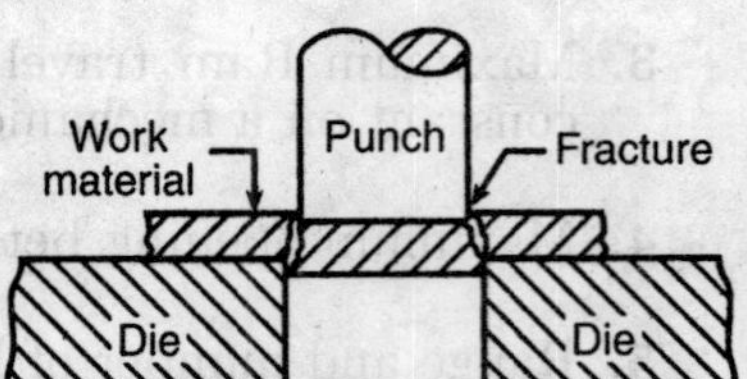

Sketch S-12.10.2

Example 1.

A 100 mm in diameter blank is to be prepared from an M.S. of 4 mm thick plate. Find the diameters of the Die and Punch.

Solution: Diameter of the Die = 100 mm. **Ans.**

Diameter of the Punch $= 100 - \frac{10}{100} \times 4 = 99.6$ mm. **Ans.**

[Considering 10% clearance on harder work material thickness]

In piercing operation, the punch will be the same diameter as the diameter of the hole and clearance is provided on Die.

Example 2.

A 50 mm hole is to pierce from a brass plate 6 mm thick. Calculate the diameters of the Die and Punch.

Solution: Diameter of the Punch 50 mm. **Ans.**

Diameter of the Die $= 50 + \frac{6}{100} \times 6 = 50.36$ mm. **Ans.**

[Considering 6% clearance on softer work material]

[*Note:* There is no hard and fast rule for clearance but it may be assumed 2 to 20% on work material thickness/side, depending on the composition of the work material and according to the design consideration.]

A12.10.3 : Angular Clearance

A small amount of draft (about $\frac{1}{4}^\circ$ to 2° per side) is provided below the die opening to eject the blanked part easily from the die cavity. This draft is known as angular clearance and is necessary to prevent back pressure caused by the blank. It depends on the work material and shape of the workpiece.

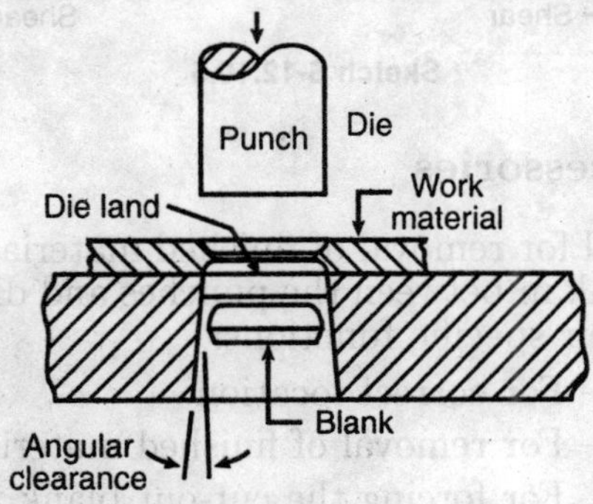

Sketch S-12.10.3

A12.10.4 : Cutting Force

The force required by the Punch to penetrate into the work material is known as cutting force. It is the product of the shearing stress of the work material and the shearing area of the workpiece.

Force, $F = f_s \times P \times t$ where, F is the cutting force in kg,

f_s is the Shearing Stress of the work material in kg/cm^2,

P is the perimeter or length of the cutting edge in cm,

t is the thickness of the work material in cm.

Example: A 100 mm square hole is to be pierced in 0.25% carbon steel plate, 2 mm thick.

Solution: Force, $F = f_s \times P \times t = 400 \times 40 \times 0.2 = 3,200$ kg. **Ans.**

[Assuming shearing stress $f_s = 400$ kg/cm^2, Perimeter, $P = 100 + 100 + 100 + 100 = 400$ mm $= 40$ cm, Metal thickness, $t = 2$ mm $= 0.2$ cm.]

A12.10.5 : Shear Angle

A small shear angle with the horizontal, provided either on the face of the punch or die, will reduce cutting forces by 30% against full shear. Full shear is equal to or greater than the work material thickness. In blanking operation, the shear angle will be provided on the die member, while in piercing operation, it will be on the punch member.

Approximate shear angle = Plate thickness +0.4 mm.

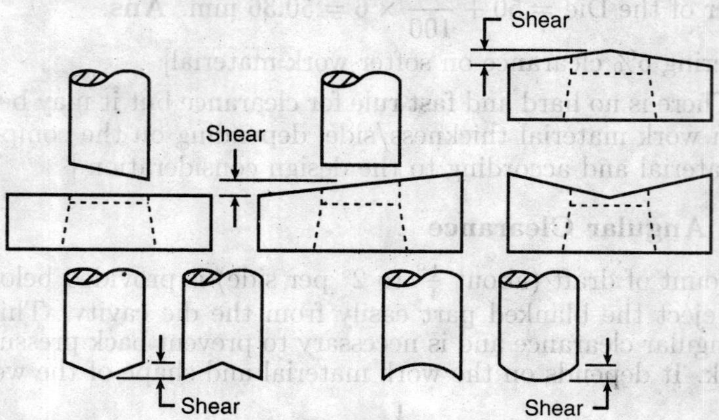

Sketch S-12.10.5

A12.11 : Press Accessories

Accessories are essential for removal of finished material, and enable the correct location of the blank in between the punches and dies. The following are the accessories with their specific functions:

1. Pilot – For correct location.
2. Stripper – For removal of finished material.
3. Knock-out – For forcing the cut-out blank from the Die.
4. Pressure Pad – For preventing the blank from bowing away from the flat face of the punch.

A12.11.1 : Pilot

The function of pilot is to locate the blank accurately with respect to the previously pierced hole for succeeding blanking and other operations. The pilots are fitted to the lower end of the punch concentrically. The pilots should fit the pierced hole with a close sliding fit.

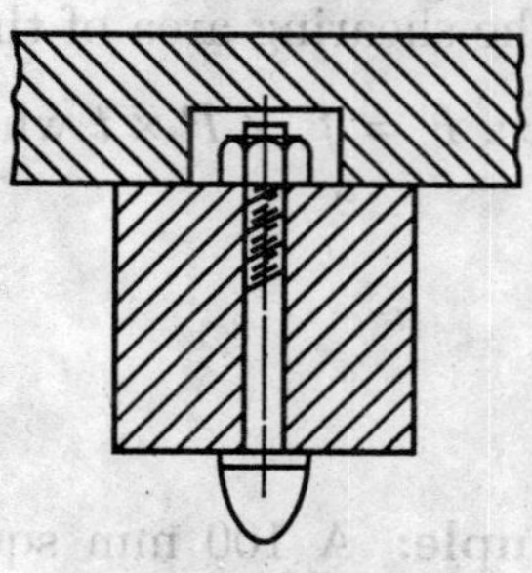

Sketch S-12.11.1

A 12.11.2 : Stripper

The function of a stripper is to discard the workpiece from the punch or the die after a blanking or piercing operation. Strippers may be classified as fixed or spring operated. Fixed strippers are generally rigidly attached to the die block or die shoe, while the spring operated strippers travel up and down on the shank of a punch; spring operated strippers are also named as pressure pad strippers, in which springs are to apply pressure to the sheet material. This type tends to hold the strip flat during the press cycle.

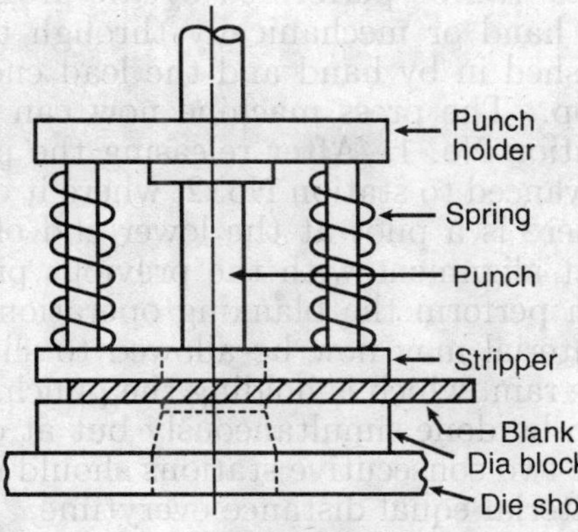

Sketch S-12.11.2

A12.11.3 : Knock-out

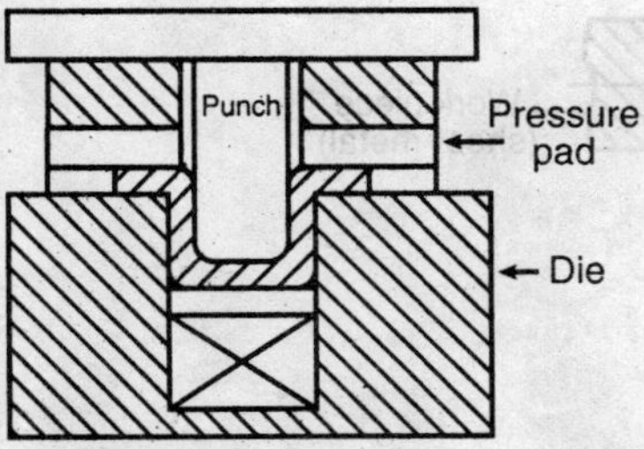

Sketch S-12.11.3

Knock-out is similar to stripper, used to force the cut-out blank from the die. Knock-out is to kick out the flanged shell after drawing operation, which cannot pass through die. Knock-out may be spring loaded or actuated by an air cylinder or cushion.

The punch is mounted on the lower shoe and knock-out is fitted on an inverted die. When the die holder descends, the sheet metal is blanked and the blank rests on the compressed spring loaded knock-out. After releasing the die holder during upward stroke, the spring loaded knock-out forces or ejects the cut-out blank from the die.

A12.11.4 : Pressure Pad

Pressure Pad is used for preventing the blank from bowing away from the flat face of the punch. During the downward movement of the punch, the blank is drawn out on the die face with sufficient pressure by the pressure pad, which results in ironing of the blank and eliminates the unevenness in the process of being drawn out to the shape required. The pressure blank serves the same function as the pressure pad by maintaining the flat flanged surface of the cup.

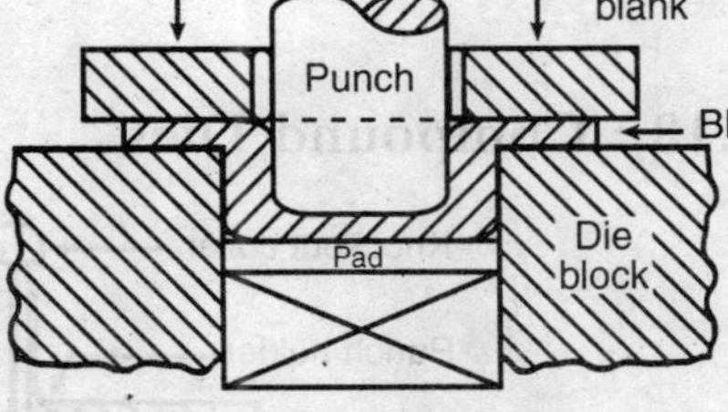

Sketch S-12.11.4

A12.12 : Progressive Dies

Progressive Dies can perform two or more press operations sequentially every time the ram descends. The workpiece is advanced through a series of stations that can perform one or more distinct die operation(s).

In the manufacture of washers, involving Piercing and Blanking operations can be performed by the progressive dies. The workpiece may be fed by hand or mechanically through the channel. The primary stop may be pushed in by hand and the lead end can then be fed into contact with the stop. The press machine now can operate to produce the pierced hole at station No. 1. After releasing the primary stop, the work material may be advanced to station No. 2, where it can contact the automatic lower die stop. There is a pilot at the lower end of the blanking punch to ensure the correct alignment with the previous pierced hole, the second hit of press now can perform the blanking operation. After releasing the die stop the work material may now be allowed to slide forward during the upward stroke of the ram, which is holding the punch. Both Piercing and Blanking operations can be done simultaneously but at different stations. The distance between the two consecutive stations should remain the same for forwarding the work material equal distance everytime.

The finished blank with a pierced hole at the end of each stroke of the press is the washer.

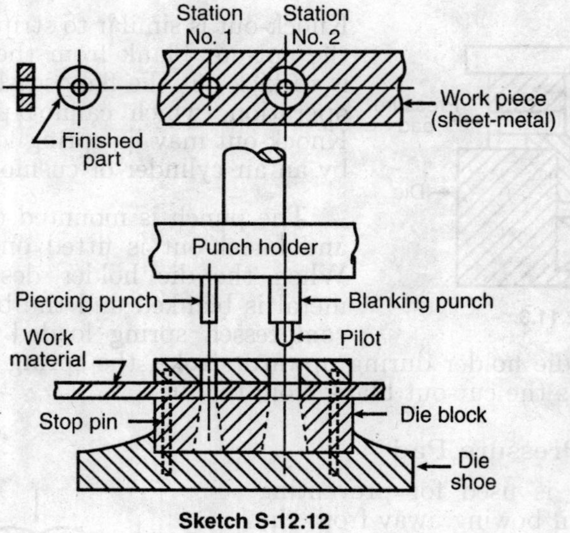

Sketch S-12.12

A12.13 : Compound Dies

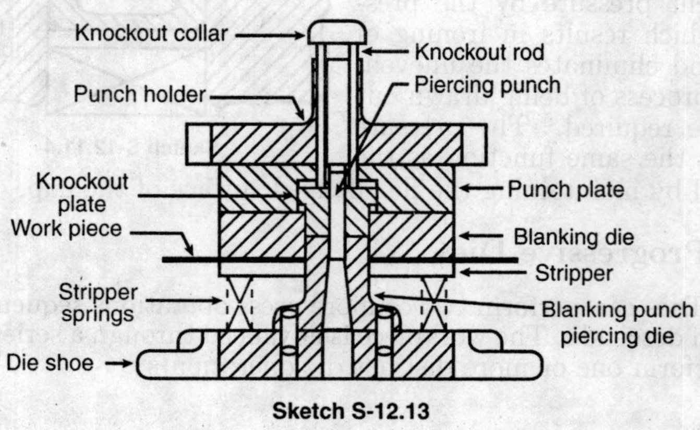

Sketch S-12.13

Compound Die can perform two or more operations in one stroke of the press at one station only. Work material may be progressively forwarded through the die, two holes for piercing and blanking may be punched each time the press slide descends. Blanking punch may act as piercing die and vise-versa. The straight cutting edges blanking die opening helps the blank not to pass through the die. The blank may be ejected by the knock-out mechanism during return stroke of the ram. Angular clearance in the piercing die helps to allow the waste material (i.e., pierced piece) to drop through the die. Knock-out plate is used to support and guide the piercing punch.

Advantages of Compound Dies over the Progressive Dies:

1. Scrap blank material can be hand fed.
2. Large parts can be blanked in a comparatively smaller press.
3. Shearing ensures the flatness of the blank.
4. Close tolerance in the pierced hole is possible.

Disadvantage:

1. Slower in operation.
2. More cost of labour to save the work material.

A12.13.1 : Comparison between Progressive Dies and Compound Dies

Sl. Nos.	Points	Progressive Dies	Compound Dies
1.	Speed of Operation	Faster	Slower
2.	Tolerances	Open	Close
3.	Size of the work suitable for operation	Longer Strips of Materials	Shorter length but large parts
4.	Saving in materials	Less (More Scraps)	More (Less Scraps)
5.	Labour cost	Less	More

A12.14 : Drawing Cup Or Shell

Blank Diameter for manufacturing the Cup by drawing operation,

$$D = \sqrt{d^2 + 4d\,h}$$

where, d is the diameter of the Cup
h is the depth of the Cup
t is the thickness of the Cup

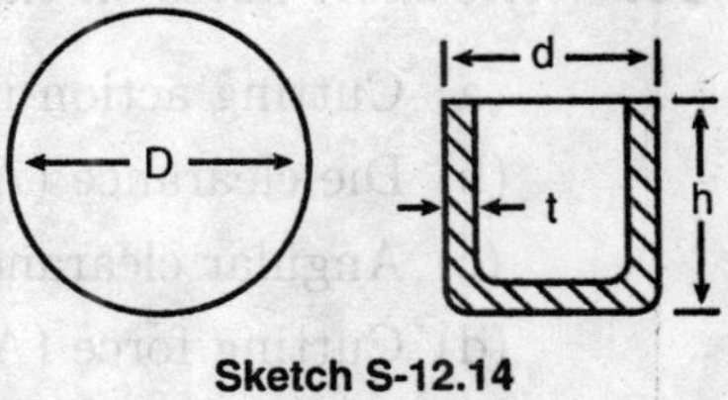

Sketch S-12.14

Example: Determine the blank diameter of a Cup of diameter 100 mm to a depth of 75 mm.

Solution: Blank diameter, $D = \sqrt{d^2 + 4dh}$

where, diam. of the Cup, $d = 100$ mm and depth of the Cup, $h = 75$ mm

$$D = \sqrt{(100)^2 + 4 \times 100 \times 75}$$

$$= 200 \text{ mm. } \textbf{Ans.}$$

A12.15 Questions

1. State the Press Machine with respect to its principle of operation, applications and operations (A12.1).
2. Classify a Press Machine (A12.2).
3. Sketch, state and explain the working of Hand Press or Fly Press or Ball Press (A12.3.1).
4. Sketch and explain the working of a Power Press (A12.3.2).
5. Sketch and state the Gap Press, Inclinable Press and Adjustable Bed Press (A12.3.3, A12.3.4 & A12.3.5).
6. Explain in brief an Arch Press, Straight Side Press and Pillar Press with necessary sketches (A12.3.6-9).
7. What do you mean by Number of Actions (A12.5)?
8. Write the functions of main parts of Power Press (A12.5).
9. What do you mean by the principles of driving mechanism (A12.6)?
10. Sketch and explain the working of a Crank Press driving mechanism (A12.6.1).
11. Sketch and explain any two types of Press driving mechanism (Any two among A12.6.2 to Al2.6.8).
12. State at least four important Press operations with necessary sketches (A12.7-[1 to 4]).
13. Specify a Press Machine (A12.9).
14. Write short notes on Press Tools (A12.10).
15. Write short notes on the following:
    (a) Cutting action in Punch and Die operations (A12.10.1)
    (b) Die clearance (A12.10.2)
    (c) Angular clearance (A12.10.3)
    (d) Cutting force (A12.10.4)
    (e) Shear angle (A12.10.5)
16. Write the functions of Press Accessories (A12.11).
17. Explain with necessary sketches the following Press Accessories:
    (a) Pilot (A12.11.1), (b) Stripper (A12.11.2), (c) Knock-out (A12.11.3), (d) Pressure Pad (A12.11.4).
18. Sketch and explain the Progressive Dies (A12.12).
19. Sketch and state the Compound Dies (A12.13).

20. Compare Progressive Dies and Compound Dies (A12.13.1).

21. State the necessary calculation to determine the blank diameter for manufacturing the cup or shell by drawing operation (A12.14).

22. Write a short note on "Preparation of Press Frames" (A12.8).

23. Compare Blanking and Piercing operations (A12.7C).

13

JIGS AND FIXTURES

A13.1 : Introduction

Jigs and Fixtures are the devices required to hold and to locate the workpieces and to guide the tools. These devices are only economical in mass production with easy loading, unloading, holding and operational facilities and ensured interchangeability of machined workpieces for quick assembly.

A13.2 : Definition Of Jigs And Fixtures

Jigs and Fixtures are the devices required for holding and locating the workpieces and guiding the tools. These devices are only economical in mass and rapid production and ensure interchangeability of finished products for achieving quick assembly and also reduce the time for loading, unloading and setting the workpieces.

Jig

A Jig is a device necessary for holding and locating the small workpieces rigidly on machine table and also for locating or guiding the tools during drilling or drilling type of operations like reaming, tapping, etc.

Fixture

A Fixture is a device necessary for holding and locating the medium and heavier sized workpieces rigidly on machine table. The device will help for easy setting a workpiece relative to the tool but does not guide the tool. Clamping, supporting and locating the workpiece is used during inspection of the workpieces or in manufacturing operations like milling, planing, grinding, turning, etc.

A13.3 : Distinction Between Jigs And Fixtures

Points	Jigs	Fixtures
Purpose	It holds and locates the workpiece and also guides the tool	It holds and positions the workpiece but does not guide the tool
Construction and clamping	Jigs are smaller in size and lighter in construction, for rapid handling the workpiece and clamping on machine is often unnecessary	Fixtures are usually massive and heavier in construction and clamping rigidly on machine table or machine bed or on cross slide is essential

Contd...

Points	Jigs	Fixtures
Suitable for operations	It is suitable for holding the workpiece and guiding tool in machining operations like drilling, reaming, tapping or similar type of operations	It is suitable for holding and locating the workpiece in machining operations like milling, turning, boring, grinding, planing, etc.
Skillness during operation	Less skill is essential	More skill is necessary as compared to jig
Fabrication time and cost	Less	More
Example	Diameter jig, Plate jig, Box jig, Channel jig, Welded jig, etc.	Air-operated chunk, may be constructed with one or more among angle plates, V-block, bolt, parallel, strips, etc.
Sketch	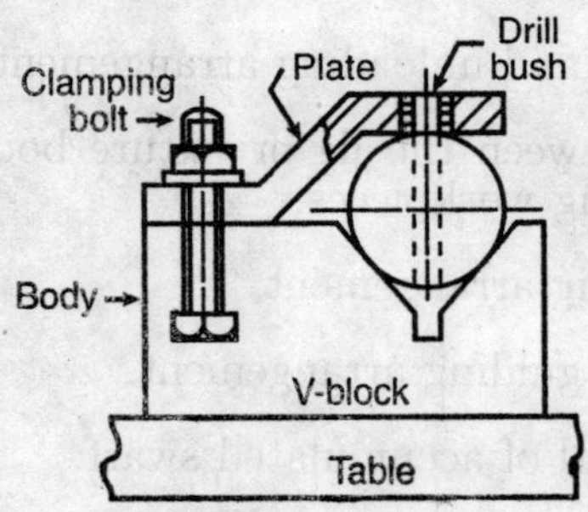 (B) Diameter jig **Sketch S-13.3(B)**	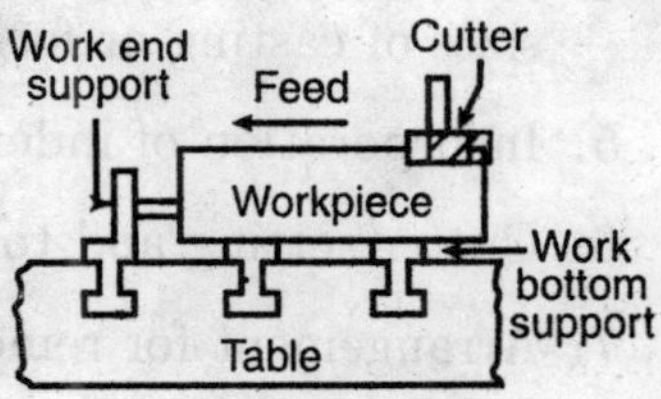 (A) Fixture **Sketch S-13.3(A)**

A13.4 : Advantages Of Using Jigs And Fixtures In Mass Production Work

The following are the points relating to the advantages of using Jigs and Fixtures in mass production work:

1. It can avoid measuring, marking, setting, etc., by locating the system.

2. It minimizes the setting time for fixing number of workpieces at a time and number of operations can be performed in one setting.

3. It can increase the operational accuracy as the work is rigidly fixed and located and the tool is guided.

4. It minimizes the machining time by increasing the depth of cut, **feed** and speed, due to better clamping rigidity of the workpiece and guiding the tool.

5. It ensures interchangeability of finished products for achieving quicker assembly.

6. It minimizes assembly cost due to interchangeability and eliminating the operations like filing, etc.
7. It minimizes the operator's fatigue due to less material handling.
8. It minimizes the labour cost by using semi-skilled worker as setting of work and tool are mechanized.
9. It minimizes the cost of inspection as the products are identical.
10. It minimizes the overall cost of the product by partly or fully automatizing the processes.

A13.5 : Basic Elements For Designing A Jig Or Fixture

The following are the dependable factors to be studied in designing a Jig or a Fixture before the actual work is started:

1. Components of the workpiece and the sequence of operations.
2. Selection of Machine Tools in relation to the operations to be performed.
3. Easy and quick loading and unloading arrangements.
4. Necessary clearance between the jig or fixture body and the variable sizes of casting or forging workpieces.
5. Incorporation of indexing arrangement.
6. Cutter setting and tool guiding arrangement.
7. Arrangement for removal of accumulated swarf.
8. Easy ejection of heavy workpieces mechanically.
9. Power devices for operating the clamping elements.
10. Locating elements on type of work and operations, number of workpieces required, degree of accuracy, etc.
11. Safety of the operator, job and machine.
12. Full proofing arrangement, to prevent the accidental error of the operator from spoiling the work.
13. Clamping and loosening the work with the fixture should be done by using power devices.
14. Fixture should be balanced by using counter-weight to prevent vibration.

A13.6 : Elements Of Jigs And Fixtures

The Elements are
(1) Body or Table (cast iron by casting or fabricated M.S. bars and plates).
(2) Clamping Device (M.S.).
(3) Locating Device.
(4) Tool Guide (tool steel or cast iron).

A13.6.1 : Body

It is to support and house the workpiece. Body may either be fabricated from M.S. bars and plates or from cast iron casting. To relieve the internal stresses, the body may be heat-treated if required. The jig body may be of various types such as Template type, Channel type, Box type, Diameter type, etc.

A13.6.2 : Clamping Device

Functions of the clamping devices are to hold the workpieces rigidly on the fixtures or on the jigs against the cutting forces. These should have the following facilities:

1. Low clamping pressure on the work to prevent distortion.
2. Easy and simple clamping to facilitate easy loading and unloading action.
3. Rigid and fool-proof clamping.
4. Case hardened the clamping faces to prevent wear.
5. Arrangements of the clamps to perform many operations in one setting.

   Clamps may be of various types such as C-clamp, studs, set screws, latch clamp, screw clamp, pivoted clamp, double action clamp, etc.

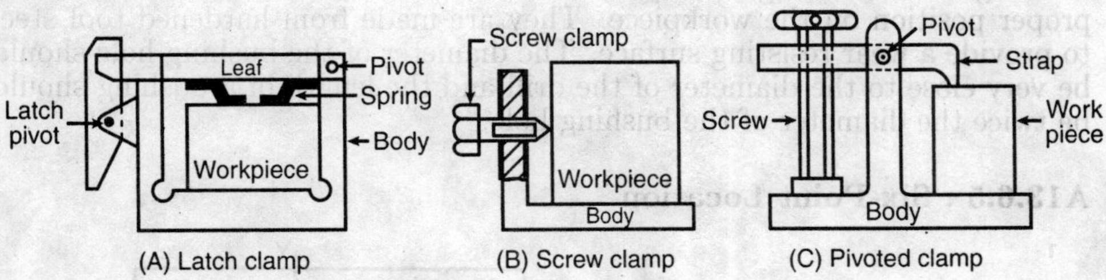

Sketch S-13.6.2 (A, B & C)

A13.6.3 : Locating Devices

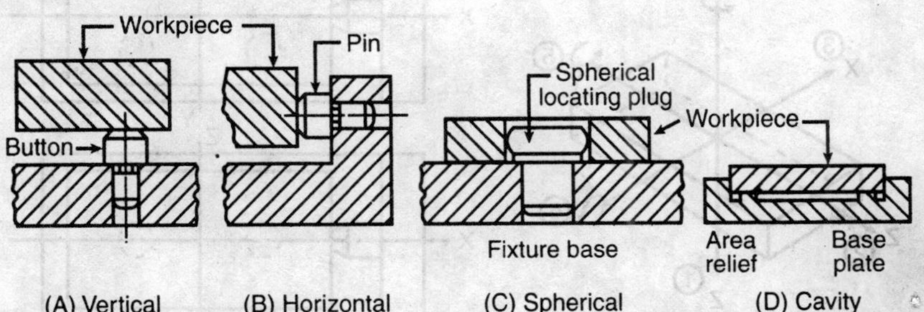

Sketch S-13.06.3 (A to D)

Correct Location is essential for the accuracy of the finished product. Locating is the dimensional and positional relationship between the workpiece and the cutting tool used on the machine. Round pin is usually used for

Horizontal Location, whereas shorter buttons are used for Vertical Location. Lock screw pins are used where the pin is subjected to excessive wear and must be frequently replaced. Large pins used in holes for locating are generally referred to as plugs with extra clearance between the workpiece and the plugs. Spherical Locating Plugs will help to prevent sticking because of misalignment. Rest pads and plates are used to support and locate the larger and heavier workpieces vertically. Pads and plates are held to the jig or fixture base plate by socket head cap screws. Nest or Cavity of a jig or fixture is a pocket into which the workpiece is placed and located. Pocket is to be machined directly into a base plate to the same contour as the workpiece.

A13.6.4 : Tool Guides Or Drill Bushings

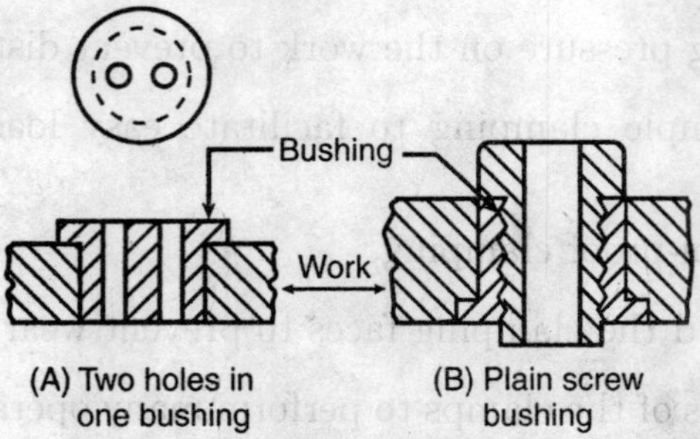

Sketch S-13.6.4 (A & B)

Bushings are used to guide drills, reamer and similar cutting tools into the proper position on the workpiece. They are made from hardened tool steel to provide a wear resisting surface. The diameter of the bushing hole should be very close to the diameter of the drill and the length of a bushing should be twice the diameter of the bushing hole.

A13.6.5 : Six-Point Location

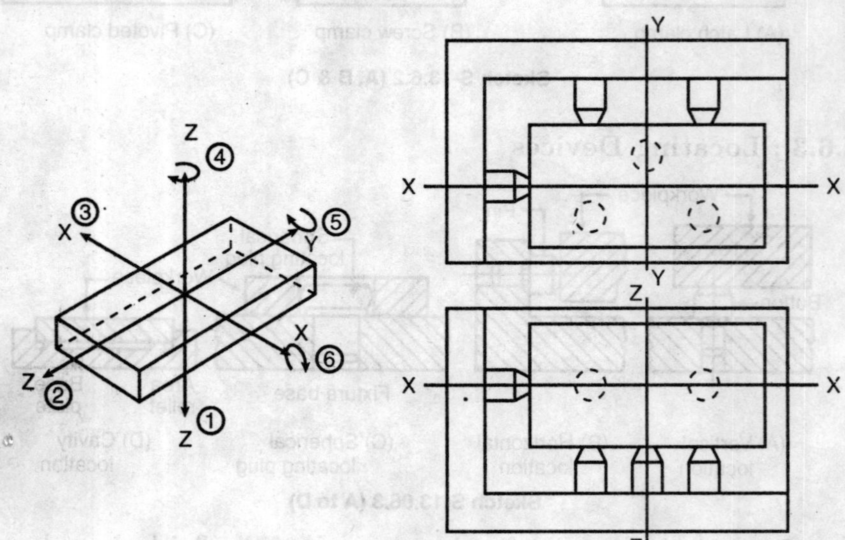

Sketch S-13.06.5

Six-Point Location Method In Jig Design

Location: It gives a relationship between Workpiece and Jig or Fixture for correct location and maintaining the accuracy of a finished job.

Principle of Location: If a rectangular block is free to move about its axis XX, YY, ZZ, it can also rotate about its axis. So degree of freedom of the body is six.

Description: Downward movement of the block along ZZ is controlled by three supporting locating points and movements along YY and XX axes are controlled by double and single locating points respectively. The rotary movements of the block about XX, YY and ZZ axes are also controlled by the bottom side and back pins. The locating points for an uneven object, like a three-legged object is determined by different arrangements, but guiding principle is same.

The workpiece is completely restrained to move along 1, 2 and 3 directions by arranging three locating points in ZZ axis, two locating points in YY axis and one one locating point in XX axis respectively. Hence the workpiece is also restrained to rotate about three axes XX, YY and ZZ, i.e., three degree of movements in 4, 5 and 6. Three free movements, i.e., upward, backward and rightward are provided for locating and unloading the workpieces and during operation can be restrained by providing clamps in Jigs or Fixtures. This principle is known as six-point location.

A13.7 : Types Of Jigs

Jigs are used to drill, ream and tap holes at much greater speeds and with greater accuracy by using a semi-skilled worker. Jigs hold the work, locate the work and also guide and control the cutting tool(s).

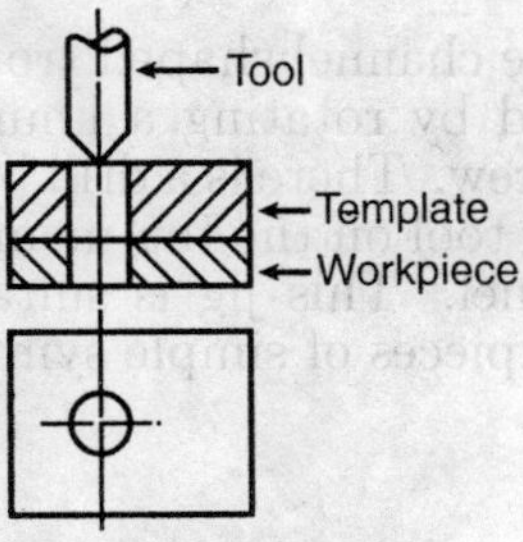

Sketch S-13.7[1]

[1] Template Jigs

This is simplest type among jigs, used for locating hole patterns on large workpieces. A template with holes at the desired position is clamped on the component to be drilled.

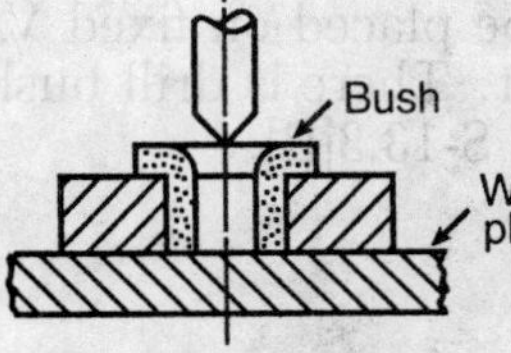

Sketch S-13.7[2]

[2] Plate Jigs

It is the modification of template jig containing bushings; forms the main structural member. As it is open in construction, also named as open jig, which facilitates part loading and removal, chips removal and clamping.

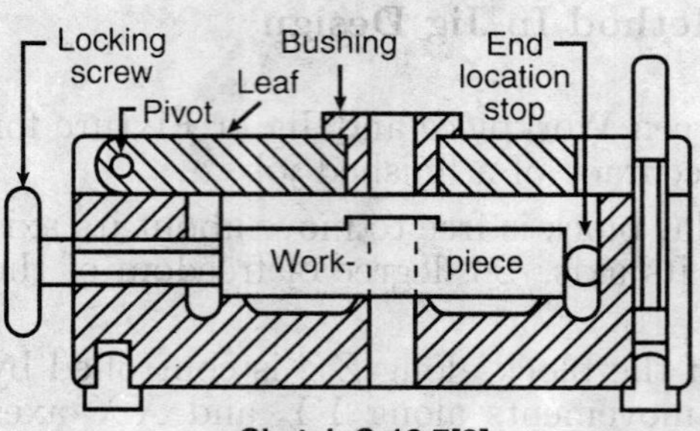

Sketch S-13.7[3]

[3] Leaf Jigs

It has a leaf or cover, which can be swung open or close to load and unload the work. After locating the workpiece inside the jig, the leaf is firmly closed and locked, and the jig is then ready for use.

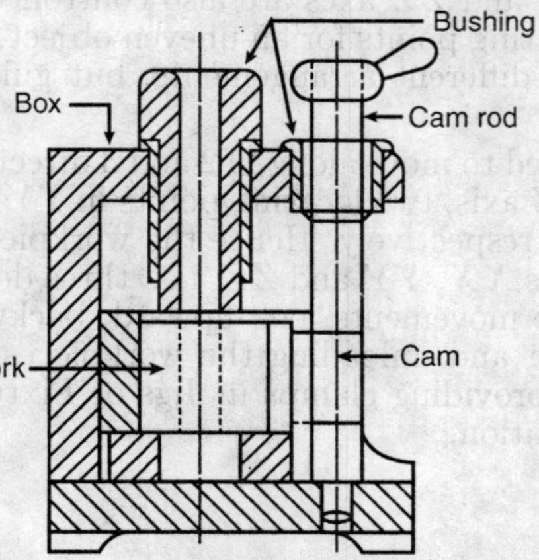

Sketch S-13.07[4]

[4] Box Jigs

It is box shaped, containing box structure, bush, cam, cam rod, etc. It contains bushings on two or more sides for the purpose of drilling holes on different sides of the part. For loading the workpiece on jig, the cam rod is taken out of the jig and the workpiece is placed in position inside the jig. Then the workpiece is locked by rotating the cam through cam rod.

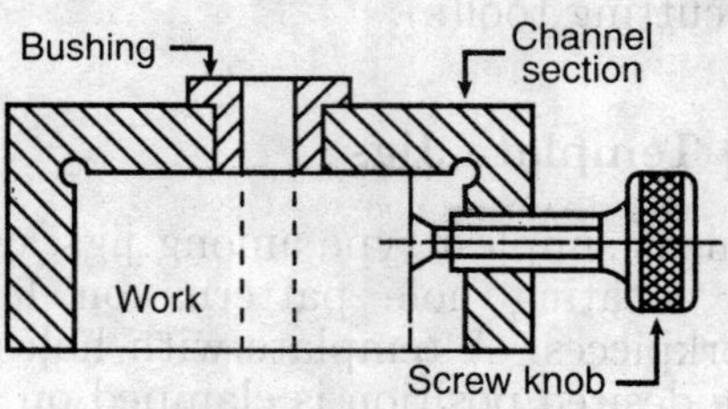

Sketch S-13.7[5]

[5] Channel Jigs

It is a simple channel shaped trough and clamped by rotating a knurled knob of a screw. There is a drill bush to guide the tool on the top member of the channel. This jig is suitable for the workpieces of simple symmetrical shape.

[6] Diameter Jigs

These types of jigs are used to drill or ream radial holes on cylindrical or spherical workpieces. The cylindrical workpieces may be placed on fixed V-block and clamped by clamping plate and clamping bolt. There is drill bush on clamping plate for guiding the tool. [Refer to sketch S-13.3[B].

A 13.8 : Types Of Fixtures

Fixture

A Fixture is a device for holding a workpiece to a machine bench in a fixed position during machining operations. It does not have special arrangement for guiding the cutting tool. A fixture is used for holding and locating the workpiece only when a large number of identical parts are to be machined to justify its cost.

Types of Fixtures

Common types of fixtures are vice, lathe, boring, broaching, milling, grinding, etc.

A13.8.1 : Vice Fixtures

Standard machine vice with the arrangement of special jaws is provided for an easy way of holding the parts for machining. Special jaws are designed according to the requirement of individual workpieces. Small castings or forgings with irregular contours can be held in standard machine vices.

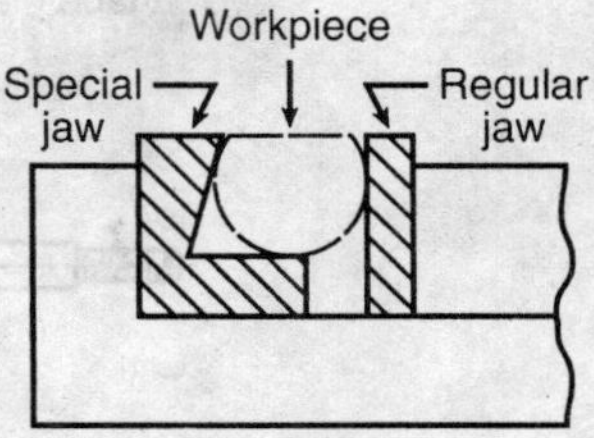

Sketch S-13.8.1

A13.8.2 : Lathe Fixtures

Most of the lathe operations can be done by using standard chucks and holding methods. But many castings and forging parts cannot easily be mounted by the standard methods and require special work holding fixtures.

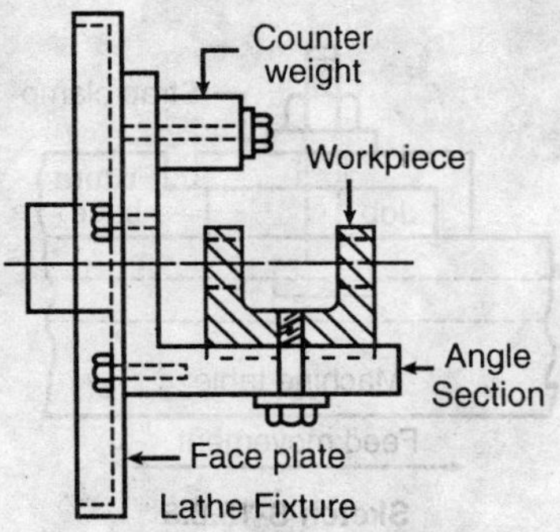

Sketch S-13.8.2

Face Plate Fixture

Face Plate Fixture is mounted to a standard lathe face plate. The flat base of fixture is bolted to the face plate. The workpiece is dynamically balanced as accurately as possible by using a suitable sliding counterweight for turning high speed operations. Accurate static balancing is satisfactory for turning low and medium speed operations.

A13.8.3 : Boring Fixtures

Boring fixtures may be of a drill dig or a mill fixture depending on the type of operation. Boring fixtures have the adjustment on the machine table or boring bar, may require additional holes or indicating surfaces for initial alignment. Single-piloted bar is guided on its leading end to prevent springing under cutting thrust and an end support is also provided for alignment by adjustment of the bearing block. Purpose of this fixture is to hold the workpiece in the correct relation to the boring bar.

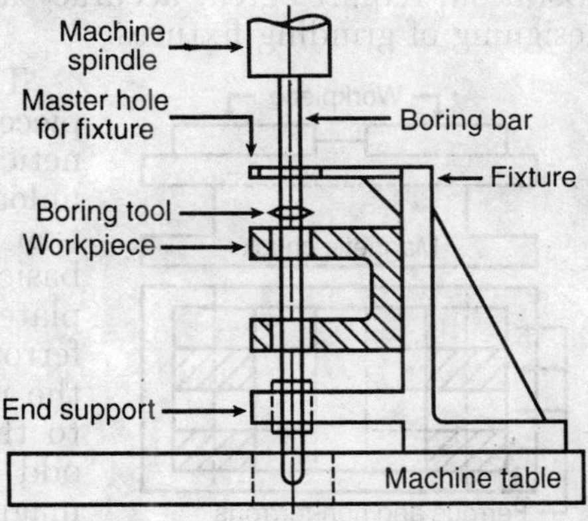

Sketch S-13.8.3

A13.8.4 : Broaching Fixtures

Simple plate, plug or horn fixtures locate the correct position of the workpiece in relation to the broach and machine face plate. The figure above shows a simple fixture used for broaching a keyway. Wire strip of desired thickness (made of tool steel and heat treated wear resistance) is used when to cut a keyway deeper than that can be achieved in one pass of the broach.

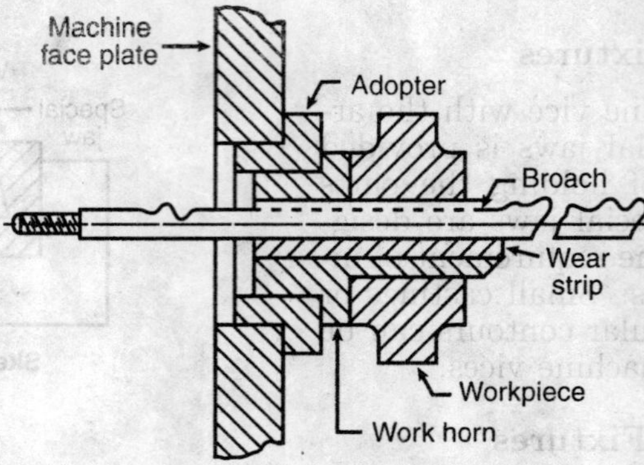

Sketch S-13.8.4

A13.8.5 : Milling Fixtures

A milling fixture holds the workpiece(s) in the correct relation to the milling cutter as the table movement carries the workpiece through the cutters. The fixture consists of fixture base, strap clamp, nuts and bolts, locating points. Strap clamp should be extremely rigid to exert the cutting forces by interrupted cutting of the milling cutter.

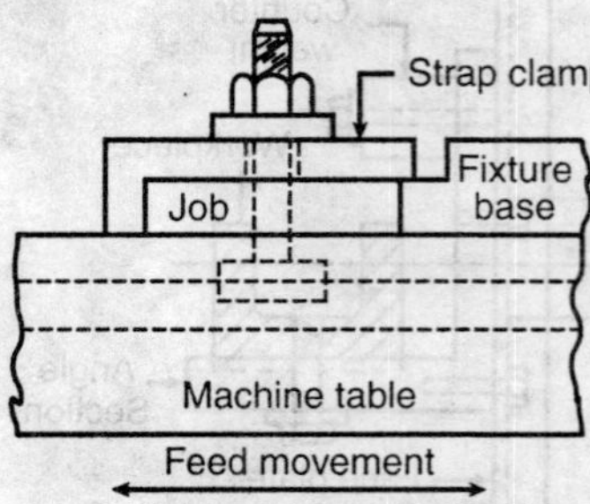

Sketch S-13.8.5

A13.8.6 : Grinding Fixtures

Lathe fixtures may be used for cylindrical grinding. Grinding is a finishing operation, requires great accuracy and close tolerance is to be considered in designing of grinding fixtures.

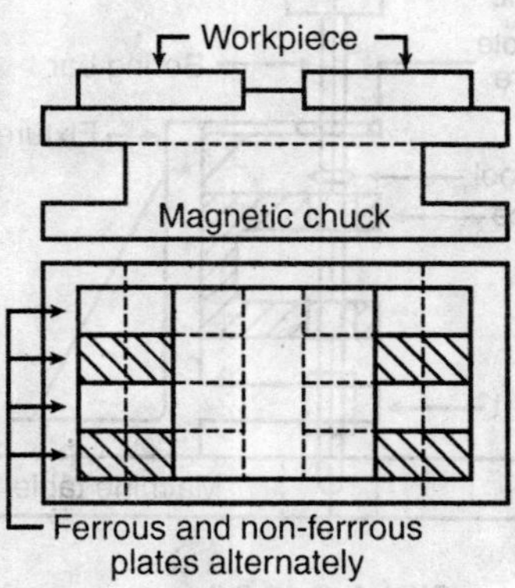

Sketch S-13.8.6

The standard method of holding the workpieces for surface grinding operations is magnetic chucking. Quick loading, mounting and unloading the workpieces and also no distortion is caused by mechanical clamping. The basic fixture is made of a number of mild steel plates separated from each other by a nonferrous spacer. The magnetic flux passes from the magnetic chuck through the steel plates to the workpiece. Plates, round bars, even odd shaped workpieces can be mounted on magnetic chucks parallels, V-blocks and magnetic chucks with apoxy tooling materials respectively.

A13.9 Questions

1. Write an 'introduction' on Jigs and Fixtures (A13.1).

2. Define Jig and Fixture (A13.2).

3. Compare Jig and Fixture (A13.3).

4. Write the advantages of using Jigs and Fixtures in mass production (A13.4).

5. Explain in brief the basic elements for designing a Jig or a Fixture (A13.5).

6. State in brief with sketches the elements of Jig and Fixture (A13.6 to A13.6.4).

7. Sketch and state in brief the Leaf Jig and Diameter Jig (A13.7 [3] and A13.7 [6]).

8. Sketch and state in brief the Box Jig and Channel Jig (A13.7 [4] and A13.7[5]).

9. Write the types of Fixtures mostly used (A13.8).

10. Sketch and state in brief on Vice Fixture and any Lathe Fixture (A13.8.1 and A13.8.2).

11. State on Broaching fixture, Milling fixture and Grinding fixture with, necessary sketches (A13.8.4, A13.8.5 and A13.8.6).

12. Sketch and explain the Six-Point Location (A13.6.5).

14

NON-TRADITIONAL MACHINING

A14.1 : Introduction

Recent developments in materials with respect to their qualities of high strength-to-weight ratio, high temperature resistance, hardness, etc., as applied in aluminium alloys, hastalloy, nitralloy, titanium, carbides, high strength temperature resistance (HSTR) alloys, etc., are being using in the aerospace, nuclear engineering industries, and other developed industries.

Conventional machining processes like operations in Lathe, Drilling, Mining, etc., are inadequate and hard to machine above mentioned materials from the standpoint of economic production, complexity, time consuming, impossibility and difficulty in machining, accuracy, surface finish, etc.

Unconventional or non-traditional machining processes have overcome the above difficulties. These non-traditional machining techniques have been invented to supplement effectively the metal cutting, casting methods, any complex shapes hard to machine, brittle materials, etc.

A14.2 : Flow Diagram Of Non-Traditional Processes

Processes	**Input energy**	**Basic mechanism**	**Source of immediate energy**	**Transfer energy medium**	**Application**
Conventional Machining	Mechanical	Shear	Cutting Tool	Physical Contact	M.S., C.I., Non-ferrous Metals etc.
Ultrasonic Machining (USM)	Mechanical	Erosion	Pneumatic or Hydraulic Pressure	High Velocity Particles	Steel, Titanium, Plastic
Electro-chemical Machining (ECM)	Electro-chemical	Ion Displacement	High Current	Electrolyte	Aluminium, Titanium, Refractories
Electrical Discharge Machining (EDM)	Thermo-electric	Vaporization	High Voltage	Electrons	Aluminium
Laser Beam Machining (LBM)	Thermo-electric	Vaporization	Amplified Light	Radiation	Aluminium, Steel, Super alloys, Titanium, Plastics, Glass.

A14.3 : Alternative Names Of Non-Traditional Machining

The Non-traditional Machining Methods can also be named as follows:

(a) New technologies.

(b) Modern machining methods.

(c) Unconventional machining methods.

A14.4 : Principle Of Non-Traditional Machining

Principle of Non-traditional Machining is to apply mechanical/chemical/ electrical/thermal energy directly to a workpiece. Almost no physical contact with the tool removes thin and uniform rate of material from the workpiece to give a desired shape or contour.

A14.5 : Advantages Of Using Non-Traditional Machining

1. Material removed which almost no physical contact between the tool and the workpiece.

2. Tool material may or may not be harder than the work material.

3. Any known material (irrespective of hardness) and any shape may be machined.

4. Superior surface finish and burr-free machining is possible.

5. Uniform material removal and rate is independent of workpiece hardness.

6. Stress-free machining and cutting forces are not depending on workpiece hardness.

7. Micro-machining is possible, etc.

A14.6 : Limitations Of Using Non-Traditional Machining

1. Machining processes are very much costly.

2. Power consumption per unit of volume of material removal per unit time is very high.

3. In using ECM and EDM methods, the work and the tool materials must be electrically conductive.

4. In using LBM method, the holes or cavities may be tapered and the depth of cut is limited.

5. USM method is not suitable to work on softer materials and the tool wear will lead to a problem.

A14.7 : Ultrasonic Machining (USM)

Principle of Machining

This type of machining can be used in machining of hard and brittle materials, whether electrically conductive or non-conductive. Ultrasonic is a process in which tool vibrates with high specific frequency (above 16 kHz) by mechanical means through hydraulic or pneumatic pressure on the work surface in between a gap of about 0.02 to 0.1 mm through which abrasive slurry flows. The tool has the desired shape of the cavity to be machined but is slightly small in size of the cavity by a few hundredths of a millimeters and made of low carbon steel or stainless steel, brazed or soldered to the tool holder.

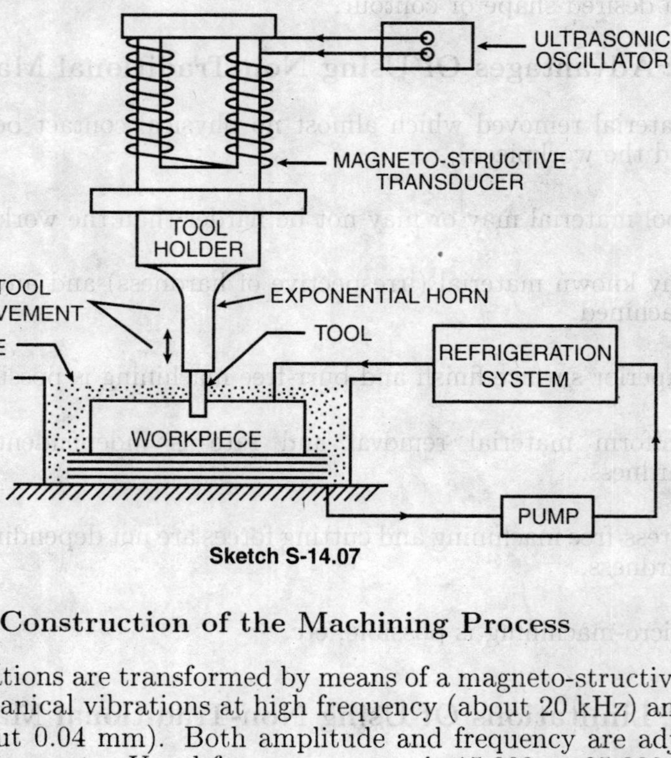

Sketch S-14.07

Working and Construction of the Machining Process

Ultrasonic vibrations are transformed by means of a magneto-structive transducer into mechanical vibrations at high frequency (about 20 kHz) and small amplitude (about 0.04 mm). Both amplitude and frequency are adjustable as per job requirements. Usual frequency range is 15,000 to 25,000 Hz and the amplitude of vibration is of 0.01 to 0.06 mm. These mechanical vibrations are transferred to the tool through different components like connecting body, tool holder, exponential horn, etc. Connecting body is brazed with transducer and a removable tool holder is fastened to the connecting body. During vibrations of the tool (as shown in the sketch) with specific frequency, an abrasive slurry (mixture of abrasive grains may be of Aluminium oxide or silicon carbide or boron carbide or diamond grains and water with 30% to 60% by volume of abrasives) is flowing through the tool-workpiece interface. The vibratory tool striking on the flow of abrasive slurry, causes thousands of microscopic abrasive grains to remove the work material by abrasion.

The machining or material removal is done by repetitive impacts of high frequency of the tool on the workpiece and the abrasive slurry in between the tool and workpiece.

The abrasive slurry heated by abrasion may be cooled to 5°C to 6°C by the refrigeration system and this is being circulated in the system by pumping.

Application

Application of ultrasonic machining is for machining profile holes, slicing, drilling, engraving, trepanning, honing, etc. Most suitable materials for machining are steel, titanium, plastic, etc.

A14.8 : Electro-Chemical Machining (ECM)

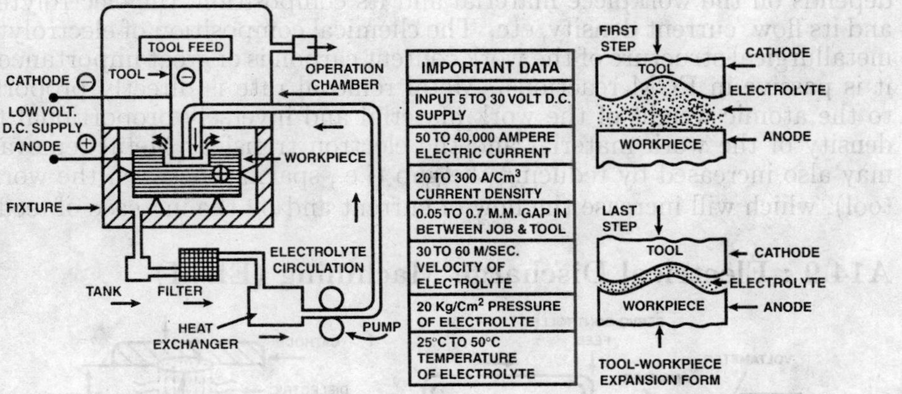

Sketch S-14.8

Principle

Electro-chemical machining (ECM) is the machining process of controlled metal removal by anodic dissolution of the positive workpiece separated from a suitably shaped negative electrode tool by flowing a thin film (about 0.5 mm) of electrolyte medium in between. Low voltage (about 20 V) and high direct current employed in the presence of electrolyte, enables a controlled removal of metal from the workpiece by anodic dissolution. Solution of sodium chloride in water is mostly used as electrolyte.

Working

The workpiece is connected to the positive terminal (Anode) and the tool is connected to negative terminal (Cathode) of the 20 volt D.C. supply. The flow of electrolyte in between the tool (Cathode) and the workpiece (Anode) completes the electrical circuit. The positively charged ions attract the tool and negatively charged ions attract the workpiece and thus initiate the flow of current in the electrolyte.

Cathode liberates hydroxyl ions (negatively charged) and free hydrogen during electrolysis process. The metal ions of anode combine with the hydroxyl ions to form insoluble metal hydroxides and the material is thus removed from the workpiece (anode). The process is to be continued till the tool reproduces its shape in the workpiece (anode). The rate of ions attraction to respective electrodes, depends on current density which varies directly with the applied voltage, concentration of electrolyte and the gap between the electrodes. The rate of metal removal depends on the rate of ions attraction.

Applications

ECM machining is to be applied only to conductive materials in specialized area where conventional machining is not feasible and in mass production work due to its high capital cost. This machining is also suitable in the aerospace industry to machine difficult and complex-shaped parts. This may apply to machine superalloys, steels, complicated contours, by designing a suitable tool.

A14.8.1 : Metal Removal Rate (MRR) in ECM Machining

MRR, i.e., metal removal rate in Electro-chemical Machining process mainly depends on the workpiece material and its composition, the electrolyte used and its flow, current density, etc. The chemical composition of electrolyte and metallurgical structure of the work content carbon is of great importance since it is passive in ECM reactions. Metal removal rate is directly proportional to the atomic weight of the work material and inversely proportional to the density of the work material and the electron transfer. Metal removal rate may also increased by reducing the gap (i.e., spacing, between the work and tool), which will increase the flow of current and by the increase of feed rate.

A14.9 : Electrical Discharge Machining (EDM)

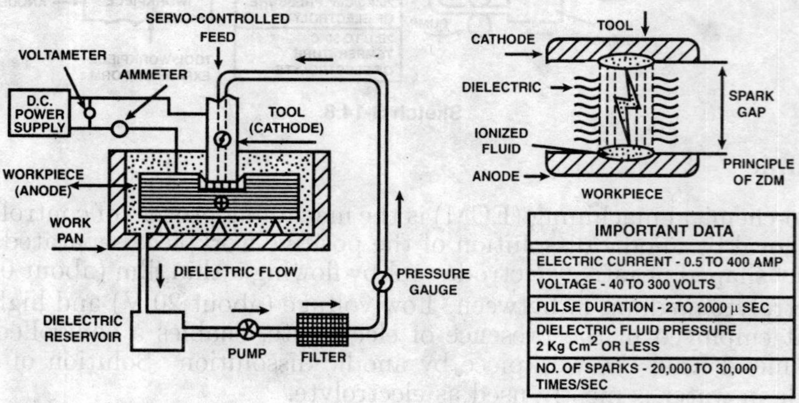

Sketch S-14.9

Principle

Electrical Discharge Machining (EDM) is the machining process of controlled erosion of electrically conductive materials by the rapid and repetitive spark discharge between the workpiece (anode) and the tool (cathode) separated by flooded dielectric fluid through the small gap (about 0.02 to 0.5 mm), and known as spark-gap.

Working

The dielectric fluid is ionized in the spark-gap under the direct pulsed application of the direct current and thus a spark-discharge is passing between the tool and the workpiece. Every spark generates sufficient heat to melt and vaporize a very small volume of metal, leaving a small crater on workpiece surface. The energy in the spark can be controlled so that the material removal rate, surface finish and tolerance can be maintained. Kerosene is mostly used as dielectric medium and is circulated by pumping at a pressure of 2 kg/cm^2 or less.

Application

This machining system is used in the manufacture and reconditioning of press tool, forging dies, irregular shaped profile, mostly used in tool rooms, moulds for injection moulding, small diameter holes in carbide or hardened steel, internal threads and internal helical gears in hardened materials. The process may also be applied to machine the metals like steels, superalloys, etc.

A14.10 : Laser Beam Machining (LBM)

Laser

The word Laser is an acronym for light (L) Amplification (A) by Stimulated (S) Emission (E) of Radiation (R).

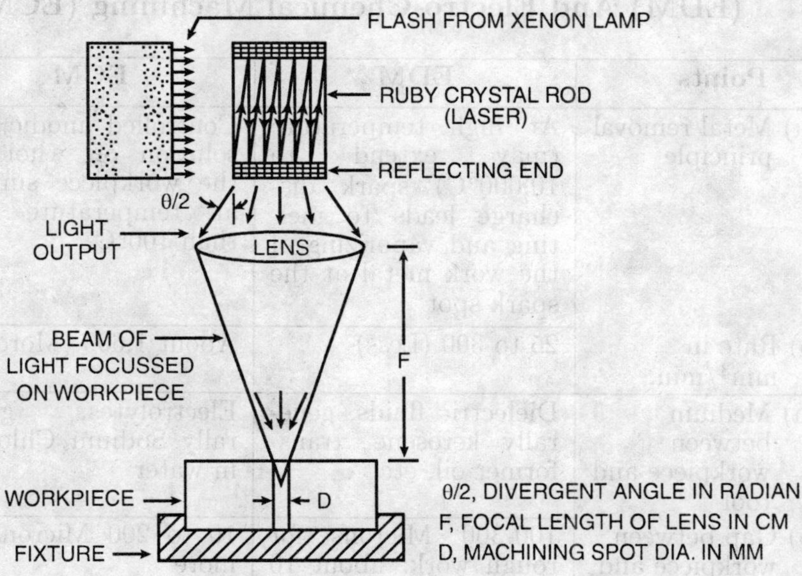

Sketch S-14.10

Principle

In Laser Beam Machining (LBM),the work material is melted and vaporized by an intense monochromatic beam of light known as Laser. The Laser beam strikes on a small area where machining is to be done; the heat generated on that area is sufficient to melt any known materials, even diamond. Laser is an electromagnetic radiation and produces monochromatic light in the form of a collimated beam.

Working

The stimulated monofrequency beam of light is radiated from the Xenon Flash Lamp and is focussed on the Ruby Crystals (Laser rod). In Laser rod, the beam of light is reflected and accelerated in the path. This accelerated beam of light is emitted as divergent (angle θ in radian) beam on a lens placed suitably in the path so that the beam of light which then converges and focuses (focal length, F) the beam of light suitably on to the workpiece at the required spot diameter (D) to be machined. A Laser beam is so powerful that it can heat, melt and vaporize any known materials, even diamond when focussed by the lens system. The temperature and energy density of the beam of light may extend to 1,00,000°K and 1,00,000 kW/cm^3 at the workpiece end.

Application

LBM machining process is very costly and found suitable only when not feasible by the other methods of machining. This system is used in machining thin and hard materials like ceramics, diamond, etc., in very small holes, in cutting complex profiles, in sheet metal trimming, in mass micromachining production, in partial cutting or engraving, etc.

A14.11 : Comparison Between Electrical Discharge Machining (EDM) And Electro-Chemical Machining (ECM)

Points	EDM	ECM
1. (a) Metal removal principle	At high temperature (may extend to 10,000°C) spark discharge leads to melting and vaporizing of the work metal at the spark spot	Controlled anodic dissolution on whole of the workpiece surface at temperature less than 100°C
(b) Rate in mm^3/min.	25 to 300 (Less)	About 1,600 (More)
2. (a) Medium between workpiece and tool	Dielectric fluids, generally kerosene, transformer oil, etc.	Electrolytes, generally Sodium Chloride in water
(b) Gap between workpiece and tool	100-300 Microns for rough work; about 10 Microns for finishing work	10 to 200 Microns or more
3 Electrical data (a) Generator	10A to 600A static Impulse Generator	100A to 40,000A static high D.C. Generator
(b) Voltage	80V to 200V D.C.	4V to 30V D.C.
(c) Current density	6 to 20A/cm^3	10–500A/cm^3
4. Tool : Cost	Less, for smaller tool cross-section	Less, for larger tool cross-section against proper power and current
Design	Easy	Difficult
Wear	Depends on machining conditions	No tool wear

Contd..

Points	EDM	ECM
5. Surface Finish	Increases with low current density and at very low metal removal rate	Increases with high current density and with higher feed rate
6. Machine (a) Type (b) Feed (c) Feed Unit	Mostly vertical type Controlled with feedback device Electro-hydraulic	Mostly portal construction Constant feed rate Electro-mechanical

A14.12 : Comparison Among USM, ECM, EDM And LBM

Points	USM	ECM	EDM	LBM
1. Application of various work materials	Steel, Titanium, Plastic	Aluminium, Titanium, Refractories	Aluminium	Aluminium, Steel, Superalloys, Titanium, Plastic, Glass
2. Application according to the shape of workpiece	Honing	Precision, Surface of revolution, Grinding	Small holes ($0.15 < D < 0.03$), Deep holes ($L/D > 20$), Double contouring	Shallow ($L/D < 20$), Deep
3. Application for Machining	Profile holes, slicing, trepanning, drilling, engraving, etc.	Only to conductive materials; machining in the aerospace industry; to machine complex shaped parts, to machine metals like superalloys, steel in tools	Only to conductive materials; machining of Press Tools and forging dies, irregular shaped profiles, injection mouldings, small diam. holes, internal threads and helical gears, and also to metals like steels, superalloys, refractories, etc.	Very small holes, thin and hard materials, complex profile, mass micro-machining production, sheet metal trimming, etc.

Contd...

Points	USM	ECM	EDM	LBM
4. Machining performance for non-conducting materials like— Ceramics, Plastics, Glass	Suitable	Not possible	Not possible	Good for ceramics and fair for glass and plastics
5. Contouring a surface	Poor	Good	—	Fair
6. Power consumed in kW/ cm^3/min.	9.00	7.20	1.80	2,700
7. Maximum metal removal rate in mm^3/min.	Medium; 400	Extra high; 16,000	Medium; 800	Very low; 4
8. Penetration rate in mm/ min.	0.5	12	12	100
9. Accuracy in mm	.005	0.01	0.01	0.01
10. Economics of processes	High metal removal efficiency, low capital cost, tooling cost and power consumption cost	Very high capital cost, very low tool wear	Higher tooling cost, metal removal efficiency and tool wear	Very low power consumption cost, very high metal removal efficiency, very low tool wear
11. Basic Mechanism of metal removal	Erosion by mechanical abrasion	Ion-displacement by Electro-chemical method	Vaporization by thermo-electric method	When Heated, melts and vaporizes instantly by thermoelectric method.
12. Immediate energy source at the job-tool interface	Pneumatic or hydraulic pressure	High Current	High Voltage	Amplified light
13. Medium for transferring energy	High velocity particles.	Electrolyte	Electrons	Radiation

A14.13 Questions

1. Write the history behind the invention of Non-traditional Machining processes (A14.1).

2. State the flow diagram of Non-traditional processes with the Conventional Machining processes (A14.2).

3. Write the alternative names of Non-traditional Machining processes (A14.3).

4. State the principles of Non-traditional Machining processes (A14.4).

5. What are the advantages of using Non-traditional Machining (A14.5)?

6. Write the limitations of using Non-traditional Machining (A14.6).

7. Sketch, label and explain the principle, working and applications of the following Non-traditional Machining processes:

   (a) Ultrasonic Machining (A14.7).

   (b) Electro-chemical Machining (A14.8).

   (c) Electrical Discharge Machining (A14.9).

   (d) Laser Beam Machining (A14.10).

8. Compare between EDM and ECM (A14.11).

9. Compare among USM, ECM, EDM and LBM (A14.12).

10. Write the factors on which MRR depends (A14.8.1).

15

FINISHING METHODS

A15.1 : Introduction

Normal machining methods like turning, milling, etc., are limited in dimensional and geometrical accuracies (precision). Geometrical errors are flatness, concentricity, cylindricity, parallelism of functional surfaces. Important functional properties are wear resistance, fatigue strength, corrosion resistance, power losses due to friction in motion. Micro-finishing processes like honing, lapping, super-finishing, etc., are essential to obtain a very high surface finish or high dimensional and geometrical accuracies. Height in micro-irregularities (H) in Microns (·0.001 mm) of different micro-machining processes are:

Precision Turning	–	1 to 12
Grinding	–	0.9 to 5
Honing	–	0.12 to 1.2
Lapping	–	0.08 to 0.25
Superfinishing	–	0.01 to 0.25

A15.2 : Honing

Definition

Honing is a bonded fine grained abrasive process in which very little material is removed and gives a suitable surface finish on a pre-machined surface with slight pressure, in a rotating and longitudinal motion.

Working

The honing tool is chucked in the spindle head of a honing machine. The work is mounted on machine table rigidly by using suitable fixture. Bonded abrasive sticks held on the outside (or inside) of a honing tool, rotate (peripheral speed 50 to 70 m/min.) continuously in one direction and simultaneously reciprocate axially. Feed-out (or feed-in) is done periodically by expansion of the sticks in machining internal surfaces (or by compression of the sticks in machining external surfaces) for each full stroke of the honing tool. It can remove 0.1 mm of material from the surface. Honing abrasive sticks remove super-fine powdered chips. Sulfurized mineral base or lard oil mixed with kerosene is needed as coolant to flush away small chips and to keep temperatures uniform.

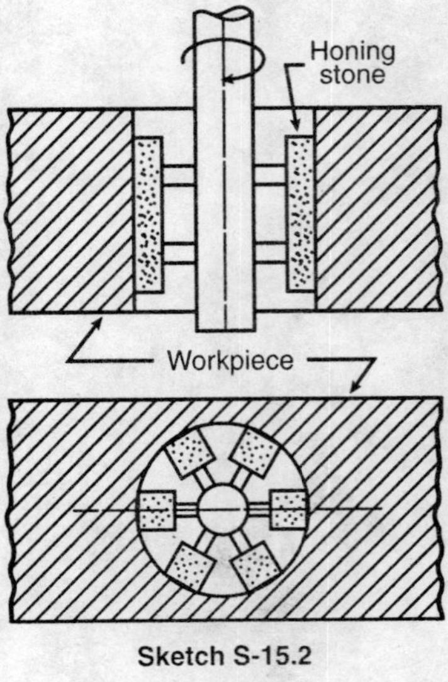

Sketch S-15.2

Application

Honing operation is employed not only to produce high surface finish, but to correct out roundness, axial distortion and slight taper in workpieces. Honing is applicable to 3 mm to 1000 mm in diameter holes and the length of the honed surface may extend to 30 m. Horizontal machines are used for honing long gun barrels and similar works.

Cutting Conditions

To obtain high surface quality and accuracy, the following are to be considered:

(a) High cutting speed—for cast iron 70 to 100 m/min. and for non-ferrous metals 150 to 400 m/min.

(b) Small cutting depth—0.03 to 0.15 mm.

(c) Small feed—0.008 to 0.08 mm/rev.

(d) Suitable coolant (Soluble-oil emulsion or kerosene).

(e) Sturdy construction of the machine.

(f) Properly fixing of workpiece with maximum rigidity.

(g) Cemented Carbide or Diamonds for the cutting edges of the tools.

A15.3 : Lapping

Lapping is precision grinding (final machining operation) with loose, dust type abrasives (Lapping compound) to produce geometrically true surfaces, improve dimensional accuracy of round and flat parts and ensure close fit between mating parts.

Lapping compound depends on the workpiece material, surface quality desired and lapping methods. Steel, cast iron, bronze, aluminium oxide powder with grain size 300 to 600 may be used for rough lapping. Chromium oxide (green colour) or polishing red or hard woods (beech, oak, etc.) are suitable for finish lapping.

Working of Lapping Methods

Lapping can be done either by hand or with lapping machines.

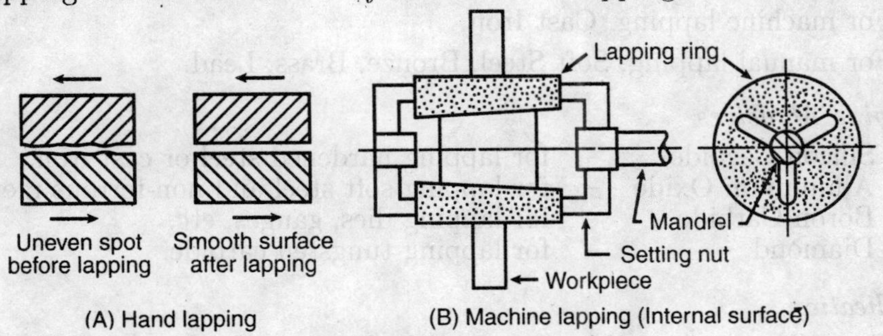

Sketch S-15.3 (A & B)

Hand Lapping

The workpiece (Pin) to be lapped, is mounted in a Lathe and requisite lapping compound is applied with a brush. The workpiece is rotated with a peripheral speed of about 20 m/min. The lapping ring is now to be shifted to and fro by hand. The bores are to be lapped with a lapping mandrel. Flat surfaces may be lapped by rubbing the parts with suitable abrasive powder over the accurately finished flat surface of a master lap. Manual lapping of flat surfaces requires operator's skill, proper pressure and lapping speed.

Machine Lapping

This machining process is suitable in production of round and flat surfaces. The workpiece to be lapped is properly clamped by a holding device and brought in between two plane lapping wheels. Lapping compound is fed through a pump or applied with a brush. An additional pressure to the workpiece may be attained by lowering upper wheel. Then the lapping wheels are put into rotation. The holding device performs an eccentric movement in addition to its rotary motion, which causes the simultaneous rolling and sliding of the round parts.

Cutting Conditions for Lapping

Speed

Most efficient lapping speed range—100 to 250 m/min.

Maximum lapping speed for fine lapping of bores—50 m/min.

Maximum cutting speed for lapping flat surface—250 m/min.

Pressure

Pressure ranges for soft materials including non-ferrous metals and loose abrasives, 0.1 to 0.3 kg/cm^2.

Pressure for hard materials, 0.7 kg/cm^2.

Uniform pressure during lapping is to be maintained.

Allowances

Allowances for lapping, 0.005 to 0.03 mm.

Harder the material, lower should be the lapping allowance.

Tolerance within, ±0.001 mm.

Lap Materials

For machine lapping, Cast Iron.

For manual lapping, Soft Steel, Bronze, Brass, Lead.

Lapping Abrasives

Silicon Carbide	–	for lapping hardened steel or cast iron.
Aluminium Oxide	–	for lapping soft steel and non-ferrous metals.
Boron Carbide	–	for lapping dies, gauges, etc.
Diamond	–	for lapping tungsten carbide.

Application

General purpose machines are used for lapping cylindrical surfaces. Single purpose machines are used for lapping valve seats, crank-shaft, crankpins and journals, camshafts, gears, optical parts etc. Vertical lapping machine is also used for both flat and cylindrical lapping.

A15.4 : Super-Finishing

Super-finishing is suitable for obtaining a surface of the highest class of finish. It is similar to honing with small allowances but differs in the large number of different abrasive stones and wheels and also the type of motions. In addition to high surface finish, super-finishing also removes chatter marks, feed spirals, grinding commas and other imperfections left by the previous grinding operations.

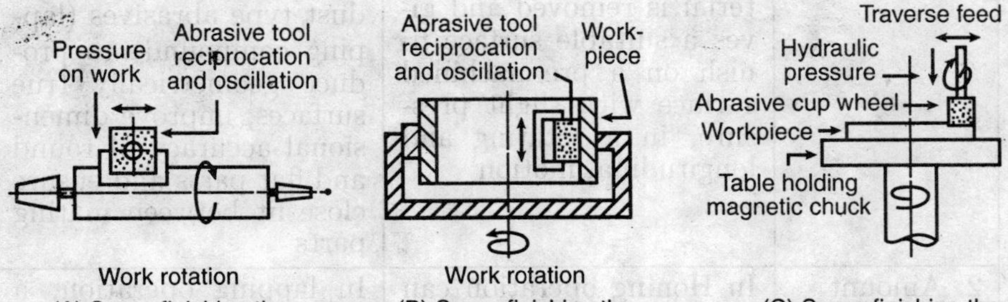

(A) Super-finishing the external cyld. surface
(B) Super-finishing the internal cyld. surface
(C) Super-finishing the flat surface

Sketch S-15.4 (A, B & C)

Working

Super finishing can be used to external and internal cylindrical, tapered, flat surfaces. External cylindrical surfaces can be super-finished by oscillating the abrasive stones at amplitude of 1.5 mm to 3 mm and frequencies of 500 to 1500 cycles per min., reciprocating axially along the rotating workpiece. External cylindrical surfaces can also be super-finished by oscillating and rotating the abrasive stones about the stationary workpiece. Internal cylindrical surfaces can be super-finished by axially oscillating and reciprocating abrasive stones applied to the rotating workpiece. Flat surfaces can be machined by rotating cup-shaped abrasive stone, with the work resting on a circular rotating table. (Abrasive stone and the circular table are rotating in opposite directions.) An additional oscillating movement and hydraulic pressure are given to the abrasive stone for traverse feeding and super-finishing.

Cutting Conditions for Super-finishing

Cutting speed	–	10 to 40 m/min.
Oscillating frequencies	–	500 to 2000 cycles/min.
Working pressure	–	1 to 4 kg/cm^2.
Allowances range	–	0.002 to 0.02 mm.

Fluids

Light mineral oil or kerosene mixed with heavier cutting oil is used as coolant to keep the stone clean and sharp.

Abrasive Stones

Aluminium oxide	–	for super-finishing of steel.
Silicon carbide	–	for super-finishing of soft steel, cast iron and non-ferrous metals.

Applications

This process is suitable to external or internal cylindrical, tapered, flat, curved, spiral surfaces, or where high class of finish is required.

A15.4C : Compare Honing & Lapping

Points	Honing	Lapping
1. Definition	Honing is a bonded fine grained abrasive process in which very little material is removed and gives a suitable surface finish on a pre-machined surface with slight pressure, in a rotating and longitudinal motion	Lapping is precision grinding (final machining operation) with loose, dust type abrasives (lapping compound) to produce geometrically true surfaces, improve dimensional accuracy of round and flat parts and ensure close fit between mating parts
2. Amount of metal removal	In Honing operation can remove 0.1 mm of material from the surface. It is applicable from 3 mm to 1000 mm in diameter holes and length of the honed surface can extend to 30 meter	In lapping operation, a very thin layer of metals 0.005 mm to 0.03 mm can be removed as final machining operation
3. Job materials to be machined	Honing operation is to be done on work materials like plastics, silver, Al, brass, C.I. to hard steel, etc.	Lapping operation is to be done on work materials like C.I. for machine lapping and soft steel bronze, brass, lead, etc., for manual lapping
4. Tool material	Honing abrasives are cemented carbide or diamond	Lapping abrasives are any of silicon carbide, aluminium oxide, boron carbide, diamond
5. Application	Honing operation is employed not only to produce highly surface finish, but to correct out roundness, axial distortion and slight taper in workpieces	General purpose machines are used for lapping cylindrical surfaces. Single purpose machines are used for lapping valve seats, crank-shaft, crankpins and journals, cam-shafts, gears, optical parts, etc.
6. Accuracy	0.12 to 1.2 Micron (0.001 mm)	0.08 to 0.25 Microns (0.001 mm)

A15.5 Questions

1. State the necessity of Finishing methods. Name some finishing methods. What do you mean by Microfinishing processes? (A 15.1).

2. What do you mean by the Honing process? Sketch, label the components and explain the working of Honing process. State also where it is to be applied (A 15.2).

3. Write the cutting conditions of Honing process (A 15.2).

4. Where Lapping is used? Write the different lapping compounds which are required for rough and finish lapping (A 15.3).

5. Sketch and explain the Hand and Machine lapping. Write also the cutting parameters necessary for lapping operations. State the purposes of different types of Lapping Machines (A 15.3).

6. Which finishing method will be suitable for highest class of finish and why? (A 15.4).

7. Sketch and explain the working of Super-finishing method. State also its application and cutting conditions (A 15.4).

8. Compare Honing and Lapping operations (A 15.4C).

16

SAFETY

A16.1 : Introduction

Working safely is first essentials of an operator. He should learn the safe way of doing the things and also the unsafe way too. The safe way is the correct way and also doing it efficiently.

Safety is an attitude for learning and practicing the safe working habits. A good and safe operator is one who is "safety conscious" always, using safe and accepted procedures. A young operator must use safe working, procedures at all times, not for a day or month or year, but for all time, if he wants to become a safe operator.

A16.2 : Accident

In every year lakhs of employees are injured in factories due to accident in mechanical, physical, electrical, radiation, chemical hazards. These accidents create a social loss, loss of life, economic loss towards earning capacity and cost, etc. The economic cost is in terms of loss of productive capacity and maintenance cost of the injured persons and their families. Number of accidents in India is nearly double the accidents occurring in USA and UK due to lower educational status of workers, more manual operations, low wages, insufficient voluntary efforts on the part of the management, etc. Intention of any industry should be to reduce the number of accidents by taking proper safety measures. An accident is a mishap. Every mishap that causes injury to the body or a member of a body should be considered as an accidental injury.

In industry, accidents are usually based on a lost time factor, i.e., the time lost from his job by the accidental person.

Injuries due to machine tools may occur either at the point of operation or by the mechanical power transmission apparatus involved in driving the machines. Injuries may also be caused by the handling materials, falls or difficulties chargeable to the job, throw of flying chips, break in the insulation of the electrical controls.

A16.3 : Causes Of Accident And Prevention

1. Operating without authority.
2. Inoperative safety devices.
3. Operating at unsafe speed.
4. Unsafe loading, placing, etc.
5. Using unsafe equipment.

6. Working on moving equipment.
7. Working at unsafe position or posture.
8. Operating without using safe protective devices.
9. Working at noisy or bad environment.
10. Unguarded or inadequately guarded.
11. Defective working conditions, slippery, corroded, cracked, etc.
12. Hazardous arrangements, layout, process, misalignment, overload, etc.
13. Unsafe lighting and ventilation (insufficient and improperly distributed).
14. Unsafe dress (improper and loose garments).
15. Unsafe planning, methods, processes, etc.
16. Unsafe design or construction.
17. Not interested to work.
18. Operating carelessly.
19. Dangerous materials with which to work.
20. Operating with lack of discipline.
21. Operating with lack of skill or knowledge.

16.4 : Hazards In Important Machine Tools

A16.4.1 : Lathe Machine

1. Loose clothing may catch the revolving work, chuck, lathe dog, etc.
2. Attempt to remove chips by hand.
3. Flying heated chips may fall on any part of the operator when machining cast iron, brass, etc.
4. Danger during lifting of heavy material by using crane to the Lathe Spindle.
5. Open stock extending beyond the Capstan Lathe may cause injury unless properly guarded.
6. Splashing of coolant or cutting oil on the floor creates a slipping hazard in case of using screw cutting and automatic lathes. For safety, operator should wear safety glasses, apron, carefully watch his job, and concentrate on the job.

A16.4.2 : Boring-Drilling Machines

1. The work may not be securely clamped.
2. Improper alignment of the tool.
3. Proper clearances may not be maintained.

4. Operator may be injured during adjusting the machine without stopping it.
5. Operator attempts to drill during holding the work in his hand.
6. Serious injuries may be caused during clearing the chips with the hands.
7. Operator's long hair may be caught by the drill bit.
8. Serious eye hazard may be caused by the flying chips.

For safety, operator should be properly dressed.

A16.4.3 : Milling Machine

1. Accidental contact with the revolving milling cutters during removing chips.
2. Serious hazards around a cutter when operator uses loose clothings.
3. Flying chips may create a hazard.

For safety, operator should be properly dressed, wearing goggles, standing on the going-away side of the cutter.

A16.4.4 : Planing and Shaping Machine

1. Severe injuries may be caused during changing stop-dogs or riding the table during the operation in the planing machine.
2. Considerable hazard during handling large work material into or out of the planing machine.
3. Work may not be securely clamped to the shaper table.
4. A shaper offers real danger from flying chips.
5. Operator may be injured if proper clearance at the end of the ram-stroke is not being maintained.

For safety, operator should roll up his sleaves, wear safety glasses, concentrate on his job, clear chips and tools on the machine table.

A16.4.5 : Grinding Machine

1. Injuries due to bursting of abrasive wheels.
2. Eye injuries from flying chips.
3. Injuries from contact with the revolving wheels, discs or belts.
4. Explosion of abrasive wheels due to improper manufacturing and installation, using them incorrectly and running at incorrect speed.
5. Health hazard due to the dust removed during grinding.

For safety, operator must use goggles, safety glass guard, have excellent work habit, and also his sleaves rolled-up.

Moreover, shop-floor should be kept clean, from waste, scraps, dust, spillage, liquid splashing, etc., safe and orderly storage of tools, parts and all the rest of the items used in manufacturing.

A16.5 : Clothing And Safety Equipments

1. Operator should wear: (A) Appropriate clothings for the jobs.
   (B) Thick soles safety shoes for heavy work.
   (C) Safety glasses, goggles or face-shield.
   (D) Shirts or jumpers with sleaves cut-off or rolled above the elbows.
   (E) Gloves during rough and sharp material machining and handling.
2. Operator should not wear rings, watches, bracelets, neckties, loose or torn clothing, etc., to guard against getting caught with moving machinery.
3. Operator should remove gloves before turning on or operating any machine.

A16.6 : Safety Guards

To maintain safety, moving parts in the shop-floor must be properly guarded. The type of guard should be selected with regard to the nature of the machine, purpose and mode of operation.

Enclosed housings are generally provided to cover the revolving parts in Grinding Machine and Planing Machine. Gear Boxes of different machines, such as Capstan and Turret Machines, are totally enclosed. Cast Iron guard are provided for the belt and pulley drives. Fencings are arranged for textile technology. Guards should be provided for all dangerous parts, wherever situated.

Fixed Guards

Sheet metal, steel sheet, metal of small mesh, sometimes cast iron are the suitable materials of robust construction for the fixed guards. Metal of small mesh is suitable when a clear view of the moving parts is essential. Tubular guard fencing is suitable for projecting stock bars in Capstan Lathes.

Automatic Guard

This is operated by the moving member of the machine in which it is fitted. Automatic guards are suitable on medium and heavy power presses.

Interlocking Guard

This ensures that the machine will not start until the dangerous parts are fully protected by the guard and will not be exposed again until the machine is stopped. Interlock guards may be used in the dangerous beater blades of the opening machine, spinning gears, hydro-extractor covers.

Trip Guard

Trip Guard is used for projecting the nips between the pair of revolving parts or between revolving and reciprocating parts, into which the material is fed by hand. Trip guard may be a rod or wire or a movable device placed in such a position relative to the nip or rolls that the operator can stop the machine by clutching the guard which cuts out the driving motor. Rubber calender rolls, mixers, plates printing machine, padding mongals, etc., are used as trip guard for quick-stopping.

Distance Guard or Positional Guard

This is suitable where close or other types of guards are not suitable. The grills or bars or rails are placed at reasonable height and distance from the trapping line or area through which arm cannot be extended over the top of the fencing to the trapping area. This type of guard may be used in Electric and Photo-electric devices, Self-contained oil circuit fluid press, Tunnel, etc.

A16.7 : General Safety Precautions

1. Before operating, the machine should have effective and properly working guards in proper place.

2. The machine should be operated by an authorized operator to do so.

3. The machine should not be stopped by hands or body.

4. Before starting, the machine, operator should see that the work and tools are clamped securely.

5. Shop-floor should be free from oil, clear of metal chips or curls and waste pieces.

6. Operator should not leave the machine until it has stopped running, even after the power is off.

7. Operator should not run the machine when the Teacher/Instructor is not in the shop-floor.

8. Operator should take the help of handling equipment for movement of long and heavy pieces of metals.

9. Machine and switch should be operated by the same operator.

10. There should not be any projecting set screws, which may cause injury by catching on sleaves or clothing.

11. Operator should not lean towards the machine.

12. Operator should concentrate on the work and not to talk unnecessarily while operating the machine.

13. Operator should have sufficient light and ventilation for performing the particular operation.

14. Operator should get first aid immediately after any injury.

15. Nobody should disturb the operator while he is operating a machine.

A16.8 : Shop-Floor Safety (House-Keeping)

1. Shop-floor should be free from oil liquid of any kind, or grease to avoid slipping hazard.

2. Shop-floor should be clean and clear at all times, to avoid accidents.

3. The tools and work should be kept in position after completion of each work, to avoid toe or foot injury.
4. All scraps should be kept in scrap boxes at all times.
5. Store-materials and finished products should be properly handled by proper handling equipments to avoid tripping.
6. Tools should be kept in safe position, when they are not in use.

A16.9 : Safety Devices On Different Machine Tools

A16.9.1 : Lathe Machines

1. Should check that the work-tool and tailstock are properly clamped before starting the machine.
2. Should use hand power for putting on or removing chucks or face plate, etc., on head-stock spindle.
3. Should protect the machine parts against damage due to falling of the chuck while putting on or removing it.
4. Chuck wrench or tool should not remain on chuck, lastly, it may cause injury during running of the machine.
5. Should not use any wrench during rotation of the work or parts.
6. Should not try to measure the work or adjust the tool while the Lathe is running.
7. Should not shift or change gears while Lathe is running.
8. Should not give heavy cuts on long and slender work.

A16.9.2 : Drilling Machines

1. Should see that the drill bit in drilling spindle is running true or concentrically.
2. Should see that the drill is properly grind to machine the proper size.
3. Should use greater speed for smaller diameter drill and vise-versa.
4. Should use less speed for drilling case iron.
5. Should not remain any wrench on drill chuck, before starting the machine.
6. Should not hold the work by hand during drilling.
7. Should run the drill at proper speed and feed otherwise it may result in breaking the drill or damage the work or bring serious injury.
8. Operator should keep his head back, to avoid injury from any moving part of the drilling machine.
9. Should change belt only when power is off.
10. Should stop the machine and then make the adjustment, if the work slips or drill stops.

11. Should scrape or file all burrs from drilled holes.

A16.9.3 : Milling Machines

1. Should use proper cutters that are correctly ground and in good condition.

2. Should see that the cutter and arbor are rigidly clamped and arbor support does not obstruct the work.

3. Should not give heavy cut or feed while using cutters in a vertical milling machine to avoid injury.

4. Should not try to tighten or to take off the arbor, while the machine is running.

5. Should check proper speed and feed, and feed the work against the direction in which the cutter is rotating.

6. Should use a brush to remove chips.

7. Should keep the hands away from the work while machining.

A16.9.4 : Shaping Machines

1. Should be sure that the parts—ram, tool head, tool, work table support, clamping screw, vice, etc., are in position and properly clamped, before starting the shapers.

2. Should use proper close-mesh wire screen over the tool to catch the chips.

3. Should check the tightness of the adjusting nut after setting the stroke length and position.

4. Should remove all wrenches from the machine, after completing the set-up.

5. Should check that the current is 'ON' before starting the machine, if magnetic chuck is used.

6. Should not try to remove the chips, while the ram is in motion.

A16.9.5 : Planing Machines

1. Should adjust the length of bed and proper speed, while the planer is idle.

2. Should check that the work clears the cross-rails and the stop-pegs and safety-dogs are in proper places, after fixing the work on machine table.

3. Should see that the feed-rod and its attachment are properly located.

4. Should not ride over a moving job, bed or platen.

5. Should not leave tools of any kind between the ways.

6. Should hold the tool with one hand or place a wooden support under it, while loosening tool holders.

A16.9.6 : Grinding Machines

1. Should use suitable holding devices or clamp for grinding short pieces.
2. Should use goggles, even if the grinder is provided with protective glass shields.
3. All wheels should be tested for soundness before use.
4. Should not adjust the tool-rest while the machine is in motion.
5. Operator should stand to one side out of line of wheel, specially when starting a new wheel.
6. Work should be fed slowly and gradually, to avoid breaking of the wheel.
7. Should use the face of the wheel only, unless the wheel is designed for grinding on the side, to avoid breaking of the wheel.
8. Should not use a loosely filled grinding wheel or run it at unsafe wheel speed.
9. Should hold the work firmly against the grinding wheel, to avoid injury to hands and fingers.

A16.10 Questions

1. Write a short note on Safety (A 16.1).
2. State the term "Accident". Write a brief note on Accident (A 16.2).
3. Write the causes of an accident and its prevention (A 16.3).
4. Explain in brief the hazards in following Machine Tools:
   (a) Lathe Machines (A 16.4.1).
   (b) Boring and Drilling Machines (A 16.4.2).
   (c) Milling Machines (A 16.4.3).
   (d) Planing and Shaping Machines (A 16.4.4).
   (e) Grinding Machines (A 16.4.5).
5. Write about the clothings and equipment that are essential to an operator to maintain safety in an industry. (A 16.5).
6. Write a brief note on different types of Safety Guards (A 16.6).
7. Write how a shop-floor is maintained to ensure safety (A 16.8).
8. What are the General Safety Precautions (A 16.7)?
9. Write the safety devices in following Machine Tools:
   (a) Lathe Machines (A 16.9.1).
   (b) Drilling Machines (A 16.9.2).
   (c) Milling Machines (A 16.9.3).
   (d) Shaping Machines (A 16.9.4).
   (e) Planing Machines (A 16.9.5).
   (f) Grinding Machines (A 16.9.6).

17

NC & CNC MACHINE TOOLS

A17.1 : Numerical Control (NC)

Intention of modern industries are to maximizing the output with keeping the input to the lowest possible level within time schedule. This requires flexible automation in Machine Tools in the form of Digital Control. Instructions in the form of programs (either in Binary or in Numeric or in Alpha-numeric) fed to the system and when read by the system to regulate various slides of the machine tool to enable the tool(s) to shape the workpiece(s) to the required shape and sizes by positional and/or continuous control. The system is known as Numeric Control (NC) system. Definition of NC as given by ELA (Electronic Industries Association) is, "A system in which actions are controlled by direct insertion of Numerical Data at some point. The system must automatically interpret at least some portion of this data."

A17.1(A) : NC Machines Over Conventional Special Purpose Machines

Intention of modern industries are to maximizing the output with keeping the input to the lowest possible level and for which quality, reliability, efficiency, market demand, product changeability minimum production time, etc., are to be taken into account. Study of detailed drawing, selection of materials, machining processes, choice of cutting speed-feed-depth of cut, choice and selection of tools, inspection of dimensions, faster production of harder and tougher materials, etc., in conventional Machine Tools (i.e., Lathe, Milling, Drilling, etc., either manually or automatically operated) cannot cover these demands.

Numerical control system can meet the market demand and can cover more than 70% of total market requirements due to its highly flexibility, most economical for producing single/batch/large number of products.

Sl. Nos.	Points	Conventional M/c	Numerical Control M/c
1.	Mass-Production	Suitable for identical parts	Suitable for Part/Batch /Mass production, scope for variation of different products without involvement of high cost
2.	Time for Design, Drawing, processing, Inspection, etc.	Comparatively more	Comparatively less

Contd..

Sl. Nos.	Points	Conventional M/c	Numerical Control M/c
3.	Process suitable to meet market at reasonable cost and time	Good	Better
4.	Possibility for changing of operation sequence or product	Not possible, Once a particular production machine is set for a particular product	Yes possible, due to its flexibility in operation sequence as well as a new product by editing the stored program
5.	Operation sequence or program possible to store	No	Yes, can be stored in software
6.	Scope for Variation of operation and product	Too costly	Less costly

A17.2 : Computer Numerical Control (CNC)

A mini-computer is incorporated with Numerical Control system to control machine tool functions from stored-in information.

It is also defined as given by ELA (Electronic Industries Association).

"The numerical control system where a dedicated stored program computer is used to perform some or all of the basic numerical control functions in accordance with control programs stored-in read/write memory (RAM) of the computer."

The program is entered into the computer through keyboard or a tape. When a part is to be machined the stored memory is called for. Any modification of a program can easily be done by editing the existing program as per requirements. CNC machines are equipped with a diagnostics software, which enables easy trouble shooting if the CNC system fails to operate and also having capability to assist in maintaining and repairing the system. Servo system mainly consists of servo motors, feed-back devices, ball screws, etc., for accurate and rapid movements in X, Y and Z axes and provides a closed loop control system.

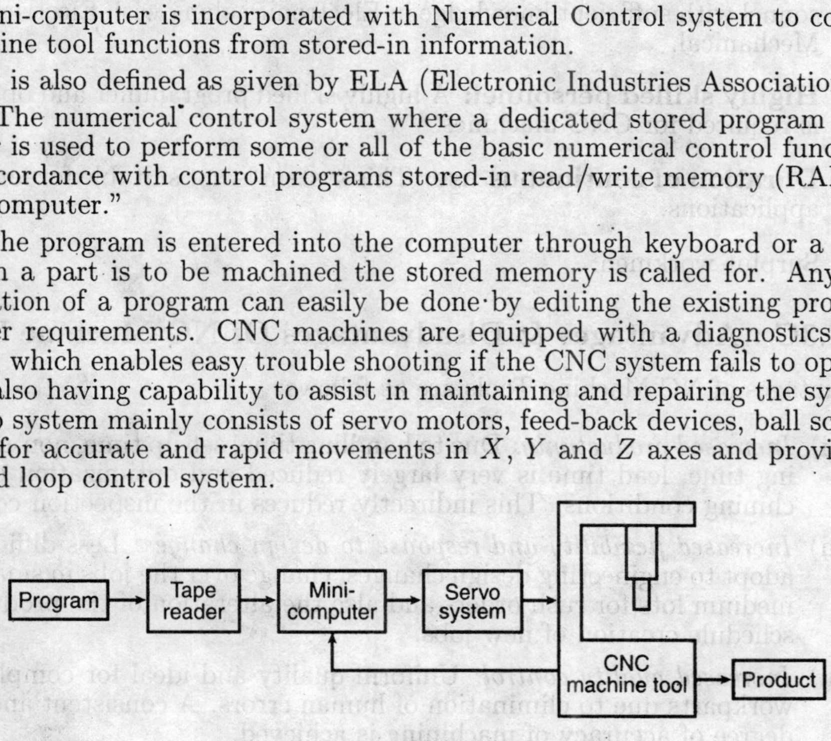

S-17.2 CNC System

A17.3A : Advantages Of CNC Machines Over NC Machines

1. **On line editing program:** Any change or mistake in cutting conditions, in dimension, etc., can easily be edited or modified by the part data stored in computer without sending the tape for re-processing.

2. **More flexibility:** Any modification in the program can be done easily without making a new tape instead of existing tape of conventional NC systems.

3. **Diagnostic:** A special software is introduced to enables easy trouble shooting if the CNC system fails to operate and also having the capability to assist in maintaining and repairing the system.

4. **Conversion of units:** Conversion of units from SI to British or from British to SI are easily convertible within the computer memory.

5. **Integration with DNC system:** In sophisticated manufacturing CNC can be integrated with Direct Numerical Control (DNC) system.

A17.3B : Disadvantages Of CNC

1. **Shop-floor conditions:** Shop-floor must be air-conditioned to run the CNC safely and smoothly.

2. **High initial investment cost:** Initial Equipment Cost and Installation Cost are much higher than Conventional NC Machines.

3. **High maintenance cost:** The complex machines requires trained personnel with sufficient knowledge in Electronics along with Electrical and Mechanical.

4. **Highly skilled personnel:** A highly skilled programmer and operator is required for CNC machines.

5. **Duration of continuous use:** This system is not suitable for long-run applications.

6. Surplus workmen.

A17.3C : Advantages & Disadvantages Of NC Machine Tools

Advantages of NC Machine Tools are as follows:

(i) *Increased productivity:* Due to handling time, set-up time, non-machining time, lead time is very largely reduced and optimization of machining conditions. This indirectly reduces in the inspection costs.

(ii) *Increased flexibility and response to design changes:* Less difficult to adopt to engineering design changes, change over the jobs in small and medium lots for rush orders and also the alteration of the production schedule creation of new jobs.

(iii) *Increased quality control:* Uniform quality and ideal for complicated workparts due to elimination of human errors. A consistent and high degree of accuracy of machining is achieved.

(iv) *Reduced handling cost:* Set-up and fitting cost, inspection cost, cost in lesser scrap, cost for elimination of special jigs and fixtures, etc., reduced handling cost.

(v) *Skillness of operator:* It is no longer important since the accuracy is dependent on program and hence reduces scrap loss and rework.

(vi) *Reduced parts inventory:* The amount of inventory control carried-out by the company is reduced owing to fewer set-up and shorter lead-times with numerical control.

(vii) *Greater utilization machine and man-power:* Better safety can be achieved to machine and operator. Semi-skilled operator can perform the operations. Optimum power utilization of machine, excellent reliability less floor-space required.

(viii) *Accurate predictions:* Programmed machining can be accurately predicted towards product cost and rate of production, no chance to deviate the actual production time with respect to the programmed time.

A17.3D : Disadvantages Of Numerical Control

Disadvantages of numerical controls are as follows:

(i) *High initial financial investment:* Cost of sophisticated and complex technology NC machine tools is much more in comparison with non-NC machine tools and therefore full utilization NC machines are essential without which whole investment will be blocked.

(ii) *Increased cost to complicated maintenance:* NC machine tools due to its complex technology requires well-trained maintenance task force to solve harder maintenance problems.

(iii) Need for replacement of personnel to high skilled personnel or well-trained retaining personnel as part programmers.

(iv) *Surplus personnel:* In case of modification of industry to increase production needs (to switch over from conventional machine tools to NC machine tool will require) less man-power for running the NC machine tools w.r.t. existing becomes a fairly large number of personnel surplus.

(v) Cost of control system is much more higher.

(vi) Operating cost/hour is higher.

(vii) Down-time of NC system is much expensive. So, it is essential for adequate training in operation and maintenance at the supplier's place before the machine is delivered.

A17.3.1 : Elements of NC Machine Tools

The main components (or elements) are:

(a) Software

(b) Machine Control Units (MCU)

(c) Machine Tools (MT)

Program(s) or set of instructions are punched on cards or magnetic tape or paper tape as per requirements of a product manufacturing this tape is known as software, which controls the movement of Numerical Control Machine Tools.

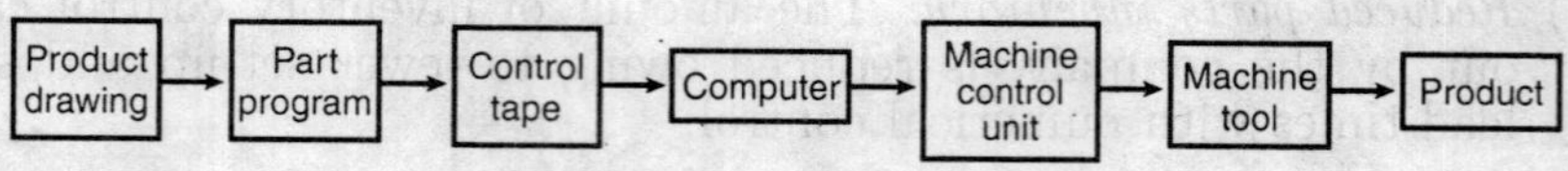

S-17.3.1

Process flow from product drawing to NC machine tools is as shown above.

The programmer plans the operation sequences with respect to product drawing and prepares a part programming in the coded blocks. These programs are punched on control tape and fed to the computer. The programs are stored in the memory of the computer. Computer controls the NC machine tool performance or movement through machine control unit where high level languages in the computer interpret to machine language for proper functioning of operations.

A17.3.2 : Types of Numerical Controls

The common types of Numerical Control Systems mostly used in machine tools are:

(a) Conventional Numerical Control (NC)

(b) Computer Numerical Control (CNC)

(c) Direct Numerical Control (DNC)

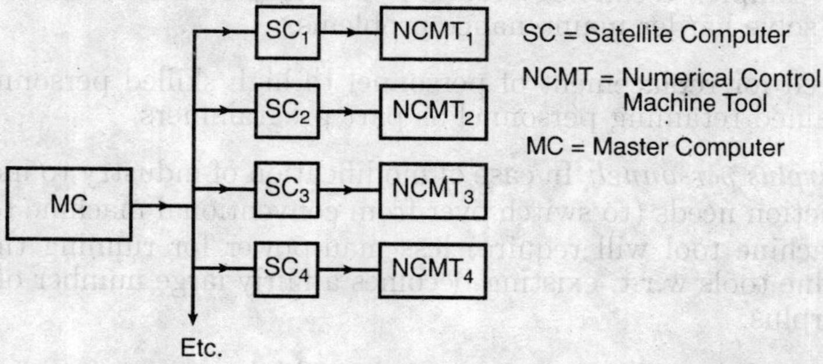

S-17.3.2

(a) Conventional Numerical Control (NC)

This system is simple and relatively cheap and is purely hardware based with integrated circuits which are permanently wired arranged on Printed Circuit Boards (PCBs).

(b) Computer Numerical Control (CNC)

This system is software based, the Machine Tools are controlled by the software of the computer. The system simplify the hardware of the conventional NC. Program is loaded in computer through keyboard or tape and stored in the memory, which can be called whenever required for machining. The program can easily be edited or modified as per requirement.

(c) Direct Numerical Control (DNC)

In this system, several NC or CNC machine tools are controlled remotely from a master computer. The processing and part processing of part program is performed a master computer (centralized), the program can easily be edited and modified according to the requirements. The master computer controls the different Numerical Control Machine tools through satellite computers. This system improves the communication between the control computer and the shop-floor:

A17.4 : Classification Of NC System

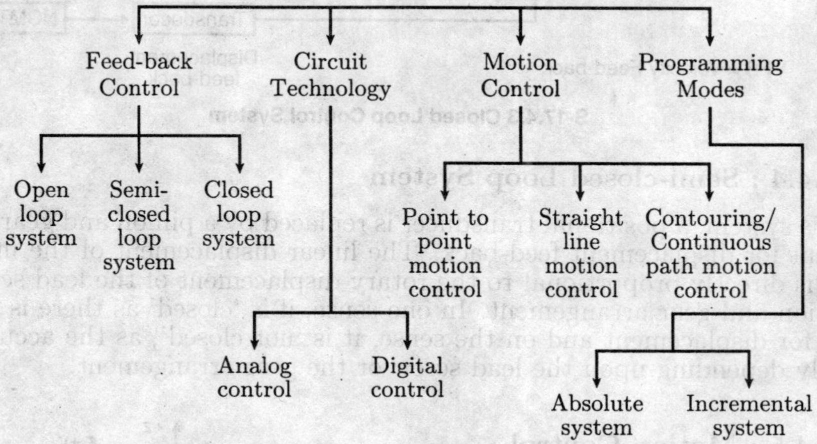

A17.4.1 : Feed Back Control

Servo mechanism consists of a group of Electro-mechanical Pneumatic or Hydraulic components which controls the position/velocity of a machine slide. The separate channel to servo system is used to the output from the data handling equipment which in turn drive the machine slides. This servo system is based on feed-back control.

A17.4.2 : Open Loop System

In this system there is no provision to ensure that the slide is actually acquired the desired position of the tool/work, it means no provision for feed-back. The final output value is not directly measured or checked against the desired value.

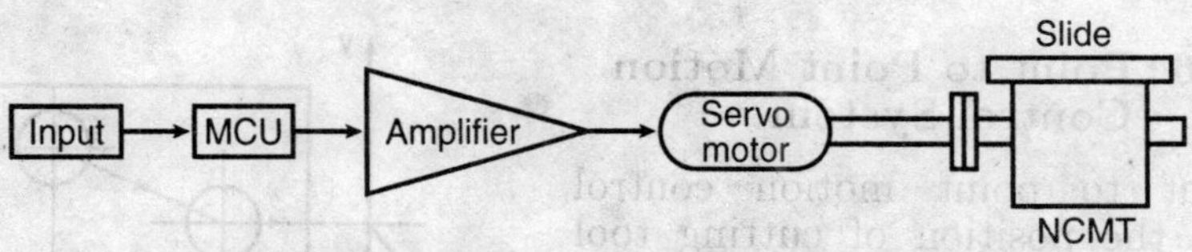

S-17.4.2 Open Loop Control System

A17.4.3 : Closed Loop System

If open loop system is provided with a feed-back device (known as transducer) which can compensate for errors in position of the tool/work. The system is known as closed loop system where transducer converts the motion of the slide to an electrical impulse. This impulse is fed back into the control unit as input signal. If the input signal and feed-back signal match (i.e., signals become of zero aptitude) the servo unit stops which indicates the desired position of the tool/work is attained. If there is any error in position, the two signals (one is input and other is feed-back) will miss-match. Another feed-back transducer (VFB) is used to control the velocity of motor in case of continuous path system.

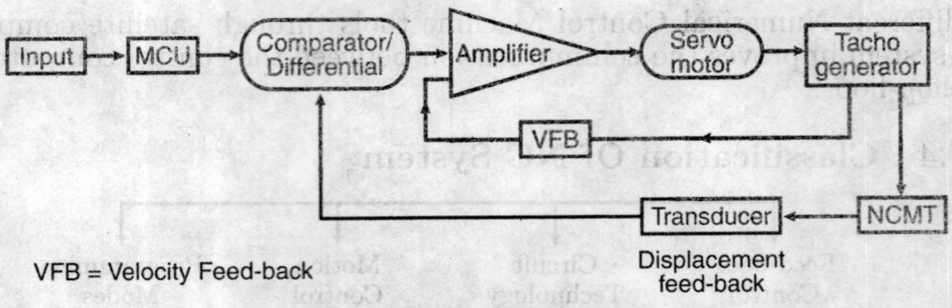

S-17.4.3 Closed Loop Control System

A17.4.4 : Semi-closed Loop System

In this system, a positional transducer is replaced by a pinion and gear arrangement for displacement feed-back. The linear displacement of the machine slide is directly proportional to the rotary displacement of the lead screw by a pinion and gear arrangement. In one sense, if is 'closed' as there is a feed-back for displacement and on the sense, it is 'not closed', as the accuracy is largely depending upon the lead screw or the gear arrangement.

A17.4.5 : Motion Control

The motion/movement of cutting tool with respect to workpiece can be achieved along the axis of motion, the main axes of motions are horizontal (X-axis), at right angle to horizontal (Y-axis) and perpendicular to both X-axis and Y-axis (Z-axis) i.e., vertical height of the tool from the workpiece.

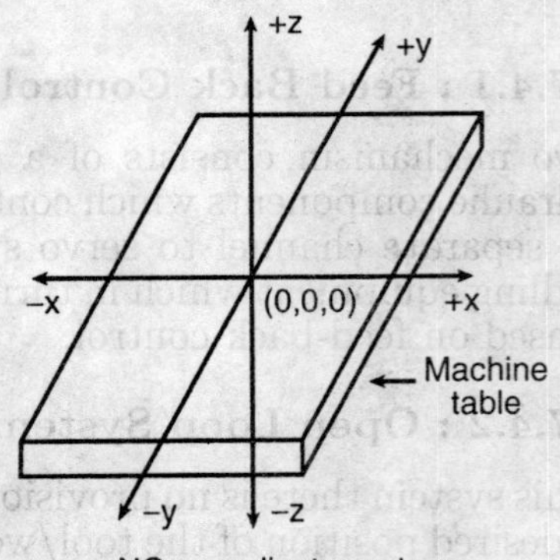

NC co-ordinate system

S-17.4.5

A17.4.6 : Point to Point Motion Control System

In point to point motion control system, the position of cutting tool changes from a particular fixed point to a point as per requirement in the shortest path and time span with respect to the workpiece. This system is cheapest and simplest in construction and suitable for drilling, tapping, boring, spot welding, etc. And the accuracy is about ±0.05 mm.

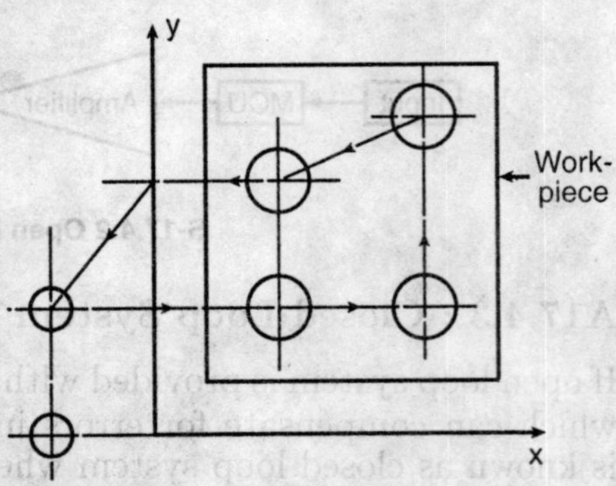

S-17.4.6

A17.4.7 : Straight Line Motion System

In this system, tool moves along a straight line path in the direction of a major co-ordinate axis along the direction of feed during milling, turning, boring operations at a controlled rate.

This type is mostly suited for plain turning or step turning, facing, face milling, horizontal and vertical milling, etc.

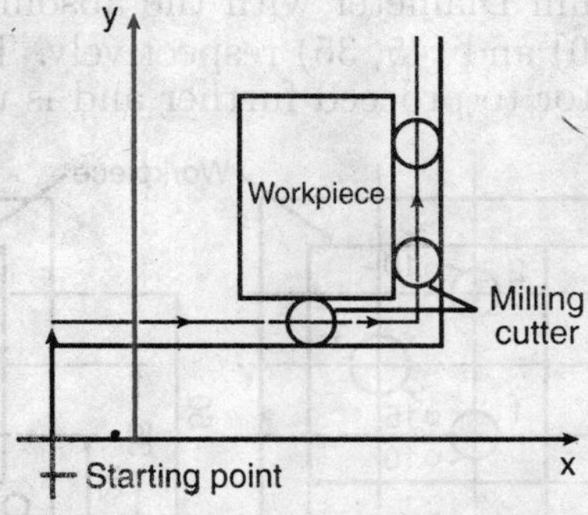

S-17.4.7 Tool moves parallel to either X-axis or Y-axis

A17.4.8 : Contouring/Continuous Path Motion Control System

The Numerical Control System, in which the position of table along with speed and tool feed are under continuous control. Each axis of motion with different speeds moves simultaneously for making conferring profiles. This systems are popular for intricated die template, flame cutting, fusion welding, etc.

A17.4.9 : Analog Control System

In this system, the table movement (and hence movement of the work) vary continuously with respect to the variation of electrical unit such as voltage. For example, the slide of a machine tool can move through 100 mm, represented by 10 volts and no movement by zero volt. Accuracy in the system can be achieved by the precision of sensors. The slide can be set to any position between two limits of slide with the variation of voltage continuously.

The accurate positioning can be achieved when the continuous input voltage is matched to the transducer output voltage. This is an ideal system for point to point positioning system.

A17.4.10 : Digital Control System

Digital Transducer is employed to convert the amount of movement of work table into convertible (in the form of a) electrical voltage pulse, where each pulse represents a basic length. Transducer produces electrical pulses as the table moves and is compared with the input signal. Number of pulses needed for a given movement of the table. Digital feedback systems are ideal for continuous path control with a finite accuracy.

A17.4.11 : Absolute System

In absolute system, all the measurements are taken from a fixed zero point or reference point with co-ordinates X = 0, Y = 0 and Z = 0, generally considered at left-hand bottom corner of the workpiece. This reference point serves as datum from all the distances are measured parallel to each axis of the system.

In Figure S-17.4.11(A) shows the original shop drawing and in Figure S-17.4.11(B) shows the absolute zero system of measuring each movement of the tool measured from (0, 0) co-ordinates. In a workpiece three holes (1, 2, 3) are to be made, out of two holes @10 mm diameter and one hole of

@15 mm Diameter with the absolute co-ordinates of the holes are (25, 20), (25, 50) and (45, 35) respectively. This system is easy and convenient for an operator to proceed further and is used in point-to-point applications.

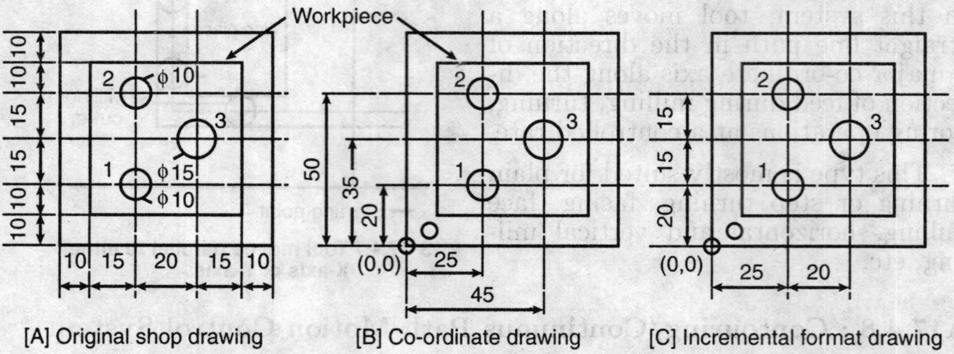

S-17.4.11 & S-17.4.12 Dimensioning of a Workpiece to indicate tool position

A17.4.12 : Incremental System

Figure S-17.4.12(C) shows the incremental formal drawing. This system needs no predetermined datum. All the measurements for the next position are measured in the form of increments of distance from the previous point at which the tool was resting. The co-ordinates of holes 1, 2 and 3 are (25, 20), (0, 30) and (20, −15) respectively. This system is generally cheaper to built.

A17.5 : Machine Control Unit (MCU)

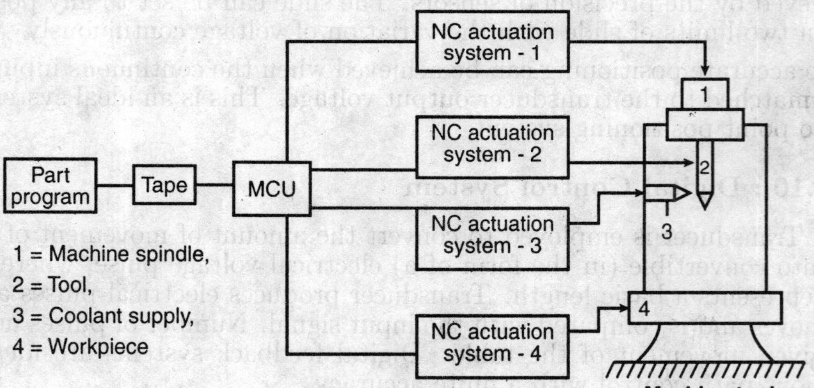

S-17.5 Machine Control Unit (MCU)

Machine Control Unit (MCU) of an NC Machine Tool is to read interprets and converts perceived input (Part Program) into appropriate movement and controls the following functions through signals for NC actuation systems:

1. To start and stop the machine spindle.
2. To vary the spindle speed and direction of rotation of the spindle.
3. To start and stop the coolant supply.
4. To change the required tool as and when required.

5. To change the desired workpiece.
6. To guide the tool tip along the desired position/path/straight line movement/rotary movement, etc.
7. To control the feed rate, etc.

A17.6 : Part Programming

A17.6.1 : Fundamentals of Part Programming

The use of NC Machine Tools requires more planning and data than conventional Machine Tools. Steps of data collections are as follows:

1. Data is to be selected with respect to sequence of operations.
2. Information in the form of dimension, shape and tool path, etc., are to be collected from shop drawing.
3. Selection of speed, feed, depth of cut, coolant required, job-tool materials, etc., are to be choiced on surface quality desired.
4. Other related data based on the machine.

Basic procedural steps are as follows:

1. Drawing of the product desired (i.e., shop drawing).
2. Preparation of NC Co-ordinate drawing (i.e., conversion of shop drawing to NC Co-ordinate drawing).
3. Process planning with respect to the sequences of operations or the route through for preparing workpiece (i.e., route sheet).
4. Preparation of part programming based on operation sequences (i.e., program manuscripts).
5. Preparation of punched tape based on program manuscripts.
6. Verification of punched tape for accuracy is by inputting the tape to the machine and by performing machining operations on a wooden block and then checked by measuring.
7. Finally the NC Machine is ready for production if the measuring the wooden job found OK but periodic inspections are necessary during running the machine to minimize rejection.

A17.6.2 : Procedure for Developing Manual Part Program

In part programming is required to consider some fundamental elements are as follows:

1. *Dimensioning:* This system generally based on either absolute or incremental.
2. *Axes drive system:* Both X and Z axis motions are obtained by permanent magnet DC stepper motors. Each linear step on the slides is 0.01 mm.

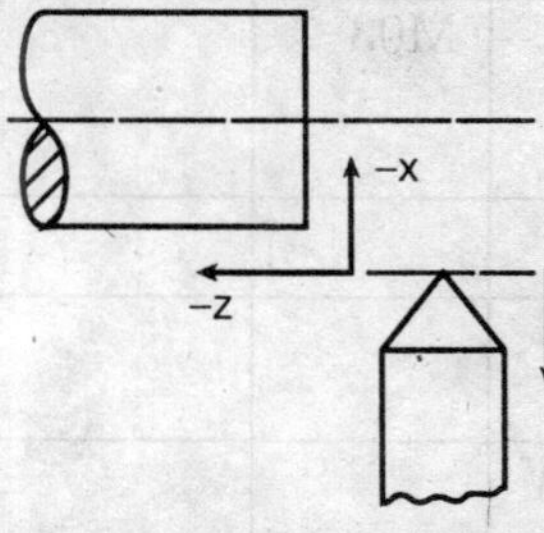

S-17.6.2

3. *Axes drive system:* Both X and Z axis motions are obtained by permanent magnet DC stepper motors. Each linear step on the slides is 0.01 mm.

4. Positive (+)sign when Tool moves away from the job.
   Negative (−) sign when Tool moves towards the job.

NC Words:

SINGLE BLOCK NC INSTRUCTIONS

N001 G01 X1.00 Z2.00 T1 F0.25 S400

N001	Indicates sequence number or line number
G01	Preparatory functions, e.g., G00—Rapid Transverse, G01, Linear Transverse, etc.
X1.00	Linear dimensioning for tool movement cross-wise.
Z2.00	Linear dimensioning for longitudinal tool movement X, Z, Y, a, b, c are the dimension data.
T1	T stands for tool selection.
F0.25	F stands for Feed route in 0.25 mm/rev., 100 mm/min.
S400	S stands for spindle speed in rpm.
M02	M stands for miscellaneous functions, i.e., M02 stands for end of program, M06 tool change, M36 feed change 1, etc.
EOB	End of Block, i.e., End of Instruction Block.

A17.6.3 : Common Codes used in Part Programming

Codes	Digits	Units	Explanations
N	N001		Line number or Sequence number
G	G00		(Rapid Traverse) Preparatory functions
X,Y,Z		mm	For primary tool motion dimension about X, Y and Z linear movement.
F	F100	mm/min.	Feed rate 100 mm/min.
S	S350	RPM	Spindle speed 350 mm/min.
T	T1	Number	Tool Number 1 (one) for Facing/Turning, etc.
M	M03		Miscellaneous functions, i.e., M03 indicates spindle start in clockwise direction.
EOB			End of Block or End of Instruction Block.
A,B,C			For angular dimensions about X, Y and Z axis respectively.
U,V,W			For secondary motion dimensions parallel to X, Y and Z axis respectively.

A17.6.4 : Problem No. 1

Two drill holes 20 mm diameter and 10 mm diameter are to drill on MS Plate of 16 mm thickness and position of holes as shown in the sketch (using absolute dimensioning).

Steps of Part Programming

1. Point A (0,0) and D are taken as reference.

2. Position of holes at B and C are (20,60) and (50,25) respectively in which centres are punched.

3. Drill bit of 20 mm diameter is fixed in the spindle.

4. The drill bit of 20 mm diameter is positioned on point 'B'.

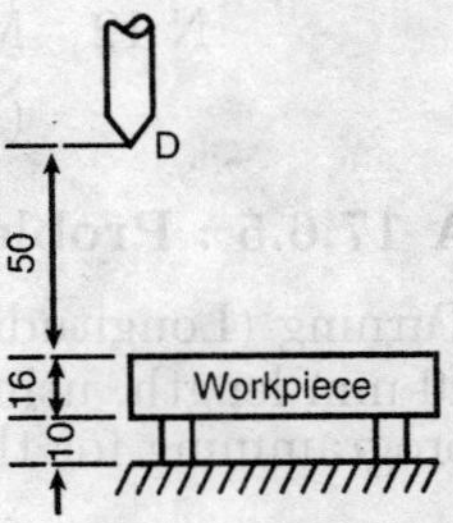

S-17.6.4

5. Rotation of drill bit is started and applied proper feed.

6. The operation of drilling is continued till full depth.

7. The drill bit is withdrawn.

8. The withdraw drill bit is replaced by 10 mm diameter.

9. Repeat the operation as laid down to 4 to 7 operations.

10. The workpiece is removed from the table.

Part Programming : Part of above steps are as follows:

1) N01 G 92 X 000 Y 000 Z 000

Operator is the move table to bring the drill bit above point A (0,0) is the control system. G 92—Absolute Preset.

2) N02 G 90 (EOB)

The distances are measured with respect to absolute system from point A. G 90 → Absolute Data.

3) N03 G 81 X 20000 Y 60000 Z 76000 S 40000

4) N04 F 100 T 01 M 03 (EOB)

5) N05

6) N06

G 81 → Drill; (20,60) (x,y); Z → 76,

S 40000 → Spindle speed 400 rpm, F 100 → Feed rate 100 mm/min.

T 01 → Tool no. 1 of 20 mm Diameter,

M 03 → Spindle Clockwise Rotation.

Tool changing; To move machine till point B comes under the tool; Drill bit rotates and moves down at reasonable speed and slow feed and continues till drill hole completes the whole depth.

7) N07 G 80 Z 0 F 0 M 05 (EOB)

G80 Cancel cycle, M 05 → Spindle rotation stop.

8)	N08	G 81	X 50000	Y 25000	Z 71000	S 40000	
9)	N 09		F 100	T 02	M 03	(EOB)	
	N 10						

Tool change over to 10 mm diameter (T02);

10)	N 11	X 000	Y 000	Z 000	F 0	M 05	(EOB)

N 12 M 02 Feed-back to original reference position (EOB)

(M 02 Stands for End of Program)

A 17.6.5 : Problem No. 2

Turning (Longitudinal and Taper) of Aluminium rod 20 mm diameter and 40 mm length, using speed 1000 rpm and feed 150 mm/min. Prepare a part programming for the above job.

(Using Incremental Dimensioning)

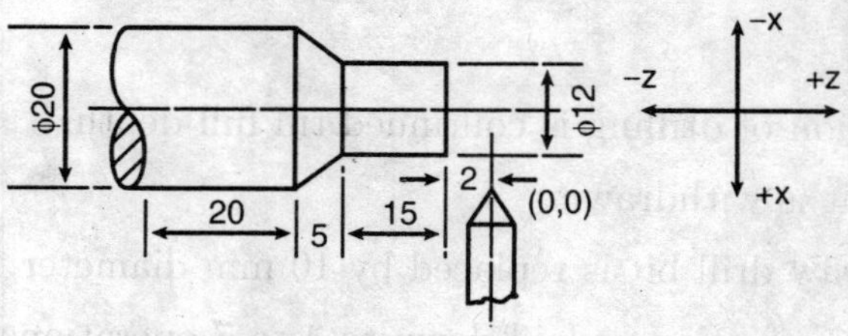

S-17.6.5

N 010	G 91	G 71	G 94	M 03	T 01	S 1000	(EOB)
N 020	G 01	X 10		F 150			(EOB)
N 030	G 00	Z 2					(EOB)
N 040	G 00	X 6					(EOB)
N 050	G 01	Z 17					(EOB)
N 060	G 01	X 4	Z 5				(EOB)
N 070	G 00	X 0	Z 22				(EOB)
N 080		M 02	(M 02 for End of Program) (EOB)				

[*Note:* G 91 → Incremental, G 71 → Metric Units, G 94 → Feed rate in mm/min., M 03 → Spindle rotates clockwise, S 1000 → Spindle speed 1000 RPM, G 01 → Linear Interpolation, F-150 → Feed 150 mm/min., G 00 → Positioning.]

Specifications of a CNC Lathe

A17.7 : Machine Specifications

1. Centres height–70 mm
2. Distance in between centres–310 mm
3. Swing over bed–100 mm
4. Swing over cross-slide–60 mm
5. Transverse of cross-slide–55 mm

Headstock

6. Hole in the Headstock spindle–16 mm
7. Spindle end use taper–Morse Taper no. 2

Tailstock

8. Spindle diameter–22 mm
9. Spindle end taper Morse Taper no. 1
10. Sleeve Travel–35 mm

Feed rate

11. Rapid Transverse–700 mm/min.
12. Programmable feed rate–1 to 700 mm/min.

Main Spindle

13. Speed Range–50 to 3200 rpm
14. Output Power–0.3 kW.

A17.8 : Questions

1. Define the terms—NC, CNC and DNC [A 17.1, A 17.2 and A 17.3.2(C)]
2. Write the elements of NC Machine Tool (A 17.3.1)
3. Explain in brief with necessary sketch the common types of Numerical Control System mostly used in Machine Tools. [A 17.3,2]
4. Differentiate between NC and CNC machines. [A 17.3A and A 17.3B]
5. what are the advantages of NC machines over conventional machines? [A 17.1(A)]
6. What are the advantages and disadvantages of CNC over NC machines? [A 17.3 A and B]
7. Give merits and demerits of NC machines [A 17.3 C and D]
8. Classify the NC System. [A 17.4]
9. Differentiate with necessary sketches the Point to Point Motion Control System and Straight Line Motion System [A 17.4.6 and A 17.4.7]
10. Differentiate Analog Control System and Digital Control System [A 17.4.9 and A 17.4.10]
11. Distinguish between Absolute System and Incremental System [A 17.4.11 and A 17.4.12]
12. Write the functions of Machine Control Unit (MCU) of a NC Machine Tool (A 17.5)
13. Sketch and label the components of Machine Control Unit [S 17.5]

18

INSTALLATION OF MACHINE TOOLS

A18.1 : Principles Of Installation Of Machine Tools/Machines

Installation of Machine Tools/Machines consists of checking up the machines for completeness, i.e., to check whether all the parts of the Machine Tools run at rated capacity with desired accuracy, Machine Tools should be balanced and with proper alignment. To achieve those objectives the following factors to be considered carefully:

(i) Site selection or location of the Machine Tool foundation

(ii) Laying of foundation bolt

(iii) Placing the machine on foundation

(iv) Aligning the machine with the driving or driven equipment

(v) Grouting on the foundation and finally

(vi) Testing for normal running at rated capacity with desired accuracy.

Thus installation of a Machine Tool will denote the process of placing the machine tool in a definite position. Say, in a bed or foundation, which is generally made of concrete for reliable operation, i.e., without much vibration or failure during working of the machine.

A18.2 : The General Requirements/Design Of Foundation Of Machine Foundation

In order to ensure that the foundation of a machine is stable and free from producing undesirable vibrations, we must know the following:

(i) Type of the Machines/Machines Tool, horizontal or vertical.

(ii) Nature of forces and moments, torque, etc.

(iii) Weight of the Machine Tool.

(iv) Size or bed plate of the Machine Tool.

(v) Speed and power of the Machine Tool.

(vi) Location of grooves, openings in the foundation for pipe lines, foundation or anchor bolts, flywheel, etc.

(vii) Location of the foundation and load bearing capacity of the soil.

A18.3 : Procedure For Machine Tools To Be Levelled On The Machine Foundation

It is essential to level the Machine Tool on the foundation block and align it with the driving or driven mechanism properly to avoid any possibility of damage to the machine or foundation block.

To level the machine tool a number of steel wedges is used. These wedges are about 5 to 8 cm long 2.5 cm to 3 cm wide, and about 3 to 6 mm thick, tapering uniformly to nothing.

If the base of the machine is a rectangle, one wedge should be driven in at each corner and one at some suitable intermediate point. For machine which stands on two legs, such as Lathe, one wedge should be driven in at each corner of the legs. It is always necessary to ensure that all the wedges are light. The machine must be levelled to horizontal position by levelling longitudinally as well as in the traverse direction. A good spirit level with an accuracy of 0.05 mm/m is to be used for testing the levelling work. For machines having long beds, such as Lathes, cross-tests should be made at each end. Sometime the machine is required to be completed with a driving machine and in such case the centre line of both the machines must be brought to the same line. Any error in levelling and alignment may cause vibration and damage to the machine.

After the levelling and alignment are done, the machine is ready for grouting.

Arrangement of wedges for levelling of Lathe Machine on foundation block.

A18.4 : Sequences For Installation Of Lathe Machine/Drilling Machine Or The Operational Steps To Be Followed For Installation Of A Lathe In A Workshop

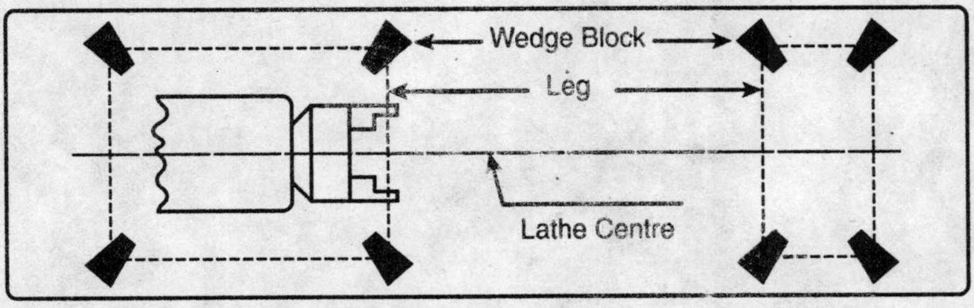

S-18.4 Lathe Machine Installation

For installing a Lathe Machine/Drilling Machine the following steps are to be followed:

(i) Reading of information including drawings generally supplied by the manufacturer of the machine.

(ii) Locating the site of installation.

(iii) Selecting suitable devices for lifting the machine.

(iv) Loading and moving the machine to be installed.

(v) Suspension of the machine for certain time by some means.

(vi) Unloading the item at the proper location.

(vii) Moving the item horizontally to place at exact location.

(viii) Making a foundation for bearing the load of the item and to minimize vibration hazards as far as possible.

(ix) Levelling and aligning the item for smooth running.

(x) Making a trial run to check satisfactory operation with desired accuracy after the foundation is set.

A18.5 : Questions On Installation Of Machine Tools

1. Write the principles of installation of Machine Tools (A 18.1).

2. State the general requirements of Machine Foundation (A 18.2).

3. Explain the procedure of levelling the Machine Foundation (A 18.3).

4. Write the sequence of operation for Installation of Lathe Machine (A 18.4).

19

OBJECTIVE TYPE QUESTIONS AND ANSWERS

Fill up the blanks:

1.1. Intension of modern industries is to create appropriate number of useful _____ at reasonable price.

1.2. _____ is a device for transforming or transferring energy.

1.3. The parts in contact with the job for doing different operations or generation of surfaces of products are known as _____.

1.4. _____ _____ used for removing metal to attend desired shape and size and operated by machine power.

1.5. In _____ _____ operations, a metal is given a desired shape and size by removing excess material in the form of chips.

1.6. In _____ _____ operation, a metal is given desired shape and size by deformation without removing any chips.

1.7. Rolling, Forging, etc., are the examples of metal _____ operation.

1.8. Turning, Drilling, Milling, Grinding, etc., are the examples of metal _____ operations.

1.9. _____ Machine Tools are to perform a great variety of machining operation with variety of jobs by using number of attachment.

1.10. _____ drive possesses low initial cost, economical in fixed charges, having overloading and variable loading facilities and one motor can run different Machine Tools at a time.

1.11. _____ tool is required in Machine Tools for giving desired shape and sizes of a product.

1.12. In _____ cutting, the cutting edge of the tool is 90 degree to the directions of relative motion between the tool and job.

1.13. In _____ cutting, feed force and cutting forces are considered.

1.14. Effective time elapsed between two successive grinding of a tool is called _____.

1.15. _____ tool life equation is the relationship between the cutting speed and tool life.

1.16. _____ is defined as the property of the material by virtue of which it may be machined easily with a cutting tool under a given set conditions.

1.17. _____ is supplied at the job-tool interface to improve the cutting condition and also tool life.

1.18. Synthetic coolants is mostly used in _____ operations.

1.19. Sodium nitrate is used with cutting fluid as _____ inhabiter.

1.20. _____ is added in compound oil cutting fluid for antiweld characteristics.

1.21. _____ oil is used as cutting fluid for machining soft steel.

1.22. Systematic naming of different single-point cutting tool angles are know _____ _____ for machining M.S., M.C.S., etc. by H.S.S. cutting tool.

1.23. Ideal condition of setting of tool should be on _____ line during operation, which will maintain both front clearance and back rake angles.

1.24. _____ is a floor of assembly of several Machine Tools like, Lathe, Powe Saw, Drilling machines, etc.

1.25. In metal cutting, the cutting tool removes a layer from the workpiec to obtain a desired shape and dimensions of the workpiece. The laye separated near the cutting edge of the tool is known as _____.

1.26. _____ chips are produced when the material ahead of the cutting edg continuously deform with fracture or rupture and flows off the tool fac by plastic deformation in the form of a ribbon.

1.27. _____ _____ chips are produced while machining typical material i which the yield strength decrease with rise in temperature like steel titanium alloys, etc., at medium cutting speed.

1.28. While turning ductile material producing long unbroken chips, the device used to break into comparatively small pieces for easy handling c chips is known as chip _____.

1.29. Wear or worn on the top flank of the tool at chip-tool interface durin machining is known as _____ wear, and wear or worn on side flank c the tool at job-tool interface is known as _____ wear.

1.30. Cast-iron during machining produces _____ chips.

1.31. Mild steel during machining with appropriate cutting condition produces _____ chips.

1.32. Negative rake angle is usually provided on _____ tool.

1.33. Binding material used in Cemented Carbide tools is _____.

1.34. Tool signature consists of _____ elements.

State whether the following statements are True or False:

1.35. Tools which are operated or controlled by a motor or any prime-move in performing different operations known as Hand Tools.

1.36. There must be two relative motions in any metal cutting.

1.37. Single purpose machine tools are designed to perform a certain specifie operations in machining a large quantitatives.

1.38. In a self-drive system, each machine tool is driven by its own motor.

1.39. In self-drive, layout is easy and less chance of accident.

1.40. Overhead crane facility is difficult in self-drive system.

1.41. Overloading is possible in individual drive.

1.42. Right-hand tool moves from right to the left-hand end of the Lathe bed.

1.43. High speed steel tool having 18% Tungsten, 4% Chromium, 1% Vanadium and rest Iron.

1.44. Cemented carbide tool having 150 m/min. cutting speed and 1500°C hot hardness.

1.45. In oblique cutting, the cutting edge of the tool is inclined (obtuse angle) at an angle greater than 90 degree to the direction of relative motion between the tool and the job.

1.46. The equation $VT_L^n = C$ is true in case of Taylor's tool life.

1.47. Machinability increases with the increase of cutting forces.

1.48. Choice of cutting fluid mainly depends on friction at job-tool interface, job material, type of operations, cutting conditions, etc.

1.49. Water is used as dry coolant.

1.50. Use of Sulphur in mixed oil is essential for tough and stringy material machining.

1.51. Mineral lard oil as cutting fluid is used to machine Hard Steel and Bronze.

1.52. Positive rake angle in single point turning tool is essential in shaping and planing machines.

1.53. Nose radius in cutting tool reduces tool life and surface finish.

1.54. Continuous chips with built-up edge are formed during machining of mild steel with low cutting speed, high feed, without using cutting fluid, etc.

1.55. Shank is that portion of the tool which does not ground to form cutting edges.

1.56. Positive back rake angle of a single point tool is the slope downward towards the shank.

1.57. Cutting fluid has no effect on the tool life.

1.58. Higher depth of cut increases rate of metal removal.

1.59. Cutting fluid is used to reduce friction and improve tool life.

1.60. In machining cast-iron, dry air is used.

1.61. Modern Machine Tools are not using self-drive.

1.62. Tool life is inversely proportional to speed.

1.63. Sodium Nitrate is not used as rust preventer with cutting fluid.

1.64. In metal cutting, chips are removed due to plastic deformation of metals.

1.65. In oblique cutting, the maximum chip thickness occurs at the middle of the generating surface.

1.66. Rate of metal removal is decreased by increasing the depth of cut in any metal cutting.

1.67. Back rake angle of a single point tool is the angle by which the face of the tool is inclined towards the shank.

1.68. In order to prevent tool against rubbing with the work, front clearance angles on tools are provided.

1.69. Positive back rake angle on tools are provided for easy flow of chips.

1.70. Crater wear is usually occurred while machining brittle materials.

1.71. Coolant or lubricant has no effect on tool life.

1.72. Tungsten increases the hot hardness of tool material.

1.73. Tool signature consists of six elements.

1.74. In machining cast-iron, cutting fluid as blow of air is required.

Choose the correct one from the alternatives given below against each question:

1.75. In machining mild steel workpiece by H.S.S. tool, the average cutting speed for general work is:

(i) 20 m/min. (ii) 30 m/min. (iii) 40 m/min. (iv) 50 m/min.

1.76. In machining cast-iron workpiece by H.S.S. tool, the average cutting speed for general work is:

(i) 25 m/min. (ii) 35 m/min. (iii) 45 m/min. (iv) 55 m/min.

1.77. In machining mild steel workpiece by Cemented Carbide tool, the average cutting speed is:

(i) 30 m/min. (ii) 60 m/min. (iii) 150 m/min. (iv) 600 m/min.

1.78. In machining M.S. workpiece by H.S.S. tool, the hot hardness of the tool is:

(i) 350°C (ii) 900°C (iii) 1000°C (iv) 1200°C

1.79. In machining M.S. workpiece by Cemented Carbide tool, the hot hardness of the tool is:

(i) 900°C (ii) 1200°C (iii) 1500°C (iv) 1650°C

1.80. Choose the machine tool designed to perform a great variety of machining operations with a variety of workpieces by using number of attachment:

(i) Single-purpose M/c Tool (ii) Product-oriented M/c Tool
(iii) Universal M/c Tool (iv) None of the above.

1.81. Choose the cutting tool material mostly used for general purpose:

(i) Medium alloy steel (ii) High speed steel
(iii) Cemented carbide (iv) Ceramic.

1.82. Choose the cutting tool material mostly used in production work:

(i) H.S.S. (ii) Cast non-ferrous stellite cobalt
(iii) Cemented carbide (iv) Ceramic.

1.83. The relation between the tool life (T) and cutting speed (V) according to the Taylor's tool life equation is:

(i) $T_L V^n = C$ (ii) $VT_L^n = C$ (iii) $V^n/T_L = C$ (iv) $T_L^n/V = C$

1.84. In any metal cutting, cutting forces at job-tool contact point is measured by:

(i) Wattmeter (ii) Dynamometer
(iii) Pyrometer (iv) Hydrometer

1.85. Cemented carbide tool tips are fixed on tool holder by:

(i) Welding (ii) Soldering (iii) Brazing (iv) Clamping

1.86. Cemented Carbide tools retain hot hardness due to use of:

(i) Chromium (ii) Tungsten
(iii) Nickel (iv) Vanadium

1.87. Main characteristics of cutting tool materials are:

(i) High toughness and low co-efficient of friction
(ii) To retain strength and hardness at elevated temperature
(iii) To withstand shock and vibration and to prevent breakage
(iv) All of the above.

1.88. The main function of chip breaker is:

(i) Break chips into segments (ii) Flashing chips from bed
(iii) To increase tool life (iv) To minimize cutting forces.

1.89. The cutting fluid mostly used for machining mild steel is:

(i) Mineral lard oil (ii) Soluble oil
(iii) Sulphur base mixed oil (iv) Chlorine base mixed oil.

1.90. Cutting fluid used for machining cast-iron is:

(i) Blow of air (ii) Water
(iii) Soluble oil (iv) Compound oil

1.91. Cutting fluid used for machining mild steel and bronze is:

(i) Soluble oil (ii) Synthetic coolant
(iii) Mineral lard oil (iv) Turpentine

1.92. In oblique cutting of metals, the cutting edge of the tool is:

(i) 90 degree to the direction of relative motion between the tool and the job
(ii) Inclined to less than 90 degree to the direction of the tool travel
(iii) Perpendicular to the workpiece
(iv) Parallel to the direction of the tool travel

1.93. Which statement is true in case of a Diamond?

(i) Hardest material (ii) Used as a tool material
(iii) Expensive (iv) Used in jewellery

1.94. The cutting edge of the right-hand tool in turning machined most efficiently when tool travels:

(i) Across bed (ii) From right to the left
(iii) From left to the right (iv) None of the above

1.95. Tool moves from Headstock end to the Tailstock end known as:

(i) Turning tool (ii) Thread cutting tool
(iii) Right-hand tool (iv) Left-hand tool

1.96. In order to decrease the cutting forces, improve strength of steel and wear resistance, alloying element generally added to steel is:

(i) Tungsten (ii) Nickel
(iii) Manganese (iv) Molybdenum

1.97. Which force is the main force in orthogonal cutting during turning operation?

(i) Feed force (ii) Cutting force
(iii) Thrust force (iv) None of the above

1.98. What are the forces considered in orthogonal cutting:

(i) Cutting and feed forces (ii) Cutting and thrust forces
(iii) Feed and thrust force (iv) Cutting feed and thrust forces

1.99. Tool life is over due to:

(i) Excessive heat generated due to friction at job-tool interface
(ii) Very poor surface finish
(iii) Sudden increase of cutting forces and hence more power consumption
(iv) All of the above

1.100. Tool life of a cutting tool can measure by:

(i) Effective operational time
(ii) Number of workpiece machined between resharpenings
(iii) Volume of material removed between resharpenings
(iv) All of the above

1.101. Factors for evaluation of machinability is:

(i) Tool life (ii) Power consumption and cutting forces
(iii) Shear angle and type of chips
(iv) Surface finish (v) All of the above

1.102. Functions of cutting fluid in a metal cutting operation are:

(i) To cool chip, tool and workpieces
(ii) To lubricate and reduce friction
(iii) To improve surface finish and decrease wear
(iv) All of the above

1.103. Most ideal property of an ideal cutting fluid is:

(i) Low viscosity (ii) Non-contaminate
(iii) High heat absorbent (iv) Low evaporation rate

1.104. Which of the following statements is wrong?

(i) Positive rake angle, increase tool life and decreases power consumptions
(ii) Negative rake angle, strengthen the tool
(iii) Zero rake angle, improves surface finish
(iv) Higher clearance angle, reduce friction and wear

1.105. The surface finish is improved by:

(i) Optimum cutting speed (ii) Greater rake angle
(iii) Moderate nose angle (iv) All of the above

1.106. Choose, which of the items will be increased for bulk of metal removal:

(i) Machine speed (ii) Feed
(iii) Tool geometry (iv) Depth of cut

1.107. Which statement favours the advantages of group drive?

(i) Group of machine tools run by a single motor
(ii) Available of variable power consumptions
(iii) Overloading is possible
(iv) All of the above.

1.108. Tool life is very much effected by:

(i) Depth of cut (ii) Tool geometry
(iii) Cutting speed (iv) Feed

1.109. Nature of curve against relationship between the cutting speed (V) and the tool life (T_L) is:

(i) Parabola (ii) Hyperbola
(iii) Straight Line (iv) None of the above

1.110. In Taylor's equation $VT_L^n = C$, value of index is closely depending on:

(i) Set conditions (ii) Work material
(iii) Tool material (iv) None of the above

1.111. Main function of an ideal cutting fluid is:

(i) As coolant (ii) As lubricant and coolant both
(iii) As lubricant (iv) As flashing action

1.112. Last element of tool geometry is:

(i) Side cutting edge angle (ii) End cutting edge angle
(iii) Trial angle (iv) Nose radius.

1.113. With same tool life, the production capacity can be increased by increasing:

(i) Feed (ii) Cutting speed
(iii) Depth of cut (iv) None of the above.

1.114. During machining cast-iron produces _____

(i) Continuous chip
(ii) Discontinuous chip
(iii) Continuous with built-up-edge chip
(iv) Ground chip.

1.115. During machining mild steel produces _____

(i) Continuous chip (ii) Discontinuous chip
(iii) Continuous with built-up-edge chip
(iv) Either (i) or (iii) (v) Either (ii) or (iii).

1.116. The factors on which tool life depends:

(i) Hardness of job material
(ii) Quality of tool material
(iii) Tool geometry
(iv) Cutting speed
(v) Cross-section of the chip
(vi) All of the above
(vii) None of the above.

1.117. Better machinable material is one which gives:

(i) Larger shear angle
(ii) Smaller shear angle
(iii) Low chip-tool contact area
(iv) High chip-tool contact area
(v) Both (i) and (iii)
(vi) Both (ii) and (iv).

1.118. With same tool life, the productivity can be increased by increasing:

(i) Cutting speed
(ii) Feed
(iii) Depth of cut
(iv) None of the above.

1.119. Machinability of the workpiece material can be predicted by the following factor:

(i) Tensile strength
(ii) Brinell hardness
(iii) Shear angle
(iv) All of the above.

Match the correct answer from group 'B' against statement given in group 'A':

1.120. Match the standard machine tool with respect to the tool movement

Group-A	*Group-B*
(a) Forward translation	(i) Drilling
(b) Reciprocating	(ii) Lathe
(c) Rotary	(iii) Milling
(d) Rotary as well as translation	(iv) Shaping
	(v) Planing.

1.121. Match the standard machine tool with respect to the job movement

Group-A	*Group-B*
(a) Forward translation	(i) Lathe
(b) Reciprocating	(ii) Milling
(c) Rotary	(iii) Planing
(d) Intermittent translatory motion	(iv) Drilling
(e) Stationary	(v) Shaping.

1.122. Match the properties of cutting tool material

Group-A	*Group-B*
(a) Nickel	(i) Increases keenness of cutting edges
(b) Tungsten	(ii) Increases strength and toughness
(c) Chromium	(iii) High elastic limit and corrosion resistance
(d) Cobalt	(iv) Fine grain structure and also increases keenness of cutting edges
(e) Molybdenum	(v) Increases hardness and strength also hot hardness and wear resistance
(f) Vanadium	(vi) Excellent hot hardness and forms stability.

1.123. Match the tool material with respect their composition

Group-A	*Group-B*
(a) Cast-non-ferrous-stellite cobalt	(i) Aluminium Oxide
(b) Ceramic	(ii) Cobalt, Tungsten and Carbon
(c) Cemented carbide	(iii) Chromium-Tungsten-Vanadium-Iron
(d) High speed steel	(iv) Chromium-Cobalt-Tungsten-Carbon-Iron

1.124. Match the cutting speed with respect to the tool materials

Group-A	*Group-B*
(a) Ceramic	(i) 30 m/min.
(b) Cemented carbide	(ii) 60 m/min.
(c) H.S.S.	(iii) 150 m/min.
(d) Cast-non-ferrous stellite cobalt	(iv) 600 m/min.

1.125. Match the hot hardness of cutting tool material w.r.t. different cutting tool materials

Group-A	*Group-B*
(a) Cemented carbide	(i) 1500°C.
(b) H.S.S.	(ii) 1200°C.
(c) Ceramic	(iii) 1000°C.
(d) Cast-non-ferrous-stellite cobalt	(iv) 900°C.

1.126. Match the application with respect to different cutting fluids normally used

Group-A	*Group-B*
(a) Soluble oil	(i) Soft steel operation
(b) Synthetic coolant	(ii) Lathe, drilling operation, etc.
(c) Compound oil	(iii) Grinding operations
(d) Sulphurised additive oil	(iv) Machining alloy steel.

1.127. Select the proper cutting fluid with respect to the job material for most of the operations

Group-A	*Group-B*
(a) Copper	(i) Blow of air
(b) Mild steel	(ii) Mineral lard oil
(c) Aluminium	(iii) Turpentine
(d) Cast-iron or brass	(iv) Kerosene
(e) Bronze or hard steel	(v) Cutting compound.

1.128. Match the statement from Gr. -B against the statement the Gr. -A

Group-A	*Group-B*
(a) Self-drive	(i) One motor is responsible for different machines
(b) Group drive	(ii) One motor is responsible for one machine
(c) Multipoint cutting tool	(iii) Twist drill
(d) Double cutting edges	(iv) Reamer

1.129. Match the statement in Gr.-B w.r.t. Gr.-A

Group-A	*Group-B*
(a) Sodium nitrate	(i) Grinding operations
(b) Solid lubricants	(ii) Soap bar or wax stick
(c) Cutting oil	(iii) Automatic machines
(d) Synthetic coolant	(iv) Rust preventer

1.130. Match the cutting fluid w.r.t. their functions

Group-A	*Group-B*
(a) As coolant	(i) Air, water
(b) As lubricant	(ii) To wash-off chips
(c) As both coolant and lubricant	(iii) Mineral oil, vegetable oil, animal oil
(d) As flashing action	(iv) Mixed oil, emulsion of oil and water.

1.131. Match the tool angles for machining M.S., M.C.S. and alloy steel by H.S.S. cutting tool

Group-A	*Group-B*
(a) Front rake angle	(i) 6 degree
(b) Side rake angle	(ii) 12 degree
(c) Front clearance angle	(iii) 9 degree
(d) Side cutting edge angle	(iv) 18 degree.

1.132. Match the cutting speed in m/min. for turning, drilling, boring, milling operations, etc., with respect to different job materials

Group-A	*Group-B*
(a) Mild steel	(i) 80 to 100
(b) Cast-iron	(ii) 50 to 70
(c) Brass	(iii) 25 to 30
(d) Stainless steel and high carbon steel	(iv) 15 to 25
(e) Aluminium	(v) 15.

1.133. Match the type of chip with respect to the job material

Group-A	*Group-B*
(a) Continuous	(i) Cast-iron
(b) Dis-continuous	(ii) Mild steel with appropriate cutting conditions
(c) Continuous with built-up edge	(iii) Mild steel without using cutting fluid and high cutting speed and feed.

1.134. Match the tool used with respect to different M/c tools

Group-A	*Group-B*
(a) Single-point cutting tool	(i) Drilling M/c
(b) Double edges cutting tool	(ii) Milling M/c, grinding M/c
(c) Multi-point cutting tool	(iii) Lathe, Shaper, Planer.

Fill up the blanks:

2.1. In _____ operation, job rotates and tool moves in forward translation.

2.2. _____ speed is necessary for operations like spinning, polishing, wood working, etc.

2.3. _____ lathe is the high speed heavy duty machine tools suitable for mass-production.

2.4. Main specifications of a centre lathe are _____ diameter over bed and _____ of bed.

2.5. _____ is the lathe part, which gives rigid supports to the tool during operations and transmit power through Feed rod and Lead screw.

2.6. All geared head is mostly suitable in _____ duty machines and where smooth running is required.

2.7. Front portion of the carriage is known as _____.

2.8. _____ of a machine tool are the essential components, without which that particular machine cannot be identified.

2.9. _____ are required for holding and supporting the working job or the cutting tool conveniently.

2.10. _____ are intendant to increase the accuracy and efficiency of the machine for specific purpose.

2.11. _____ centre is in tailstock end for supporting heavy job rotating at high speed.

2.12. _____ chuck is suitable for holding of repetition or heavy and large jobs on lathe machine spindle.

2.13. _____ are used to prevent bending of a slender job due to its own weight.

2.14. _____ is used for holding and rotating the hollow jobs on machine spindle.

2.15. In _____ operation, the job rotates and the tool is fed longitudinally by giving the desired depth of cut.

2.16. Lathe bed is made of _____ iron.

2.17. Lead screw of a lathe has _____ start thread.

State whether the following statements are True or False:

2.18. Operations like turning, facing, drilling, etc., are required moderate speed.

2.19. Operations like knurling, thread cutting, etc., are required high speed.

2.20. Horizontal lathe machines are those in which spindle axis is horizontal.

2.21. In automatic lathe, job will be fed from one side and finished product will come-out from the other side of the machine during running without the help of a operator.

2.22. In automatic lathe, all the operations are performed by the machine itself without the participation of the operator.

2.23. Bed of a lathe machine is made from mild steel.

2.24. Morse tapers are used in live and dead centres.

2.25. In stepped pulley drive, spindle speeds are varied in Arithmetic Progression.

2.26. In all geared head drive, spindle speeds are varied in Geometric Progression.

2.27. Full-proof arrangement is a safety arrangement against simultaneous engagement of half nuts and longitudinal or cross feed.

2.28. Chucks, steady rests, driving plate, carriers are the examples of lathe parts.

2.29. Taper turning attachment, dividing head attachment, feed stop, etc., are the accessories of lathe machine.

2.30. If the centre is mounted on headstock and rotated with it is known as live centre.

2.31. Ball centre is used during taper turning by off-setting the tailstock to minimise wear and strain.

2.32. Three-jaws-self-centering chuck is used to hold slender jobs with smaller diameter.

2.33. Collet chuck is used for production work to hold the bar type of jobs up to 65 mm diameter.

2.34. Magnetic chuck is only used in non-magnetic material jobs for light and thin sheet operation by grinding.

2.35. Face plate is a circular disc or plate with several rectangular slots placed radially for clamping the job by bolts, angle plate, clamps, etc.

2.36. Gang mandrel is suitable to drive the jobs having larger diameter holes more than 100 mm.

2.37. Steady rests are placed with an interval of 120 to 150 cm of the job length to prevent sag or deflection.

2.38. Follower rest has 3 jaws support 120° apart and makes the job to move concentric.

2.39. Catch plate imparts a positive drive between lathe spindle and job.

2.40. Universal chuck is mainly used for irregular jobs and giving better gripping.

2.41. Centering operation is the operation of producing conical holes on job face(s) to make the bearing support by the lathe centre(s).

2.42. Knurling is the process of embossing, producing a roughened surface on a smooth surface of a cylindrical job to provide effective gripping.

2.43. Reaming, a single-point cutting tool will not follow the previous drilling hole and will trace its own path.

2.44. In taper turning by swivelling the compound rest, the job rotates on lathe axis and tool moves in angular path.

2.45. In taper-turning by taper turning attachment, job rotates on lathe axis and tool travels on resultant path.

2.46. In taper-turning by a form tool, job rotates on lathe axis and tool moves cross-wise direction, perpendicular to the lathe axis.

2.47. Lead is equal to the product of pitch and number of start.

2.48. Change gear ratio in thread cutting operation is equal to the ratio of the product of pitch of the job and number of start to the pitch of the lead screw.

2.49. Half-nut is used to engage the lathe carriage to the lead screw for screw cutting.

2.50. When back gears are used in a centre lathe, the spindle speed reduces considerable.

2.51. The carriage and tailstock of a lathe is guided on different spindles.

2.52. The lathe spindle is usually made hollow and front portion is provided with internal morse taper.

2.53. The tool should be set properly at the centre height of the job.

2.54. When the tool is set below the centre height of the job the clearance angle is decreased.

2.55. The elements of hand reamer are body, neck, shank and body also consists of chamfer, starting taper, sizing section and back taper.

2.56. Swing diameter over bed is less than the swing diameter over carriage.

2.57. Half nut is also known as lock nut to transmit the lead screw power to tool end during thread cutting.

2.58. In turning operation, the spindle speed decreases with decrease in diameter of the work.

2.59. An operation of enlarging the end of a hole conically is known as counter-sinking.

Choose the correct one from the alternatives given below against each question:

2.60. In an Engine Lathe, the cutting tool is fed against workpiece axis in:

(i) Longitudinal direction only
(ii) Cross direction only
(iii) Both longitudinal and cross directions
(iv) Vertical direction only.

2.61. Two main specifications of a centre lathe is:

(i) Swing diameter over bed and length of bed
(ii) Swing diameter over bed and carriage
(iii) Speed and feed ranges
(iv) Swing diameter over bed and maximum length of the job can accommodate in between centres.

2.62. A Centre Lathe is specified by:

(i) The swing diameter over bed (ii) The length of bed
(iii) The diameter of the hole in H.S. spindle
(iv) All of the above.

2.63. Slow speed in head-stock spindle is necessary in:

(i) Spinning (ii) Turning
(iii) Thread cutting (iv) Knurling
(v) Both (i) and (ii) (vi) Both (iii) and (iv).

2.64. Standard taper used in Lathe centres is:

(i) Brown and Sharp (ii) Morse
(iii) Jarno (iv) Metric taper.

2.65. A Lathe machine having four-steps cone pulley drive with one back gear will have:

(i) Eight direct speeds (ii) Four direct speeds
(iii) Eight indirect speeds (iv) Four direct and four indirect speeds.

2.66. When a turning tool travels parallel to the Lathe axis, the movement of tool is known as:

(i) Longitudinal feed (ii) Cross feed
(iii) Angular feed (iv) None of the above.

2.67. The included angle of Lathe centres for general work is:

(i) 25° (ii) 40° (iii) 60° (iv) 75°.

2.68. The different H.S. spindle speed of all geared-head drive on a lathe form:

(i) Harmonical progression (ii) Arithmetical progression
(iii) Geometrical progression (iv) None of the above.

2.69. An operation of embossing, producing a roughened diamond-shaped pattern on a smooth surface of a cylindrical job to provide effective gripping is known as:

(i) Grooving (ii) Knurling
(iii) Turning (iv) All of the above.

2.70. The cutting edge width of a parting off tool varies from

(i) 4–16 mm (ii) 8–25 mm (iii) 15–35 mm (iv) 3–8 mm.

Match the correct answer from group 'B' against statement given in group 'A':

2.71. Match the functions of Lathe parts

Group-A	*Group-B*
(a) Bed	(i) Main source of power, transmits power to feed rod and lead screw.
(b) Headstock	(ii) Sufficiently rigid and good damping qualities
(c) Tailstock	(iii) Rigid supports, moves and controls the cutting tool
(d) Carriage	(iv) Holds other end of the job, performing drilling, boring, reaming, taper for long jobs.
(e) Feed rod	(v) Transmit power for thread cutting operation
(f) Lead screw	(vi) Transmit power to the tool end for automatic longitudinal and cross feeding.

2.72. Match the statement of different transmission system in Lathe machine

Group-A	*Group-B*
(a) Stepped pulley	(i) Device used to change direction of rotation of headstock-spindle to feed rod and lead screw.
(b) All geared head drive	(ii) Number of spindle speed increases in arithmetic progression.
(c) Back gear drive	(iii) Number of speed increases in geometric progression
(d) Tumbler gear	(iv) Reduces speed and number of speeds will be doubled
(e) Apron mechanism	(v) Device by which the rotary movement of feed rod can be converted to longitudinal and transverse motion of the tool.

2.73. Match the statement in Gr.-B w.r.t. Gr.-A

Group-A	*Group-B*
(a) Parts	(i) Additional equipments such as feed stop.
(b) Accessories	(ii) Holding and supporting the workpiece such as chucks, steady rest, mandrel, etc.
(c) Attachment	(iii) Essential components such as tailstock bed, etc.

2.74. Match the centre with respect to their application

Group-A	*Group-B*
(a) Half Centre	(i) Mostly used in production work and expensive
(b) Ball Centre	(ii) For supporting heavy job rotating at high speed
(c) Pipe Centre	(iii) To allow more space for the tool when turning down the end
(d) Tipped Centre	(iv) Minimizes wear and stain during taper turning by tailstock set-over method
(e) Frictionless Centre	(v) For holding hollow jobs such as shell, pipe, etc.

2.75. Match the functions of different chucks used in Lathe machine

Group-A	*Group-B*
(a) 2-jaw chuck	(i) Used to hold magnetic material for light and thin sheet operation
(b) 4-jaw independent chuck	(ii) Used to hold heavy and irregular-shaped work rigidly
(c) 3-jaw concentric or self centering chuck	(iii) Used for holding heavy and large jobs of production work.
(d) Magnetic chuck	(iv) Used for holding regular -shaped job
(e) Air or oil operated chuck	(v) Used to hold small job and small stock removal like polishing, etc.

2.76. Choose mandrel with respect to their functions

Group-A	*Group-B*
(a) Step Mandrel	(i) Suitable for the jobs having different hole diameter
(b) Gang Mandrel	(ii) Used to drive the jobs having internal threads
(c) Collar Mandrel	(iii) Suitable for production work for holding number of identical jobs at a time
(d) Cone Mandrel	(iv) Suitable for to drive the jobs having larger diameter holes more than 100 mm
(e) Screw Mandrel	(v) Used to drive different jobs of different sizes of holes at a time

2.77. Match the different Lathe operations

Group-A	*Group-B*
(a) Reaming	(i) Operation of making the ends of a job to produce a flat surface square with axis of operation
(b) Boring	(ii) Operation of producing a cylindrical surface of a job by removing excess material
(c) Straight turning	(iii) The operation of enlarging an existing hole by turning inside with some form tool
(d) Facing	(iv) Operation of sizing or finishing a drilled hole to the required size by means of a tool.

2.78. Match the different taper-turning methods in Lathe Machine with respect to their principles of taper-turning

Group-A	*Group-B*
(a) Taper-turning by swivelling the compound rest	(i) Job rotates on lathe axis and tool moves in guided angular path
(b) Taper-turning by off-setting the tail-stock	(ii) Job rotates on lathe axis and tool moves cross-wise direction
(c) Taper-turning by taper turning attachment	(iii) Job rotates on lathe axis and tool moves in angular path
(d) Taper-turning by a form tool	(iv) Job rotates at an angle to the lathe axis and tool travels longitudinally to the lathe axis
(e) Taper-turning by combination feeds	(v) Job rotates on lathe axis and tool travels on resultant path.

2.79. Match the different taper-turning methods with respect to their applications

Group-A	*Group-B*
(a) By swivelling the compound rest	(i) Longer jobs of steep angle in mass production
(b) By off-setting the tailstock	(ii) Mass production for chamfering on bolts, nuts, bushes, etc.
(c) By taper-turning attachment	(iii) Longer jobs and small tool angle of taper
(d) By a form tool	(iv) By hand feeds for making the ball of a hammer
(e) By combination feeds	(v) Shorter length and steeper angle.

2.80. Match the units with respect to different cutting conditions used in Lathe operations

Group-A	*Group-B*
(a) Cutting speed	(i) mm
(b) Feed	(ii) mm/rev.
(c) Depth of cut	(iii) m/min.
(d) R.P.M.	(iv) Revolution/min.

State whether the following statements are True or False:

3.1. In Capstan Lathe, hexagonal turret rests directly on a saddle, which can move both the ways, i.e., along bedways and transverse directions.

3.2. In Turret Lathe, the hexagonal turret rests on a short slide or ram, which slides on saddle. The saddle can move on lathe bedways.

3.3. In bar feeding arrangement, the bar in Capstan Lathe is pushed forward immediately after releasing the collet without stopping the Lathe machine.

3.4. Collets are used for gripping small bars in Capstan and Turret Lathes of mass production work.

Fill up the blanks:

4.1. ______ is being done both for surface finish and also for enlargement of the hole.

4.2. Systematic naming of different elements of a twist drill and showing of different angles are known as twist drill ______ and ______.

4.3. The operation of producing a large hole by removing metal along the circumference of a hollow cutting tool is known as ______.

4.4. The operation of finishing relatively large holes such as automobiles cylinders by means of relatively slow moving abrasives is known as ______.

4.5. Machining time in minute is ratio of total travel of drill to the product of spindle speed and the ______.

4.6. The tool used to withdraw a drill from the sleeve is called ______.

4.7. One or more ______(s) is used as a media to hold the twist drill into the spindle.

4.8. Only one ______ is used as a media to hold the twist drill into the spindle.

4.9. ______ holder is used to compensate for non-alignment of drill spindle with work and permits self-alignment of the tool.

State whether the following statements are True or False:

4.10. In drilling operation, job is fixed and the drill bit is rotating and feeding vertical translatory movement.

4.11. In super-sensitive and sensitive drilling machine, drill spindle is running at very low speed and automatic feed with diameter over 16 mm.

4.12. In Ratchet drilling machine, drill bit is hand operated, lever length limited range 300 to 600 mm rod and length of feed range 60 to 85 mm.

4.13. Drift is used for taking out the socket or sleeve or twist drill or drill chuck from the spindle by driving it into the keyslot.

4.14. Spot-facing is an operation of smoothening and squaring around the top surface of the hole.

4.15. Reaming speed is generally less than the drilling speed.

4.16. A short taper lead is to be selected on machine reamer for tough and soft material or blind holes.

4.17. Over-all length of machine-jig reamer consists of guide length, cutting edge length and shank length.

4.18. The hole drilled for tapping should be smaller than the tap size by the depth of thread.

4.19. Point angle of a twist drill for general purpose is 118°.

4.20. Helix angle of a drill for drilling Aluminium is more than 30°.

Choose the correct one from the alternatives given below against each question:

4.21. The metal is removed in drilling machine by:

(i) Extrusion (ii) Shearing
(iii) Shearing and Extrusion (iv) Shearing and Compression.

4.22. Name the machine tool in which work is usually rotated while the drill bit is fed into the work:

(i) Radial drilling M/c (ii) Multi-spindle drilling M/c
(iii) Deep hole drilling M/c (iv) Gang drilling M/c.

4.23. The operation of producing a large hole by removing metal along the circumference of a hollow cutting tool is known as:

(i) Counter-boring (ii) Counter-sinking
(iii) Trepanning (iv) Spot-facing.

4.24. A 50 mm Drilling Machine means that the machine can drill a hole:

(i) In 50 mm thick plate
(ii) Having cross-section area of drill hole of 50 mm
(iii) Of maximum diameter 50 mm
(iv) None of the above.

4.25. The tool necessary to withdraw a drill bit from the drilling machine spindle or sleeve is known as:

(i) Drill drawer (ii) Drill puller
(iii) Drill withdrawal (iv) Drift.

4.26. The twist drill bit is mostly made of:

(i) High carbon steel (ii) High speed steel
(iii) Cemented carbide (iv) Ceramic.

4.27. The value of depth of cut in drilling with respect to drill diameter is:

(i) Double (ii) Equal
(iii) One-half (iv) One-fourth.

4.28. Tap size drill is _____ the outside diameter of the thread on the tap:

(i) Greater than (ii) Smaller than
(iii) Equal to (iv) None of the above.

Match the correct answer from group B against statement given in group A:

4.29. Match the different drilling machines with respect to their operation and suitability

Group-A	*Group-B*
(a) Flexible shaft drilling M/c	(i) A drilling M/c operated by compressed air, mostly used in mines, roads, etc., are to crack
(b) Electric drilling M/c	(ii) Manually operated drilling M/c is mostly suitable where electric power is not available
(c) Ratchet drilling M/c	(iii) Operated by hand power, pneumatic power or electric power (AC or DC) suitable for high speed non-productive jobs
(d) Air drilling M/c	(iv) Operated by electric power, and can be moved elsewhere
(e) Portable drilling	(v) Similar to portable electric drilling M/c but the shaft is long and flexible.

4.30. Match the different drilling machines with respect to their main functions

Group-A	*Group-B*
(a) Radial drilling M/c	(i) For drilling very long holes of smaller diameter and suitable in production work
(b) Gang drilling M/c	(ii) Number of identical holes is drilled at a time and suitable for production work
(c) Multi-spindle drilling M/c	(iii) For medium to large and heavy jobs to drill at any angle and mostly suitable in tool room and maintenance work
(d) Deep hole drilling M/c	(iv) Series of operations are performed by minimum movement of job on common work table and mostly suitable for production work.

4.31. Match the approximate range of angles in degree of a twist drill

Group-A	*Group-B*
(a) Rake or Helix angle	(i) 80° to 150°
(b) Chiesel edge angle	(ii) 8° to 12°
(c) Lip clearance angle	(iii) 120° to 135°
(d) Point angle	(iv) 30°

4.32. Match the statements of different angles of a twist drill with respect to their name

Group-A	*Group-B*
(a) Point angle	(i) The obtuse angle, included between chiesel edge and the lip viewed from the end of a drill.
(b) Chiesel Edge angle	(ii) The angle included in the conical part of a twist drill.
(c) Helix or Rake angle	(iii) The angle formed by the flank and a plane at a right angle to the drill axis.
(d) Lip clearance angle	(iv) The angle formed by the leading edge of the land with a plane having the axis of the drill.

Fill up the blanks:

5.1. _____ is the extension of milling machine spindle on which all the various cutters are mounted.

5.2. ______ machine in which the metal is removed from a workpiece to obtain the required shape by a multi-point rotating cutter fixed on machine spindle.

5.3. ______ is mounted on the upper part of the milling machine column and received power from the motor, can be rotated at different speeds.

State whether the following statements are True or False:

5.4. In universal milling machine table can have the swivelling arrangement about a vertical axis.

5.5. In omniversal milling machine has the arrangements of knee swivelling in addition to the movement of the table as in case of universal milling machine.

5.6. Plain milling machine is less rigid than universal milling machine and so not suitable for production purpose.

Choose the correct one from the alternatives given below against each question:

5.7. In plain milling machine table can be moved against a rotating cutter mounted on arbor:

(i) Vertically (ii) Longitudinally
(iii) Cross-wise (iv) All of the above.

5.8. Milling cutter is mounted on:

(i) Spindle (ii) Tool post (iii) Arbor (iv) Table

5.9. In which of the following milling machines is most suitable for production work?

(i) Universal milling machine
(ii) Triplex type (Fixed Bed) milling machine
(iii) Omniversal milling machine
(iv) Plain milling machine.

5.10. In which of the following milling machines is most suitable for versatility of works?

(i) Universal milling machine
(ii) Fixed bed type milling machine
(iii) Omniversal milling machine
(iv) Plain milling machine.

5.11. Select the machine tool in which the forward movement of the workpiece in longitudinal, cross or vertical direction fed against the rotating cutter:

(i) Lathe (ii) Drilling (iii) Broaching (iv) Milling.

5.12. In which milling machines is most suitable in Tool-Room Operations:

(i) Plain milling machine
(ii) Universal milling machine
(iii) Omniversal milling machine
(iv) Fixed bed type milling machine.

5.13. Milling machine is mainly specified by:

(i) Manufacture type and code number
(ii) Diameter of arbor and table size with movements
(iii) Range and number of spindle speeds and table feeds
(iv) Net weight, floor space and kW of the motor.

5.14. Select the milling machine most suitable for helical gear cutting:

(i) Plain milling machine (ii) Hand milling machine
(iii) Universal milling machine (iv) None of the above.

5.15. Select the milling machine cutter holding device:

(i) Arbor (ii) The spring collet
(iii) Screw on arbor (iv) All of the above.

5.16. The space between the back of one tooth and face of the subsequent tooth in a plain milling cutter is known as:

(i) Face (ii) Fillet
(iii) Gash (iv) None of the above.

5.17. The curved surface at the bottom of the chip space which joins the back of tooth to the face of the next tooth is known as:

(i) Gash space (ii) Chip space
(iii) Fillet (iv) None of the above.

5.18. The process of removing metal by a milling cutter, which is rotated against the direction of the feed of the workpieces:

(i) Face milling (ii) Up milling
(iii) End milling (iv) Down milling.

5.19. The process of removing metal by a milling cutter, which is rotated in the same direction as the feed of the workpiece:

(i) Face milling (ii) Conventional milling
(iii) Up milling (iv) Climb milling.

5.20. Cutting forces in down milling operation _____ per tooth movement of the cutter.

(i) Is uniform (ii) Is minimum
(iii) Decreases from maximum to zero (iv) Increases from zero to maximum.

5.21. Cutting forces in up milling operation _____ per tooth movement of the cutter:

(i) Is uniform
(ii) Varies from maximum to zero
(iii) Varies from zero to maximum
(iv) Is minimum.

5.22. For milling 13 teeth on a gear blank, the index crank to move

(i) 3 complete turns and 1 hole on 39 hole circles of I.P. no. III
(ii) 3 complete turns and 3 holes on 39 hole circles of I.P. no. III
(iii) 3 complete turns and 6 holes on 39 hole circles of I.P. no. III
(iv) 3 complete turns and 9 holes on 39 hole circles of I.P. no. III.

5.23. For milling 49 teeth on a gear blank, the index crank to move

(i) 10 holes on 49 hole circles of I.P. no. III
(ii) 20 holes on 49 hole circles of I.P. no. III
(iii) 30 holes on 49 hole circles of I.P. no. III
(iv) 40 holes on 49 hole circles of I.P. no. III.

Match the correct answer from group 'B' against statement given in group 'A':

5.24. Match the statements in Gr. 'B' with Gr. 'A' of milling operations

Group-A	*Group-B*
(a) Conventional milling	(i) Up-milling
(b) Climb milling	(ii) Same direction
(c) Job-Tool motion in up milling	(iii) Opposite direction
(d) Job Tool motion in down milling	(iv) Down milling.

5.25. Match only one answer from Gr.-B with respect to statement is Gr.-A of milling operation

Group-A	*Group-B*
(a) Cutting forces varies from zero to maximum	(i) Down milling
(b) Difficult to use of cutting fluid in milling process	(ii) Conventional milling
(c) Cutter forces sufficiently on the job to press downwards	(iii) Up milling
(d) Better surface finish if the table is free from back lash error	(iv) Climb milling.

Fill up the blanks:

6.1. _____ machine is used to machine 900 mm cube maximum size among reciprocating type of machine tools.

6.2. _____ maximum size of a block can be machined by a shaping machine.

State whether the following statements are True or False:

6.3. In shaping operation, job reciprocates and tool giving in linear motion.

6.4. The overhang portion of the shaper tool should be kept as minimum as possible.

6.5. In universal shaping machine, in addition to the two movements as in plain type machine, the table can be swivelled and tilted to permit machining at any angle.

6.6. In draw cut type shaping machine, the metal is removed from the job during the forward motion of the ram.

6.7. In plain type shaping machine, the table holding job permits only two movements either in horizontal or in vertical for giving desired feed.

6.8. For production work, a horizontal, plain and push cut type shaper is commonly used.

6.9. For tool room and maintenance works, a horizontal tool travel universally and push cut type shaper is mostly suitable.

6.10. Draw cut type machine is mostly used as conventional cutting for standard and universal type horizontal shaper.

6.11. In travelling tool head type machine, the ram holding the tool during reciprocation, can move cross-wise for required feeding while the job will remain stationary.

6.12. Cutting-return ratio is fixed in a particular planning machine.

6.13. Cutting-return ratio in case of a shaper can be changed independently.

Choose the correct one from the alternatives given below against each question:

6.14. Choose the machine tool in which tool reciprocates in horizontal axis and job feeds in intermittent translatory motion

(i) Slotting machine (ii) Shaping machine
(iii) Planing machine (iv) None of the above.

6.15. Choose the machine tool in which job reciprocates in horizontal axis and tool feeds in intermittent translatory motion

(i) Slotting machine (ii) Shaping machine
(iii) Planing machine (iv) None of the above.

6.16. Choose the machine tool in which tool reciprocates in vertical axis and job feeds in intermittent translatory motion

(i) Slotting machine (ii) Shaping machine
(iii) Planing machine (iv) None of the above.

6.17. In an universal planer metal is removed during

(i) Forward strokes
(ii) Return strokes
(iii) Both forward and Return strokes
(iv) None of the above.

6.18. Choose the Machine Tool, in which the ram holding the tool travels in a vertical direction can be adjusted 10° on either side on its vertical axis for machining an inclined surface, the addition to its vertical direction

(i) Shaper
(ii) Slotter
(iii) Both of them
(iv) None of them.

6.19. Select the machine in which three surfaces can be machined at a time

(i) Open housing planer
(ii) Double housing planer
(iii) Pit planer
(iv) Edge or plate planer.

Match the correct answer from group-B against statement given in group A:-

6.20. Match the statement in Gr.-B w.r.t. Gr.-A of reciprocating type of machines in relation to their construction and rigidity

Group-A	*Group-B*
(a) Shaping machine	(i) Heavier in construction and more rigid
(b) Slotting machine	(ii) Lighter in construction and more rigid than vertical shaper
(c) Planing machine	(iii) Lighter in construction and less rigid.

6.21. Match the cost relation of the respective machine tools in Gr.-B w.r.t. Gr.-A

Group-A	*Group-B*
(a) Shaping machine	(i) Cost most
(b) Slotting machine	(ii) Cost more
(c) Planing machine	(iii) Cost moderate.

6.22. Match the special features of different planing machines

Group-A	*Group-B*
(a) Open side planer	(i) Carriage holding the cutting tool moves to and fro past the stationary workpiece
(b) Double housing planer	(ii) Two parts of the table is reciprocated separately or together

(c) Pit planer	(iii) Having only one column on one side of the base
(d) Edge or Plate planer	(iv) Having two columns at two sides of the base
(e) Divided table planer	(v) Table stationary and column carrying the cross-rail holding tool head(s) can reciprocate on horizontal rails situated on both sides of the table.

6.23. Match special features of different shaping machines

Group-A	*Group-B*
(a) Horizontal shaper	(i) Metal is removed from the job during the forward motion of ram
(b) Vertical shaper	(ii) The table can also be swivelled and tilted to permit machining at an angle
(c) Travelling tool head shaper	(iii) Ram holding the tool during reciprocation can move cross-wise
(d) Universal shaper	(iv) Ram holding the tool travels in horizontal axis
(e) Push cut type shaper	(v) Ram holding the tool travels in vertical directions.

6.24. Match the main specifications in relation to different reciprocating machine tools

Group-A	*Group-B*
(a) Shaper	(i) Maximum length of stroke and diameter of the table
(b) Slotter	(ii) Maximum length stroke of tool table movement horizontally and vertically
(c) Planer	(iii) Maximum length of stroke and width of the table and distance from the table to rail.

Fill up the blank:

7.1. The main dimensions of a _____ are the maximum length of stroke of the ram and the diameter of the table.

Fill up the blanks:

8.1. _____ machine is used to machine 3 m × 3 m × 18.25 m rectangular block maximum size among reciprocating type of machine tools.

8.2. _____ maximum size of a block can be machined by a planing machine.

Fill up the blanks:

9.1. _____ is a process of removing materials in the form of chips from a workpiece by mechanical action of many small abrasive particles bonded together in a wheel.

9.2. An _____ is a hard material which can be used to cut or wear away by abrasion to other materials.

9.3. _____ is the degree of hardness of the bond and is designated by the letter of alphabet.

9.4. _____ indicates the size of the cutting teeth or the size of the abrasive particles used in making a grinding wheel.

9.5. The relative spacing of abrasive grain is known as _____ and denoted by the number of cutting edges per unit area of wheel face.

9.6. Degree of hardness caused by the _____ is called the grade of the wheel.

9.7. _____ bond is suitable for maximum metal removal and 75% of the total use.

9.8. _____ bond is used for good surface finish and roll finish.

9.9. _____ bond is used for general purpose and diameter of grinding wheel is limited to 915 mm.

9.10. _____ bond is used for diameter up to 1525 mm and small amount of metal removal as knife, tool, etc.

9.11. _____ bond is for extremely thin sheets, best finish, and for serious bending at high surface speed.

9.12. _____ a wheel is the operation to create a true and concentric surface with the wheel spindle axis.

9.13. _____ a wheel is the operation to remove glaze or other adhesion particles and to expose fresh cutting particles.

9.14. Commercial _____ tool is used for turning and dressing the newly manufactured grinding wheel.

State whether the following statements are True or False:

9.15. A grinding wheel is designated or specified by a line with letters and figures which show the characteristics of a wheel.

9.16. Silicon carbide abrasive is suitable for materials of high speed steel, tool steel, steel casting.

9.17. Aluminium oxide abrasive is used for materials of cemented carbide.

9.18. In selecting grade, harder wheel is used for softer materials.

9.19. In selecting grade, softer wheel is used for harder materials.

9.20. Dense structure is used for finishing cuts of brittle and hard materials.

9.21. Open structure is used for heavy cuts of soft and ductile materials.

9.22. Centreless grinder is suitable for production purpose.

9.23. Centre type or cylindrical grinder is suitable for general purpose.

9.24. Centre type or cylindrical grinder is a continuous process of cylindrical surface grinding.

9.25. Centreless type of grinder is intermittent process of cylindrical surface grinding.

9.26. The same diamond pointed tool is used for both the operations of truing and dressing but not for same purpose.

Choose the correct one from the alternatives given below against each question:

9.27. For production of finishing of flat and plane surfaces on a workpiece the method of grinding is

(i) External cylindrical grinding (ii) Internal cylindrical grinding
(iii) Surface grinding (iv) Universal type tool grinding.

9.28. Principle of operation of finishing flat surfaces by surface grinders

(i) Grinding by horizontal spindle on reciprocating table
(ii) Grinding by horizontal spindle on rotary table
(iii) Grinding by vertical spindle on reciprocating table.
(iv) Grinding by vertical spindle on rotary table
(v) All of the above.

9.29. Select the grinder mostly used to sharpen and reconditioning various types of cutters, die, jig, fixture, light surface, cylindrical and internal surfaces etc.

(i) Surface grinder (ii) Cylindrical grinder
(iii) Tool and cutter grinder (iv) None of the above.

9.30. Select the grinder in which work is not secured between centres or in a chuck or on any fixture, but the workpiece is supported by using two wheels and a work-rest

(i) External cylindrical grinder
(ii) External surface by centreless grinder
(iii) Surface grinder
(iv) Internal surface by centreless grinder.

9.31. Which factor(s) effect(s) the selection of the grinding wheel?

(i) Abrasive (ii) Grain size (iii) Grade (iv) Structure
(v) Bonding (vi) All of the above (vii) None of the above.

9.32. Which factor(s) effect(s) the rejection of the grinding wheel?

(1) Wheel speed (ii) Work speed
(iii) Work material (iv) Area of contact
(v) None of the above (vi) All of the above.

9.33. Newly manufactured wheels taken from kilns needs to shape and dimensions by

(i) Turning (ii) Dressing
(iii) All of the above (iv) None of the above.

9.34. Degree of hardness of the bond in a grinding wheel is designated by

(i) Brinell hardness number (ii) Rockwell hardness number
(iii) Vickers pyramid number (iv) Letters of alphabet.

9.35. The choice of grain size depends on

(i) The nature of grinding operation
(ii) The work material
(iii) The material removal rate
(iv) Surface finish desired
(v) All of the above.

9.36. The process to expose fresh cutting action by removing glaze or adhesion particles is called by

(i) Clearing (ii) Dressing (iii) Turning (iv) Facing.

9.37. The process to create a true and concentric surface of the grinding wheel with the wheel spindle axis is called by

(i) Clearing (ii) Dressing (iii) Truing (iv) Facing.

Match the correct answer from group 'B' against statement given in group 'A':

9.38. Choose the grain size w.r.t. different tool materials

Group-A	*Group-B*
(a) Tool steel	(i) 80
(b) High speed steel	(ii) 46
(c) Tungsten carbide	(iii) 36

9.39. Match the statement for selecting of bonding in Gr.-B w.r.t. Gr.-A

Group-A	*Group-B*
(a) Vitrified	(i) For precision grinding cams, rolls for high finish and high speed snagging of castings
(b) Silicate	(ii) For wheel diameter up to 1525 mm and small amount of metal removal
(c) Rubber	(iii) For general work and wheel diameter up to 915 mm and maximum metal removal.
(d) Synthetic resinoid	(iv) For extremely thin sheets, best finish and for serious bending at high surface speed.

Fill up the blanks:

10.1. ______ is the operation of machining metal by means of a tool having a number of teeth gradually increasing in size.

10.2. Main specifications of a broaching machine are maximum ______ and maximum ______ developed by the slide.

10.3. Machining time in broaching operation is the ratio of length of broach ______ to the ______.

10.4. ______ surfaces or ______ holes cannot be broached.

10.5. High accuracy and surface finish of about ______ mm is possible in broaching operations.

10.6. Broaching is ______ cutting edge(s) tool.

10.7. Cost of broach tool is very ______.

10.8. Semi-finishing teeth of a broach removes ______ metal.

10.9. Roughing teeth of a broach removes ______ metal.

10.10. Finishing teeth of a broach removes ______ metal.

10.11. Broach is a tool provided with multi-cutting edges, generally made of ______.

10.12. Sufficient ______ must be used to perform cooling of workpiece and broach tool, to reduce friction and flashing of the chips.

10.13. Rise per tooth in roughing teeth of a broach is approximately ______ mm and cutting speed about ______ m/min.

10.14. Hook or rake or face angle is about ______ degree and back-off or clearance angle of about ______ degree of a internal broach tool.

State whether the following statements are True or False:

10.15. Roughing, semi-finishing and finishing cuts are completed in one pass of the broach.

10.16. Obstruction surfaces or blind holes can be broached.

10.17. Broach is a single-purpose tool, requires different broaches for different shapes and sizes.

10.18. Broach has a single cutting edge and removes large amount of metal.

10.19. Height of the successive teeth of a broach is gradually decreasing in order.

10.20. Broaching operation is not suitable for mass production.

10.21. Broaching machine is designed to manufacture workpieces at a high rate and with satisfactory accuracy and finish.

10.22. Main specifications of a broaching machine are the maximum force developed by the slide and the maximum length of stroke.

10.23. During broaching operation, tool pulled or pushed while the job provides intermittent linear feed.

10.24. In machining by broaching, the machining time is the ratio of cutting speed to the length of broach travel.

Choose the correct one from the alternatives given below against each question:

10.25. The broaching operation in which the work is moved continuously past the stationary broach is called

(i) Push broaching (ii) Pull broaching
(iii) Continuous broaching (iv) Surface broaching.

10.26. The process of machining metal for finishing in one stroke either by pulling or pushing a cutting tool is known as

(i) Shaping (ii) Planing
(iii) Forming (iv) Broaching.

10.27. The height of roughing teeth of a broach is

(i) In progressively increasing (ii) In progressively decreasing
(iii) No change through-out (iv) None of the above.

10.28. The operation performed by finishing teeth of a broach is

(i) Removed maximum metal (ii) Removed minimum metal
(iii) Performed burnishing operation (iv) None of the above.

10.29. Broaching operation can be used

(i) For mass production only
(ii) For completing the machining in one pass only
(iii) For high accuracy and surface finish
(iv) All of the above.

Match the correct answer from group B against statement given in group A:

10.30. Match the metal removal rate of a broach w.r.t. type of cutting

Group-A	*Group-B*
(a) Roughing teeth	(i) Perform burnishing operation
(b) Semi-finishing teeth	(ii) Removes maximum metal
(c) Finishing teeth	(iii) Removes minimum metal.

Fill-up the blanks:

11.1. The boring machine is one of the most ——— machine tools.

11.2. Vertical boring machine is similar in job-tool motion as in vertical ———.

State whether the following statements are True or False:

11.3. Boring Machine is designed for machining of large and heavy workpieces in mass production work of machine frame, cylinder machine housing etc.

11.4. In horizontal boring machine, the spindle holding the boring tool rotates in high precision bearings while workpiece mounted on table remains stationary.

11.5. In vertical boring machine, the tool rotates on horizontal table about a vertical axis and the job remains stationary.

11.6. Horizontal Boring machine requires less floor space than vertical boring machine.

11.7. The main specification of a vertical boring machine is the maximum diameter of the work that can be accommodated on the machine table.

11.8. In horizontal boring machine, tool head swivelling is possible.

11.9. In vertical boring machine work swivelling is possible.

11.10. In vertical boring machine, there is a turret type tool indexing arrangement.

Choose the correct one from the alternatives given below against each question:

11.11. The main specification(s) of a horizontal boring machine is(are)

(i) Maximum size of boring spindle diameter
(ii) Maximum spindle travel in horizontal and vertical directions
(iii) Maximum travel of table in longitudinal and crosswise directions
(iv) All of the above.

11.12. Which boring machine having the highest accuracy?

(i) Precision boring machine (ii) Jig boring machine
(iii) Deep hole boring machine (iv) Vertical boring machine.

11.13. Select the vertical boring machine operation(s)

(i) Horizontal flat surface
(ii) Internal and External taper turning
(iii) Grooving
(iv) Cylindrical turning
(v) All of the above.

11.14. Select the operation(s) can be done by horizontal boring machine

(i) Machining vertical flat surfaces (ii) Drilling operation
(iii) Counter-boring (iv) All of the above
(v) None of the above.

11.15. Number of cutting edge(s) can be used as cutter of a boring bar is

(i) Single-point cutter (ii) Double-point cutter
(iii) Multi-point cutter (iv) All of the above
(v) None of the above.

11.16. Material for cutter mostly suitable for production work is

(i) High speed steel (ii) Cemented carbide tips
(iii) Cast non-ferrous-stellite cobalt (iv) All of the above.

11.17. Maximum spindle diameter of a horizontal boring machine is

(i) 120 mm (ii) 220 mm (iii) 320 mm (iv) 400 mm.

11.18. Maximum diameter of machine table of a vertical boring machine is

(i) 600 mm (ii) 900 mm (iii) 6000 mm (iv) 9000 mm

11.19. Boring tool is mounted on

(i) Boring head (ii) Boring bar
(iii) All of the above (iv) None of the above.

Match the correct answer from group 'B' against statement given in group 'A':

11.20. Match the three principal types of horizontal boring machine w.r.t. their range of spindle diameter

Group-A	*Group-B*
(a) Floor type	(i) 50 to 125 mm
(b) Medium size	(ii) 100 to 200 mm
(c) Table type	(iii) 125 to 320 mm.

Fill up the blanks:

12.1. Press is a metal _____ machine tool used for most cold working operations.

12.2. Press tools are _____ and _____.

12.3. In press bottom end of the ram holding the tool is ____ and top face of the bed holding the other tool is ____.

12.4. During press operation, ____ of the two fly balls is equal to the useful work of the ram.

12.5. In ____ press, the ram is driven by the crank-shaft and connecting rod and the crank-shaft gets power either by ____ or ____ means.

12.6. The operation of cutting out to some desired shape from a metal strip or plate the cutting out portion is known as ____.

12.7. The operation of making a hole of any desired shape in the blank portion is known as ____.

12.8. The operation of forming of shapes on thick metal blank required severe compressive loads to cause metal now in cold state is known as ____.

12.9. The main specifications of a press machine are maximum ____ and ____ travel.

12.10. ____ is essential to make a fracture and also the space between the punch cutting edge and die cutting edge.

12.11. In blanking operation, the clearance is provided on the ____.

12.12. In piercing operation, the clearance is provided on the ____.

12.13. In blanking operation, diameter of the punch is ____ than the diameter of die.

12.14. In piercing operation, diameter of die is ____ than the diameter of the punch.

State whether the following statements are True or False:

12.15. Progressive dies are faster operation than compound dies.

12.16. Compound dies are open in tolerances than progressive dies.

12.17. In progressive dies operations create more scraps than in compound dies operations.

12.18. Labour cost in compound dies is less than the progressive dies.

12.19. The operations to impart a raised or depressed fancy design on the surface of the blank material are known as embossing.

12.20. In bending operation, the sheet metal takes the shape of a die.

Choose the correct one from the alternatives given below against each question:

12.21. Presses are widely used to mass-production in the field of

(i) Automotive (ii) Kitchen appliances
(iii) Electrical goods (iv) Air-craft
(v) Telephones (vi) All of the above
(vii) None of the above.

12.22. The squeezing operation of forming of shapes on thick metal blank for producing medals, etc., requires severe compressive loads to cause metal flow in cold state is called

(i) Blanking (ii) Piercing (iii) Coining (iv) Latching.

12.23. The operation of cutting out to some desired shape from a metal strip or plate, the cutting out portion is called

(i) Blanking (ii) Piercing (iii) Coining (iv) Latching.

12.24. The operation of making a hole of any desired shape in the blank, the blank portion is called

(i) Blanking (ii) Piercing (iii) Coining (iv) Latching.

12.25. If a desired shape is partially punched along a line in the work material and then bending down the non-cut portion, the combined action is called

(i) Blanking (ii) Piercing (iii) Coining (iv) Latching.

12.26. _____ dies can perform two or more press operations sequentially every time the ram descends.

(i) Compound Die (ii) Simple Die
(iii) Progressive Die (iv) Combination Die.

12.27. _____ die can perform two or more operations in one stroke of the press at one station only.

(i) Compound Die (ii) Simple Die
(iii) Progressive Die (iv) Combination Die.

12.28. Piercing and Blanking operations can perform simultaneously in a

(i) Compound Die (ii) Simple Die
(iii) Progressive Die (iv) Combination Die.

12.29. Both cutting and non-cutting operations can perform at one station of the press in every stroke of the ram by

(i) Compound Die (ii) Simple Die
(iii) Progressive Die (iv) Combination Die.

Match the correct answer from group 'B' against statement given in group 'A':

12.30. Match the press accessories with respect to their functions

Group-A	*Group-B*
(a) Pilot	(i) Forcing the cut-out blank from the die
(b) Stripper	(ii) For correct location
(c) Knock-out	(iii) For preventing the blank from bowing away from the flat face of the punch.
(d) Pressure Pad	(iv) For removal of finished material.

Fill up the blanks:

13.1. _____ is the dimensional and positional relationship between the workpiece and the cutting tool.

13.2. _____ can hold and locate the workpiece and also guide the tool.

13.3. _____ can hold and position the workpiece but does not guide the tool.

13.4. Functions of _____ devices are to hold the workpieces rigidly on the fixtures or on the jigs.

13.5. _____ jigs are used to drill or ream radial holes on cylindrical or spherical workpieces.

State whether the following statements are True or False:

13.6. Jig is a device which can hold and locate the medium and heavier sized workpieces.

13.7. Fixture is a device necessary for holding and locating the small workpieces and also guiding the tool(s).

13.8. Less skill is essential during operation with jig w.r.t. fixture.

13.9. Fixtures are usually massive and heavier in construction and clamping rigidly on machine table.

13.10. Air-operated chuck can be constructed among angle plate, V-block, parallel, strips, etc., are the examples of jig.

13.11. Elements of fixture, are body, clamping and locating devices.

13.12. Elements of jig are body, clamping and locating devices and also tool guide.

13.13. Drill bush on clamping plate is used to guide the tool.

Choose the correct one from the alternatives given below against each question:

13.14. Advantages of using jigs and fixtures are mainly

(i) To ensure interchangeability of finished products and minimizes the cost of inspection
(ii) To increase rate of production and operational accuracy
(iii) To avoid marking, setting, locating automatically and holding quickly
(iv) To minimize cost of assembly, operator's fatigue, labour cost, overall cost of product
(v) All of the above
(vi) None of the above.

Fill up the blanks:

14.1. Non-traditional machining methods can also be named as _____ machining methods.

14.2. _____ is a process in which tool vibrates at very high specific frequency and at small amplitude in non-traditional machining processes.

14.3. Abrasive slurry is the mixture of _____ grains and water.

14.4. Electro-chemical machining process operates on at _____ voltage and direct current.

14.5. Electrical discharge machining process operates on at _____ voltage and _____ direct current.

14.6. _____ among non-traditional machining processes is found suitable only when not feasible by the other methods of machining.

14.7. In LBM process the temperature and energy of density of the beam of light may extend to _____ K and _____ kW/cm^3 at the workpiece end.

14.8. In ECM and EDM processes, the workpiece and tool material must be electrically _____.

14.9. MRR in ECM process, increased by _____ the gap between the workpiece and tool and _____ the flow of current.

State whether the following statements are True or False:

14.10. Materials suitable for non-traditional machining, with respect to their qualities of high strength-to-weight ratios, high temperature resistance, hardness, etc.

14.11. Materials like hastalloy, Nitralloy, Titanium, HSTS, etc., can be easily machined by non-traditional machining.

14.12. In NTM, tool material must be harder than workpiece material.

14.13. In NTM, superior surface finish and burr free machining is possible.

14.14. Micro-machining in NTM is not possible.

14.15. Power consumption and machining cost in NTM is moderate.

14.16. In LBM process, the holes or cavities may be tapered and the depth of cut is limited.

14.17. Ultrasonic vibrations are transformed by means of a magneto-structive transducer into mechanical vibrations.

14.18. In EDM process, every spark generates not sufficient heat to melt and vaporise a very small volume of metal from the workpiece.

14.19. In laser rod, the beam of light is reflected and accelerated in the path.

14.20. Dielectric is generally mixture of sodium chloride in water.

14.21. Electrolyte fluids are generally kerosene, transformer oil, etc.

14.22. In LBM process, contouring a surface can generate very well.

14.23. In USM process, accuracy up to 0.005 mm is possible.

14.24. Power consumed in LBM process is about 2700 kW/cm^3/min.

Choose the correct one from the alternatives given below against each question:

14.25. In LBM process, contouring a surface is

(i) Poor (ii) Good (iii) Fair (iv) Bad.

14.26. Maximum metal removal rate in ECM process is

(i) Medium (ii) Low (iii) High (iv) Extra high.

14.27. In EDM process, penetration rate in mm/min. is about

(i) 0.5 (ii) 12 (iii) 100 (iv) 50

14.28. Accuracy in USM process in mm is about

(i) 0.005 (ii) 0.01 (iii) 0.001 (iv) 0.05.

14.29. In EDM process, power consumed in kW/cm^3/min. is about

(i) 9.00 (ii) 7.20 (iii) 1.80 (iv) 2700.

14.30. Alternative name of the non-traditional machining processes is any of the following

(i) New technology (ii) Modern machining methods
(iii) Unconventional machining method
(iv) All of the above (v) None of the above.

14.31. Limitation(s) of using NTM is(are) any of the following

(i) Very much costly machining
(ii) High power consumption per unit volume per unit time
(iii) In ECM and EDM method, work material must be conductive
(iv) All of the above
(v) None of the above.

Match the correct answer from group 'B' against statement given in group 'A':

14.32. Match the different non-traditional machining processes w.r.t. their basic mechanism of metal removal

Group-A	*Group-B*
(a) USM	(i) Vaporization by thermo -electric method
(b) ECM	(ii) Ion-displacement by electro-chemical method
(c) EDM	(iii) Erosion by mechanical abrasion
(d) LBM	(iv) When heated, melts and vaporizes instantly by thermo-electric method.

14.33. Match the different NTM processes w.r.t. their immediate energy source at the job-tool interface

Group-A	*Group-B*
(a) USM	(i) Amplified light
(b) ECM	(ii) High voltage
(c) EDM	(iii) High current
(d) LBM	(iv) Pneumatic or Hydraulic pressure.

14.34. Match the different NTM processes w.r.t. their medium for transferring energy

Group-A	*Group-B*
(a) USM	(i) Electrolyte
(b) ECM	(ii) Electrons
(c) EDM	(iii) Radiation
(d) LBM	(iv) High velocity particles.

Fill up the blanks:

15.1. _____ is a bonded fine-grained abrasive process in which very little material is removed from a cylinder bore in an I.C. Engine while abrasive tool is rotating and vertical linear feeding.

15.2. _____ is suitable for obtaining a surface of the highest class of finish (about 0.05 micron).

15.3. _____ is the precision grinding with loose, dust type abrasives to produce geometrically true surface (about 0.1 micron).

Match the correct answer from group 'B' against statement given in group 'A':

15.4. Match the different types of micro-machining processes w.r.t. their accuracies in microns (1 micron = 0.001)

Group-A	*Group-B*
(a) Superfinishing	(i) 1 to 12
(b) Honing	(ii) 0.9 to 5
(c) Grinding	(iii) 0.12 to 1.2
(d) Precision turning	(iv) 0.08 to 0.25
(e) Lapping	(v) 0.01 to 0.25

15.5. Match the range of cutting speed w.r.t. different finishing methods

Group-A	*Group-B*
(a) Honing	(i) 10 to 40 m/min.
(b) Lapping	(ii) 70 to 100 m/min.
(c) Superfinishing	(iii) 100 to 250 m/min.

Fill up the blanks:

16.1. _____ is an attitude for learning and practicing the safe working habits.

16.2. A good and safe operator is one who is _____ conscious always, using safe and accepted procedures.

16.3. Working _____ is the first essential of an operator.

16.4. An _____ is a mishap.

16.5. To maintain safety, moving parts in the shop-floor must be properly _____.

16.6. Operator should get first _____ immediately after any injury.

Fill up the blanks:

17.1. Set of _____ are punched on cards or magnetic tape as per requirements of a product manufacturing, this tape is known as _____, which controls the movement of _____ control Machine Tools.

17.2. The Machine Tools are controlled by the software of the computer and can easily be edited or modified as per requirement is known as _____ control Machine Tools.

17.3. _____ computer controls the different numerical control machine tools through stellite computers.

17.4. _____ computer is incorporated with numerical control system to control machine tool functions from stored-in information.

17.5. In on _____ editing program, any change or mistake in cutting condition, in dimension, etc., can easily be edited or modified by the part data stored in computer.

17.6. In sophisticated manufacturing CNC can be integrated with _____ numerical control system.

17.7. In _____ loop system is provided with a feed-back device (known as transducer) which can compensate errors in position of the tool/work.

17.8. In _____ loop system is not provided with a feed-back device.

17.9. In _____ motion control system, the position of cutting tool changes from a particular fixed point to a point as per requirement in the shortest path and time.

17.10. In _____ control system, the table movement vary continuously with respect to the variation of electrical voltage.

17.11. _____ transducer is employed to convert the amount of movement of work table into convertible electrical voltage pulse, each pulse represents a basic length.

17.12. In _____ system, all the measurements are taken from a fixed zero point or reference point.

17.13. In _____ system, needs no predetermined dactum. All the measurements are measured from the previous point at which the tool was resting.

17.14. _____ control unit of an N.C. Machine Tool is to read, interpret and convert perceived input (part program) into appropriate movements and controls through signals for N.C. actuation system.

Answers Sheet Of Objective Questions

1.1. Products	1.2. Machine	1.3. Tools	1.4. Machine Tools
1.5. Metal cutting		1.6. Metal forming	
1.7. Forming	1.8. Cutting	1.9. Standard/Versatile/General	
1.10. Group	1.11. Cutting	1.12. Orthogonal	1.13. Orthogonal
1.14. Tool-life	1.15. Tailor's	1.16. Machinability	
1.17. Cutting-fluid		1.18. Grinding	1.19. Rust
1.20. Chloride	1.21. Cutting/Compound/Mixed oil		
1.22. Tool geometry/Tool signature		1.23. Centre	1.24. Machine-shop
1.25. Chip	1.26. Continuous	1.27. Non-homogeneous	
1.28. Braker	1.29. Crater & flank		1.30. Discontinuous
1.31. Continuous	1.32. Shaper/Planer/Slotter		1.33. Cobalt
1.34. Seven	1.35. F	1.36. T	1.37. T
1.38. T	1.39. T	1.40. F	1.41. F
1.42. T	1.43. T	1.44. T	1.45. F
1.46. T	1.47. F	1.48. T	1.49. F
1.50. T	1.51. T	1.52. F	1.53. F
1.54. T	1.55. T	1.56. T	1.57. F
1.58. T	1.59. T	1.60. T	1.61. F
1.62. T	1.63. F	1.64. T	1.65. F

1.66. F	1.67. T	1.68. T	1.69. T
1.70. F	1.71. F	1.72. T	1.73. F
1.74. T	1.75. (ii)	1.76. (ii)	
1.77. (iii)	1.78. (ii)	1.79. (iii)	1.80. (iii)
1.81. (ii)	1.82. (iii)	1.83. (ii)	1.84. (iii)
1.85. (iii)	1.86. (ii)	1.87. (iv)	1.88. (i)
1.89. (iii)	1.90. (i)	1.91. (iii)	1.92. (ii)
1.93. (i)	1.94. (ii)	1.95. (iv)	1.96. (iii)
1.97. (ii)	1.98. (i)	1.99. (iv)	1.100. (iv)
1.101. (v)	1.102. (iv)	1.103. (iii)	1.104. (iii)
1.105. (iv)	1.106. (iv)	1.107. (iv)	1.108. (iii)
1.109. (iii)	1.110. (iii)	1.111. (ii)	1.112. (iv)
1.113. (iii)	1.114. (ii)	1.115. (iv)	1.116. (vi)
1.117. (v)	1.118. (iii)	1.119. (iv)	1.120. ii-iv-iii-i
1.121. ii-iii-i-v-iv		1.122. ii-vi-iii-v-i-iv	
1.123. iv-i-ii-iii	1.124. iv-iii-i-ii	1.125. i-iv-ii-iii	1.126. ii-iii-i-iv
1.127. iii-v-iv-i-ii	1.128. ii-i-iv-iii	1.129. iv-ii-iii-i	1.130. i-iii-iv-ii
1.131. iii-ii-i-iv	1.132. iii-iv-ii-v-i	1.133. ii-i-iii	1.134. iii-i-ii

2.1. Turning	2.2. High	2.3. Automatic	2.4. Swing, Length
2.5. Carriage	2.6. Heavy	2.7. Apron	2.8. Parts
2.9. Accessories	2.10. Attachment	2.11. Frictionless	2.12. Air-operated
2.13. Rests	2.14. Mandrel	2.15. Turning	2.16. Chilled cast
2.17. Single	2.18. T	2.19. F	2.20. T
2.21. T	2.22. T	2.23. F	2.24. T
2.25. T	2.26. T	2.27. T	2.28. F
2.29. F	2.30. T	2.31. T	2.32. F
2.33. T	2.34. F	2.35. T	2.36. F
2.37. T	2.38. F	2.39. T	2.40. F
2.41. T	2.42. T	2.43. F	2.44. T
2.45. F	2.46. T	2.47. T	2.48. T
2.49. T	2.50. T	2.51. F	2.52. T
2.53. T	2.54. F	2.55. T	2.56. F
2.57. T	2.58. F	2.59. T	2.60. (iii)
2.61. (i)	2.62. (iv)	2.63. (vi)	2.64. (ii)
2.65. (iv)	2.66. (i)	2.67. (iii)	2.68. (iii)
2.69. (ii)	2.70. (iv)	2.71. ii-i-iv-iii-vi-v	
2.72. ii-iii-iv-i-v	2.73. iii-ii-i	2.74. iii-iv-v-vi-ii	2.75. v-ii-iv-i-iii
2.76. v-iii-iv-i-ii	2.77. iv-iii-ii-i	2.78. iii-iv-i-ii-v	2.79. v-iii-i-ii-iv
2.80. iii-ii-i-iv			

3.1. F	3.2. F	3.3. T	3.4. T

4.1. Boring	4.2. Nomenclature and Angles	4.3. Trepanning
4.4. Honing	4.5. Feed per Revolution	4.6. Drift

4.7. Sleeve 4.8. Socket 4.9. Floating 4.10. T
4.11. F 4.12. T 4.13. T 4.14. T
4.15. F 4.16. T 4.17. T 4.18. F
4.19. T 4.20. T 4.21. (iii) 4.22. (iii)
4.23. (iii) 4.24. (iii) 4.25. (iv) 4.26. (ii)
4.27. (iii) 4.28. (ii) 4.29. v-iv-ii-i-iii 4.30. iii-iv-ii-i
4.31. iv-iii-ii-i 4.32. ii-i-iv-iii

5.1. Arbor 5.2. Milling 5.3. Spindle 5.4. T
5.5. T 5.6. F 5.7. (iv) 5.8. (iii)
5.9. (ii) 5.10. (iii) 5.11. (iv) 5.12. (iii)
5.13. (ii) 5.14. (iii) 5.15. (iv) 5.16. (iii)
5.17. (iv) 5.18. (ii) 5.19. (iv) 5.20. (iii)
5.21. (iii) 5.22. (ii) 5.23. (iv) 5.24. i-iv-iii-ii
5.25. ii-iii-i-iv.

6.1. Shaping 6.2. 900 mm^3 6.3. F 6.4. T
6.5. T 6.6. F 6.7. T 6.8. T
6.9. T 6.10. F 6.11. T 6.12. F
6.13. F 6.14. (ii) 6.15. (iii) 6.16. (i)
6.17. (iii) 6.18. (i) 6.19. (ii) 6.20. iii-ii-i
6.21. ii-iii-i 6.22. iii-iv-v-i-ii 6.23. iv-v-iii-ii-i 6.24. ii-i-iii.

7.1. Slotter

8.1. Planing 8.2. 3 m × 3 m × 18.25 m

9.1. Grinding 9.2. Abrasive 9.3. Grade 9.4. Grit or grain
9.5. Structure 9.6. Bond 9.7. Vitrified 9.8. Rubber
9.9. Vitrified 9.10. Silicate 9.11. Rubber 9.12. Truing
9.13. Dressing 9.14. Diamond 9.15. T 9.16. F
9.17. F 9.18. T 9.19. T 9.20. F
9.21. F 9.22. T 9.23. T 9.24. F
9.25. F 9.26. T 9.27. (iii) 9.28. (v)
9.29. (iii) 9.30. (ii) 9.31. (iv) 9.32. (v)
9.33. (iii) 9.34. (iv) 9.35. (v) 9.36. (ii)
9.37. (iii) 9.38. iii-ii-i 9.39. iii-ii-iv-i.

10.1. Broaching 10.2. Length of stroke and force
10.3. Travel and Cutting speed 10.4. Obstruction and blind
10.5. 0.001 10.6. Multi 10.7. High 10.8. Minimum
10.9. Maximum 10.10. No 10.11. High Speed Steel
10.12. Coolant 10.13. (Any value between 0.02 to 0.12 mm and 2 to 10 m/min.)
10.14. Any value between 12° to 15° and 1° to 4° 10.15. T
10.16. F 10.17. T 10.18. F 10.19. F

10.20. F   10.21. T   10.22. T   10.23. T
10.24. F   10.25. (iii)   10.26. (iv)   10.27. (i)
10.28. (iii)   10.29. (iv)   10.30. ii-iii-i.

11.1. Versatile   11.2. Lathe   11.3. T   11.4. T
11.5. F   11.6. F   11.7. T   11.8. F
11.9. F   11.10. T   11.11. (iv)   11.12. (ii)
11.13. (v)   11.14. (iv)   11.15. (iv)   11.16. (ii)
11.17. (iii)   11.18. (iii)   11.19. (iii)   11.20. iii-ii-i

12.1. Forming   12.2. Die and Punch   12.3. Punch, Die
12.4. Kinetic Energy   12.5. Power, Mechanical, Hydraulic
12.6. Blanking   12.7. Piercing   12.8. Coining   12.9. Force, Ram
12.10. Clearance   12.11. Punch   12.12. Die   12.13. Less
12.14. More   12.15. T   12.16. F   12.17. T
12.18. F   12.19. T   12.20. F   12.21. (iv)
12.22. (iii)   12.23. (ii)   12.24. (i)   12.25. (iv)
12.26. (iii)   12.27. (i)   12.28. (i)   12.29. (iv)
12.30. ii-iv-i-iii.

13.1. Locating   13.2. Jig   13.3. Fixture   13.4. Clamping
13.5. Diameter   13.6. F   13.7. F   13.8. T
13.9. T   13.10. F   13.11. T   13.12. T
13.13. T   13.14. (v)

14.1. Modern or Unconventional   14.2. Ultrasonic   14.3. Boron Carbide
14.4. Low V and High I   14.5. High V and Low I
14.6. LBM   14.7. 100000 K and 100000 kW/cm^3 14.8. Conductive
14.9. Reducing, Increasing   14.10. T   14.11. T
14.12. F   14.13. T   14.14. F   14.15. F
14.16. T   14.17. T   14.18. F   14.19. T
14.20. F   14.21. F   14.22. T   14.23. T
14.24. T   14.25. (iii)   14.26. (iv)   14.27. (ii)
14.28. (i)   14.29. (iii)   14.30. (iv)   14.31. (iv)
14.32. iii-ii-i-iv   14.33. iv-iii-ii-i   14.34. iv-i-ii-iii.

15.1. Horning   15.2. Super-Finishing   15.3. Lapping
15.4. v-iii-ii-i-iv   15.5. ii-iii-i.

16.1. Safety   16.2. Safety   16.3. Safety   16.4. Accident
16.5. Guarded   16.6. Aid.

17.1. Instruction, Software, Numerical
17.2. Computer Numerical   17.3. Master   17.4. Mini
17.5. Line   17.6. Direct   17.7. Closed   17.8. Open
17.9. Point-to-Point   17.10. Analog   17.11. Digital
17.12. Absolute   17.13. Incremental   17.14. Machine.

20

MISCELLANEOUS OBJECTIVE TYPE QUESTIONS

A20 : State whether the following statements are true or False:

1. If the turning tool is held in such a way that its cutting edge is above the centre of rotation of work piece, the front rake angle of the tool will be decreased.
2. Addition of cobalt in high speed tool increases keenness and stability of the cutting edge.
3. As regards belt made of leather, the grain side of the belt is the driving side.
4. Capstan and turret lathes as special purpose Lathes.
5. In speed Lathe as the speed of the lathe is very high it can be suitably used for high feed and high depth of cut.
6. A parting tool is a purely end cutting tool.
7. A facing tool removes metal by side cutting edges and so no top rake is necessary for a facing tool.
8. The angle formed by the flank and a plane at right angle to the twist drill axis is known as rake angle of the drill.
9. In a draw type of shape, the cutting pressure acts towards the column which allows to take deep cut on the job.
10. The cutting and return speed of a crank type of shape is almost uniform throughout the length of the stroke.
11. For machining chilled cast iron surface by shaping tool comparatively high feed and lower depths of cut is preferable.
12. An oblique single point cutting tool will remove more metal in the same life than an orthogonal cutting tool, other factors being the same.
13. Larger and heavier chucking work is being machined in Capstan Lathe.
14. The end formed by the intersection of the flute surface and the body clearance of a twist drill is known as the face of the drill.
15. Vanadium in the cutting tool of H.S.S. improves hot hardness and makes the cutting tool more wear resistant.

16. Cutting and return speed of a planer are almost uniform throughout the stroke.

17. Odd threads are those in which the products of pitch of the lead screw in mm and the number of teeth of the worm gear of the thread chasing dial must be in exact multiple of pitch of the job to be cut.

18. While machining mild steel bar with HSS cutting tool the average cutting speed should be restricted to 50 metres/mim.

19. Half centres are used in lathe operation for reducing wear and strain.

20. A parting tool is a purely end cutting tool.

21. Automatic screw machines, capstan lathe and turret lathe are turned as special purpose lathes.

22. In capstan and turret lathes, the external threads are usually cut by thread chasers.

23. When a number of single spindle drilling machine column are placed side by side on a common base and have a common work table, the machine is known as multi-spindle drilling machine.

24. When the taper on the tool shank of a drill is larger than the spindle, taper sleeves are used to hold the tools.

25. If the cutting edges of a turning tool is held below the centre of the work to be turned, the rake angle is reduced and clearance angle is increased.

26. In case of Capstan Lathe, the hexagonal turret is mounted on the saddle which slides directly on the bed.

27. In a centre lathe, half centres are generally used for turning operation.

28. Tungsten in the H.S.S. tool provides keenness of the cutting edge of the tool.

29. In a centre lathe the tumbler gear is used to change the direction of rotation of the main spindle.

30. In U.S.M. process the vibratory tool striking on the flow of abrasive slurry, causes thousands of microscopic abrasive grains to remove the work material by mechanical shear erosion.

31. In E.C.M. process the workpiece is connected to the negative supply and the tool or the cathode which is connected to the positive terminal is advanced towards the workpiece through electrolyte that completes the circuit.

32. The essential property of a dielectric fluid in E.D.M. process in that they must not breakdown electrically in the shortest possible time when the breakdown voltage has been reached.

33. In L.B.M. process, there is no direct contact between the tool and the workpiece.

34. Push type of broaching tool can be made longer than the pull type of broaching tool as the push type of broaching tool is subjected to compression.

35. Soft, tough and ductile materials and heavy cuts require dense structure grinding wheel.

36. Progressive dies as faster operation than compound dies.

37. The process of removing metal by milling cutter, which is rotated against the direction of the workpiece is known as down milling operation.

38. Concave milling cutters have teeth curved outwards on the circumferential surface to form the contour a semicircle.

39. Honing operation is more precision than lapping operation.

40. E.C.M process is based of Michael Faraday's Classical Laws of Electrolysis requiring basically the two elements, an electrolyte, a gap and a source of A.C. power of sufficient capacity.

41. E.D.M. is the process of metal removal based on the principle of erosion of metals by a constant electric spark discharge between the electrode tool and the work.

42. Laser in an electromagnetic radiation which produces monochromatic light in the form of an almost collimated beam.

43. A broach is a multiple edge cutting tool that has successively same cutting edges along the length of the tool.

44. Dressing is the process of changing the shape of the grinding wheel as it becomes worn from the original shape owing to breaking away of the abrasive and bond.

45. In the rack cutter generating system both internal and external teeth on a gear blank can be cut.

46. The tooth profile of involute gears changes with the number of teeth on the gear for the same module.

47. The convex milling cutters have teeth curved outwards on the circumferential surface to form the contour of a semicircle.

48. Centreless type of grinder is an intermittent process of cylindrical surface grinding.

49. In practice a set 9 cutters are used to cut all gears ranging teeth from 12 teeth to rack for the same module.

50. In a broaching operation both roughing, semi-finishing and finishing cuts are completed in one pass of the tool.

51. The hardness of grinding wheel is specified by Brinell hardness number.

52. The thickness of chip in down milling process is minimum when the tool begins its cut and it increases to the maximum when the cut terminates.

53. Any solid material which can be melted without decomposition can lee cut with laser beam machining process.

54. In blanking operation of press work, the clearance is provided on the punch.

55. Aluminium oxide is best suited abrasive material for grinding wheel because it is very hard.

56. In grinding, curved, convex and concave surfaces, surface grinders are employed.

57. The essential requirements of a dielectric fluid used in E.D.M. process are that they must remain electrically non-conductive until the required breakdown voltage has reached.

58. Numerical control cannot by applied to all of the machine tools.

59. A high rate of production is secured on hobbing machine because the cutting action is continuous in one direction.

60. A broach is a multiple edge catting tool that has successively same cutting edges along the length of the tool.

61. Silicon carbide grinding wheel is used for grinding materials of low tensile strength.

62. The thickness of the chip in up milling process minimum at the beginning of the cut and it reaches to maximum when the cut terminates.

63. Electric discharge machining is a process of metal removal based on the principle of erosion of metals by a constant electric spark discharged between the electrode tool (usually cathode) and the work (anode).

64. The spindle will rotate 12° by one complete turn of the index crank of universal dividing head.

65. Cycloidal gear cannot be produced by generating methods of producing gears.

66. A dense structure grinding wheel is required for hard and brittle materials and where finishing cuts are required.

67. Buffing operation is followed by polishing operation.

68. Internal gears cannot be cut by pinion cutter generating process.

69. All involute gear cutters have equal pressure angle.

70. the V-belt made of fabric and vulcanized rubber with cotton cord runs in 60°-V-groove turned in the rim of the pulley.

71. The minimum diameter of the pulleys for transmitting power should be at least 24-times the thickness of the belt used to rim on the pulley.

72. The crowning of pulleys, in most case should not exceed 30 mm on the diameter per meter of width and the width of the pulley should be one fourth greater than the width of the belt used.

73. cobalt is added to the high speed steel to increase the hot hardness and wear resistance of the tool.

74. stellite is the trade name of nonferrous cast alloy composed of Principally of carbon mixed with other elements such as tungsten and cobalt to form carbides.

75. the speed lathe consists of a bed, a head stock a tail stock and a Tool post mounted on the carriage.

76. Capstan and turret lathes are special purpose lathes.

77. The guide ways of the lathe bed may be flat or V-ways having included angle of 90°.

78. Thrust bearings are provided to the lathe spindle for taking up the radial loading.

79. As the heat conductivity of Ceramics are low, Ceramics tools must be used with coolants.

80. The tool is set in slightly advanced position than the two jaws of a follower rest during the rotation of the job.

81. If the cutting edge of a turning tool is held in such a way that its cutting edge is above the centre of the work piece the rake angle will be increased where as the clearance angle will be decreased.

82. Reaming is more accurate process than boring.

83. In Indian Standard System, twist drills are designated by the series to which they belong, the diameter, the I.S. number and the material of the drill.

84. The drilling tool may be removed from the spindle of the drilling machine by pressing a lapered sleeve into the slotted hole of the spindle.

85. The taps used serially for cutting external threads are taper, bottom and plug.

86. In vertical boring machine the work is rotated on a vertical axis and the tool is fed longitudinally in a vertical plane.

Fill in blanks with suitable word/words.

87. The edge formed by the intersection of the surface left by the provision of secondary clearance and the flute of the reamer is known as _____ of the reamer.

88. The average cutting speed for machining cast iron work piece with H.S.S. tool is _____ m/min.

89. The side cutting edge angle _____ the tool life for the same depth of cut.

90. A shaft has to be cut with a taper of 1 : 3. The tangent of the angle of swivelling the compound rest will be _____.

91. The trade name of a non-ferrous cast alloy composed of cobalt, chromium and tungsten is known as ______.

92. Conditions favour to the formation of ______ chips include low cutting speed, low rake angle, high feed & depth of cut.

93. In case of capstan lathe the hexagonal turret is mounted on ______ which slides on the saddle.

94. The sleeve with the tool may be removed from the spindle by forcing a ______ within the slot of the spindle.

95. Three surfaces can be machined at a time by ______ planer.

96. The central portion of the drill situated between the roots of the flutes and extended from the point towards the shank is called ______ of the drill.

97. The bed of a lathe machine is made massive with sufficient depth and width to absorb ______.

98. The tool head of a planing machine is mounted on the ______ by the saddle.

99. In ______ shaper the tool is set in the reversed direction to that of a standard shaper.

100. In case of capstan lathe the hexagonal turret is mounted on a ______.

101. The end thrust on the job along the axis of rotation of the job due to axial feeding action of tool on the job is being taken up by ______ in case of lathe machine.

102. Cutting forces or power at the cutting tool may be measured by ______.

103. To get large velocity ratio ______ gearing is used.

104. ______ is the operation of smoothing and squaring the surface around a hole for the seat for a nut or the head of a screw.

105. The operation is producing a large hole by removing metal along the circumference of a hollow cutting tool is known as ______.

106. The operation of enlarging the end of a hole cylindrically and co-axially to a certain depth is called ______ operation.

107. The holding device with adjustable jaws set radially to hold straight shank drill are known as ______.

108. A Lathe machine part which carries the saddle, cross-slide, compound rest, tool post and apron is known as ______.

109. For setting washer over a casting surface we generally do ______ operation.

110. No top rake angle is required in a ______ tool as it removes metal by its side cutting edges.

111. Larger and heavier chucking works are usually handled on a _____ Lathe.

112. The metal column that separates the drill from the flutes and runs the entire length between the flunks is known as _____ of the drill.

113. In lathe operation while machining cast iron with H.S.S. tool the average cutting speed should be restricted _____ metres/min.

114. Orthogonal cutting is mainly confined to _____ turning tool.

115. When the table is stationary and the column carrying the cross-rail reciprocates on massive horizontal rails mounted on both sides of the table, the planer is known as _____ planer.

116. When the cutting face of a grinding wheel has particles of metal being ground adhering to it and the pores of the opening of the wheel face having been filled up with metal, the wheel is said to be _____.

117. When the milling cutter is rotated about an axis perpendicular to the work surface, the operation is called _____ operation.

118. To locate a work having odd irregular profiles _____ locator is used.

119. To obtain less velocity ratio _____ gearing is used.

120. For cutting 34 teeth of involute gear cutter no. _____ is to be used.

121. The machining time to broach a hole of a MS job having length of the broach travel 1 metre and cutting speed of 2 m/min, is _____ minutes.

122. Recently a new development in U.S.M. has taken place in which the tool impregnated with _____ is used and no slurry is used.

123. _____ gearing is used when the power is to be transmitted to two shafts with their axis-non-intersecting and in different planes.

124. _____ gears can not be produced by generating method of gear cutting.

125. The operation of cutting a sheet metal through part of its length and then bending the cutting portion is known as _____ operation.

126. One complete turn of the crank of the universal dividing head will cause the spindle and work to rotate through _____ degree.

127. To cut involute teeth of 60 teeth spur gear of certain module on a gear blank cutter no. _____ in the sets of cutter is used.

128. _____ grinder is capable of grinding irregular, curved, convex and concave surfaces.

129. The _____ is the extension of machine spindle on which cutters are securely mounted and rotated.

130. _____ milling machine has the arrangement of knee swivelling in addition to movements of tables as in case of a universal milling machine.

131. As the principle of generating gears system is based upon _____ system, cycloidal gears cannot be produced by generating methods .

132. _____ is precision grinding process that is used to produce geometrically true surfaces, correct mirror surface imperfections, improve dimensional accuracy with loose dust type abrasives.

133. The tool in case of U.S.M. process is made of _____.

134. Laser is an electromagnetic _____.

135. The operation of machining several surfaces of a workpiece simultaneously by feeding the table against a number of cutters mounted on the arbor is known as _____.

136. The _____ is the operation of production of a flat surface which may be vertical, horizontal or at an angle in reference to the table surface.

137. When the particles of the metal to be ground adhere with the cutting face of grinding wheel during operation, the wheel is said to be _____.

138. For low tensile strength materials, the wheels made of _____ is recommended.

139. To locate a work having add, irregular profiles _____ locator is used.

140. In a _____ die with both cutting and non-cutting operations are accomplished at one station of the press in every stroke of the ram.

141. the operation of forming the edges of a component into a roll by bending the sheet metal in order to strengthen the edges and to provide smoothness to its surface is known as _____ operation.

142. The most important part of the laser apparatus is the laser _____.

143. The process of removing metal by a milling cutter which is rotated in the same direction of the travel of the workpiece is known as _____.

144. For cutting gear teeth of involute profile of same module, the number of cutters required to cut any gear from a pinion of 12 teeth to a rack are _____.

145. The operation of production of flat vertical surfaces on both sides of a workpiece by using two side milling cutters mounted on the same arbor is known as _____.

146. The operation performed by a milling cutter to produce a flat machined surface perpendicular to the axis of rotation of the cutter is known as _____.

147. When two shafts, the axes of which intersect, are to be connected by gearing, the wheels are known as _____.

148. The process of changing shape of the grinding wheel as it becomes worn from the original shape is known as _____.

149. The relative spacing of abrasive grains in a grinding wheel is referred to as the _____ of the wheel.

150. The main function of the _____ is to discard the workpiece from the punch or the die after the end of the cutting.

151. Low cutting speed, low rake angle, high feed, lack of cutting fluid and large depth of cut tend to promote the formation of _____ chips.

152. The r.p.m. of a 60 teeth Worm Wheel being driven by a worm shaft whose lead is equal to 3 pitch and revolving at 300 r.p.m. is _____.

153. Conditions tending to promote the formation of _____ chips include low cutting speed, low rake angle, high feed lack of cutting fluid and large depth of cut.

154. While turning a M.S. bar with a single point H.S.S. Cutting tool, the following observations are made:

Depth of Cut – 1 mm

Feed – 1 mm/rev.

Cutting speed – 30 metre/min.

Time to Tool Failure – 30 minutes.

The Tool life as per volume of metal removed until tool fails _____ mm^3/min.

155. A clutch is a form of connection for transmitting motion between a driving shaft and a driven shaft member on the _____ axis.

156. The saddle is mounted on the _____.

157. For general purpose work the included angle of the lathe centre is usually _____ degree.

158. _____ is needed to be fitted in the tail stock to face the end of a bar.

159. To grip various work pieces with different hole diameters within a limit _____ mandrel is used.

160. The spindle end of a milling machine has a taper 1 : 4. The tan of the setting of the compound rest is _____.

161. Tool signature consists of _____ elements.

162. The turret of a capstan lathe is mounted on a _____ which slides on the saddle.

163. Reaming operation can not correct the _____ of the hole made by drilling.

164. When a number of single spindle drilling machine columns are placed side by side on a common base and have a common work table. The machine is known as _____ drilling machine.

165. The angle included between the two lips projected upon a plane parallel to the drill axis and parallel to the two cutting lips is known as _____ angle of a twist drill.

166. The size of a horizontal boring machine is specified by the _____ of its _____ in mm.

167. The operation of Producing a large hole by removing metal along the circumference of a hollow cutting tool is known as _____.

Answers of Miscellaneous Objective Type Questions

1. F	2. F	3. T	4. F
5. F	6. T	7. T	8. F
9. T	10. F	11. F	12. T
13. F	14. F	15. F	16. T
17. F	18. F	19. F	20. T
21. F	22. T	23. F	24. F
25. T	26. F	27. F	28. F
29. F	30. T	31. F	32. F
33. T	34. F	35. F	36. F
37. F	38. F	39. F	40. F
41. T	42. T	43. F	44. F
45. F	46. T	47. T	48. F
49. F	50. T	51. F	52. F
53. T	54. T	55. F	56. T
57. T	58. T	59. T	60. F
61. T	62. T	63. T	64. F
65. T	66. F	67. F	68. T
69. T	70. F	71. T	72. F
73. T	74. F	75. F	76. F
77. F	78. F	79. F	80. T
81. T	82. F	83. T	84. F
85. F	86. T	87. Heel	88. 18 to 20

89. Increases 90. $\tan^{-1}\frac{1}{6}$ or 0.1666 rad 91. stellite
92. Discontimous or Segmental 93. Ram or Short-Slide
94. Drift 95. Double Housing 96. Web
97. Shock and Vibration 98. Cross-rail 99. Draw type
100. Sub-bed 101. Thrust Bearing/Roller Bearing
102. Force Dynamometer 103. Worm and Worm wheel
104. Spot-facing 105. Trepanning 106. Counter-boring
107. Drill chuck 108. Carriage 109. Spot-facing
110. Thread-cutting 111. Turret 112. Web
113. 20 114. Parting-off/grooving 115. Pit
116. loaded 117. Face-milling 118. Outside pin
119. Compound 120. 4 121. 1/2 122. Diamond dust
123. Spiral 124. Cycloidal 125. Latching 126. 9
127. 2 128. Surface 129. Arbor 130. Omniversal
131. Involate 132. Truing 133. Low-carbon steel/stainless steel

134. radiation 135. Gang Milling 136. End Milling
137. Loaded 138. Silicon Carbide 139. Outside pin
140. Combination 141. Curling 142. Crystal/Rod
143. Down Milling 144. 8 145. Straddle Milling
146. Face Milling 147. Bevel gearing
148. Truing 149. Structure 150. Stripper 151. Dis-continuous
152. 15 153. Built-up-edge 154. 9×10^5
155. same 156. carriage 157. 60 158. Half-centre
159. Expansion 160. $\frac{1}{8}$ or 0.125 161. 7 162. Sub-bed
163. Location 164. Gang 165. Point 166. Spindle, diameter
167. Trepanning.

INDEX FOR COMPARATIVE STUDY

INDEX TO VOLUME I

U

INDEX TO VOLUME II

N

O